Statistical Evaluations in Exploration for Mineral Deposits

Springer

Berlin
Heidelberg
New York
Barcelona
Budapest
Hong Kong
London
Milan
Paris
Santa Clara
Singapore
Tokyo

Friedrich-Wilhelm Wellmer

Statistical Evaluations in Exploration for Mineral Deposits

Translated by D. Large

With 120 Figures and 74 Tables

 Springer

Prof. Dr. Friedrich-Wilhelm Wellmer

Bundesanstalten für Geowissenschaften und Rohstoffe (BGR)
Alfred-Bentz-Haus
Stilleweg 2

D-30655 Hannover, Germany

Translator:

Dr. Duncan Large
Paracelsusstraße 40

D-38116 Braunschweig, Germany

ISBN-13: 978-3-642-64325-5   e-ISBN-13: 978-3-642-60262-7
DOI: 10.1007/978-3-642-60262-7

The book was originally published under the title „Rechnen für Lagerstättenforscher und Rohstoffkundler. Lagerstättenstatistik, Explorationsstatistik einschließlich geostatistischer Methoden". (Ellen Pilger Verlag 1989)

Library of Congress Cataloging-in-Publication Data

Rechnen für Lagerstättenforscher und Rohstoffkundler. English. Statistical evaluations in exploration for mineral deposits / by Friedrich-Wilhelm Wellmer. Includes bibliographical references and index. ISBN-13: 978-3-642-64325-5   1. Prospecting--Statistical methods.   I. Wellmer, Friederich II. Title. TN270.R42513 1996   622'.1'072--dc20

Springer-Verlag Berlin Heidelberg 1998
Softcover reprint of the hardcover 1st edition 1998

Cover Design: design & production, Heidelberg
Dataconversion: U. Hellinger, Heiligkreuzsteinach

SPIN: 10063759   32/3136 – 5 4 3 2 1 0

Contents

Preface

This textbook, *Statistical Evaluations in Exploration for Mineral Deposits,* is a translation of the German textbook *Rechnen für Lagerstättenkundler und Rohstoffwirtschaftler Teil II, Lagerstättenstatistik, Explorationsstatistik einschließlich geostatischer Methoden.*

The English translation of Teil I (Part I) with the title, *Economic Evaluations in Exploration,* was also published by the Springer-Verlag, and ist frequently referred to in this textbook, *Statistical Evaluations in Exploration for Mineral Deposits.* Exploration progress in stages, and after each stage a decision has to be made to proceed to the next. The same is true when a new exploration project is proposed, or a decision has to be made to join other partners in an exploration project (farm-in). These yes/no decisions involve the statistical evaluation of exploration data, and an economic appraisal of such an exploration project max, at a later stage, result in a profitable mining operation. Both textbooks, *Statistical Evaluations in Exploration* and *Economic Evaluations in Exploration for Mineral Deposits,* have a similar aim – illustrating by numerous examples the methods used to arrivat at the correct yes/no decision.

This book was not intended to be a normal textbook an statistics or geostatistics, and little theory is provided. Most calculations are explained by examples disguised to conceal their source, frequently from the mineral exploration experience of the author, who worked worldwide for a German mining company. The textbook is intended as a practical vade mecum for a geologist left to his own resources to derive the right decision when he has to appraise on his own (and frequently quickly) a mineral property or an exploration project; but it also addresses geologists, universities and research institutions who teach or are involved in mineral economic evaluations, recommendations or decisions.

Typical for a task to evaluate exploration data statistically is the problem of dealing with very inhomogeneous data sets. For example, there might be a property with some percussion drill data, some data from core drilling, some with good core recoveries, some with low core recoveries, a soil geochemical survey, some data from chip sampling in trenches and a bulk sample in an exploration pit. One cannot afford to disregard low quality data sets. The competition for good exploration projects is high, and therefore the maximum information value has to be derived from all data available. Such practical problems are dealt with in the textbook.

The author would like to thank Mr. N. Champigny (Santiago de Chile), Dr. R. Braun (Cologne) and M. Günther (Hannover) for critically reviewing the manuscript, but shortcomings are, of course, the responsibility of the author. He also thanks G. Bardossy (Budapest) for making available statistical data of bauxite deposits, D. Krige (Johannesburg) for data from the Hartebeestfontein mine, and H. Siemes (Aix-la-Chapelle) for the (H-1)-table. My special thanks are due to Dr. D. Large for translating the German text.

Friedrich-W. Wellmer

The Most Important Notations and Abbreviations

1. Latin letters

a   1. the geostatistical range in a transitive type variogram
         2. in the equation of a line $y=a{\cdot}x + b$: a is the inclination of the line; for the line of regression a is one of the regression coefficients

b   1. average of all the weighting factors
         2. in the equation of a line $y=a{\cdot}x + b$: b is the value where the line intersects the y-axis; for the line of regression b is one of the regression coefficents

b_n annuity present value factor

C   coefficient of variation

C_o nugget effect

C_1 (or C) ⎫
                     ⎬ the sill (or pitch) value in a transitive type variogram
$C_1 + C_o$ (or C) ⎭

$\text{Cov}_{x,y}$ the covariance between the value x and y

d   distance or diameter

e   natural number 2.7183

EMV   expected monetary value

ev   expected value per discovery

Fi   filter value

f_i frequency in one interval of a histogram

Fu   upgrading factor

G   grade

g   outlier threshold

GT   accumulation value=product out of grade G and thickness T

G_w weight of sample

h	1. lag in the calculation of a variogram
	2. mode, most frequent value in a frequency distribution
Hp	plotting percentage or plot frequency
i	1. one of the values in a range of samples like x_i
	2. rate of interest
[K]	kriging matrix
ki	confidence interval
L	length of a target to be searched for
L, l	linear equivalent
L_B	length of a drill core to be analysed
l	1. length of a block
	2. confidence interval for the line of regression
m	median
M_w	true thickness
NPV	net present value
NSR	net smelter return
n	number of samples
p	probability of success
p_i	probability in an interval i
Q	metal content of an ore deposit
q	probability of failure
R	range (between maximum and minimum values)
Ri	amount of risk capital invested
r	coefficient of correlation
r^2	coefficient of determination

r, r^2 : of a regression analysis

r	radius of a target with a form of a circle
S	surface area of an ore deposit
S, T	distances of survey grids (dimension of a grid cell)
S_f	coefficient of skewness
Sh	Sheppard correction factor
Si	significance level
s	standard deviation of the samples
s^2	variance of the samples
s_X	standard deviation of the mean of a sample
$s_x{}^2$	the variance of the mean of a sample
SW	s-value (scale of the transformed y-axis in a probability grid, see Chap. 5.1.2.)
T	thickness
t	Student's t-factor
th	threshold value
t_{si}	Sichel's t-estimator
U	size of uncertainty
V or v	volume, (mostly ore deposit or block volume), in combination: V=ore deposit volume; v=block volume
VCC	volume-variance comparison line
W_o	Wainstein-factor for the calculation of the upper interval of confidence for a lognormal distribution
W_u	Wainstein-factor for the calculation of the lower interval of confidence for a lognormal distribution
x_A	outlier

x_c cut-off grade

x_i one of the values in a series of x-values; for values categorized the value in the center of a category (or interval)

$x_{i,h}$ value in a distance (lag) h from value x_i

$\overline{x}$ mean (arithmetic) of all the samples x_i

y_i one of the values in a series of y-values

$\overline{y}$ mean (arithmetric) of all the samples y_i

2. *Greek letters and other symbols with the exception of variances* σ^2

α 1. mean of a lognormal distribution
         2 inclination of a drill hole
         3. inclination of the variogram of the de Wijs type
         4. one of the axes of an ellipse (the other one β)

β 1. dip angle of a geological unit like stratum, ore deposit etc.
         2. standard deviation of a lognormal distribution
         3. one of the axes of an ellipse (the other one α)

β^2 variance of a lognormal distribution

γ 1. geometric mean of a lognormal distribution (ln γ = m = median value)
         2. variogram-value

Δ width of an interval in a histogram

ε recovery in a beneficiation plant

θ correction factor in a three parameter lognormal distribution (θ is one of the parameters)

λ the vector λ

λ_i kriging factor

μ 1. mean of the population
         2. Lagrange parameter

ρ density

σ^2 variance: for the different variances see under (3) below

σ	standard deviation of the population
$\hat{\sigma}^2$	best estimate for σ^2
χ^2	chi-square factor in the chi-square test
∞	infinity
∂x	partial derivation after x
ϕ	area or area segment under the normal distribution

3. *Variances σ^2 or standard deviations σ*

Combination $\sigma^2_{x/y}$ hereby x is the support
y the universe within which the sample is taken (see Chap. 1)

example: $\sigma^2_{(o/D)}$ is the variance of all samples (e.g. drill core samples, considered as points/O) within the whole ore deposit/D (see illustration below)

σ^2_D	dispersion variance
σ^2_e	extension variance (variance of a sample extended to the block e.g.)
σ^2_G	variance of the grade of a deposit
σ^2_{GT}	variance of the accumulation value GT
σ^2_K	kriging variance
σ^2_Q	variance for the metal content Q of a deposit
σ^2_S	variance for the surface S of an ore deposit
σ^2_{xi}	variance of all values x_i
σ_{xi+h}	variance of all values in a distance (lag) h of x_i

Introduction to Some Fundamental Statistical Concepts

1.1
General Definitions

A stack of spheres may be used as an example for the evaluation of data (Fig. 1), and in statistical terms the total number of individual spheres is then known as the *population*. One sphere extracted at random from the stack is referred to as a *sample*. In the example shown, each sphere is a natural sample unit, or individual. For instance, in order to determine the *mean* diameter of all the spheres, measurements are usually made on a random selection of the spheres, rather than every individual sphere, and the mean is then *estimated* from this sample of spheres (Fig. 1).

NORMAL STATISTICS

population   sample   mean
                      x_i $\overline{x}$

EXPLORATION STATISTICS

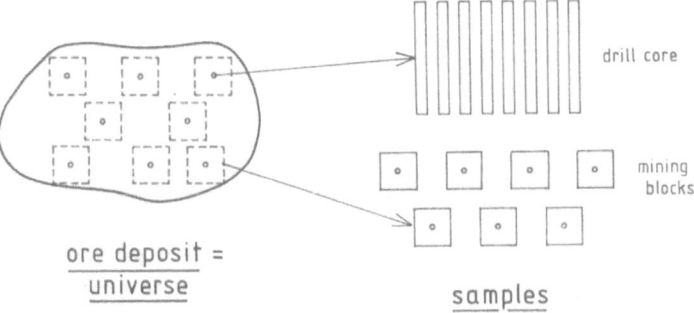

ore deposit =
universe   samples

Fig. 1. Comparison of terms used in normal statistical studies with those used in exploration statistics

A comparable sample unit is not available for the sampling of a mineral deposit or for a regional geochemical programme, and thus the sample units must be artificially determined by the sampler. Hazen (1967) introduced the general term *universe*, which includes the total mass of material within the area of interest and the source of all those data that might be of interest to the sampling project. The universe can, for example, be the whole of a mineral deposit or only that part of a deposit down to a particular depth. Similarly, it can be an area for geochemical sampling. The sample is then a small, definable, part of the universe that can be analysed or otherwise examined. The sample can be, for example, a 0.2-kg sample, a 1-m length of NQ drill core (see Economic Evaluations in Exploration, Appendix C 2), or a bulk 10-t sample (Fig. 1).

The population is therefore defined as the number of all possible samples of *one* specific type that can be extracted from the universe. The random selection of samples with the same specification from the population is subsequently referred to as the *sample size* and not just as a sample, which is the usual statistical terminology. The phrase sample is not really applicable, since reference might be made not only to drill core samples, but also to a random selection of mining blocks.

The universe can therefore include quite different populations that might have dissimilar statistical parameters. The variations in grade within a population of 1-m core sections will definitely differ from those derived from a population of 10-t bulk samples or 10 x 10 x 5-m ore blocks (Fig. 1). It is a common mistake simply to compare the grade variation of one population (samples) with that of another (mining blocks), or even to include core samples together with dump samples in

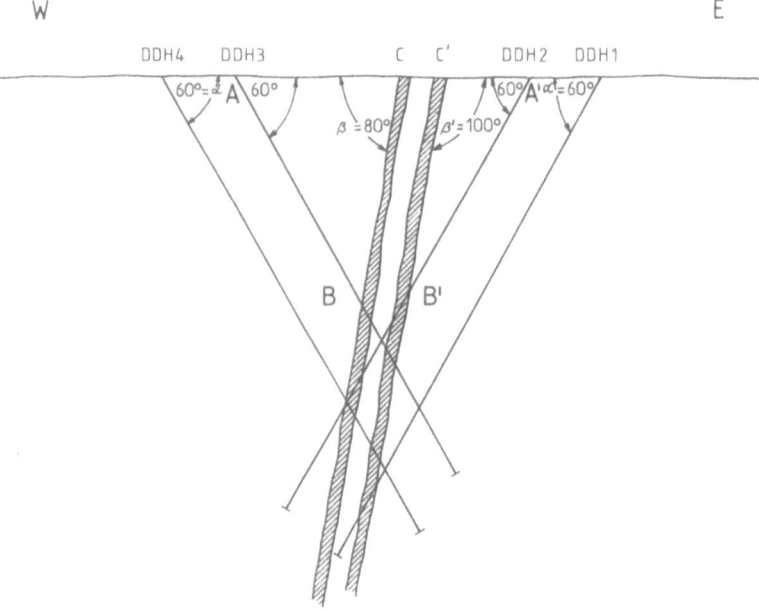

Fig. 2. Section through a Sb-vein zone

the same statistical analysis. It is therefore important to evaluate a specific sample population for variations of grade. If comparisons are made with other types of sample, then at least the specifications should be similar. This is known as *support* of the samples, and is discussed below.

Assignment: Two relatively thick Sb veins, each 2 - 3 m wide, are intersected by four drill holes (Fig. 2).

It was initially assumed, on the basis of information obtained from other veins in the vicinity, that these veins would be dipping towards the east, and hence two holes were drilled from the east towards the west. The second hole indicated that the veins actually dipped to the west and consequently the subsequent holes were drilled from the west towards the east. The core was split into 50-cm sections for analysis. The results from all the holes, including the first two down-dip holes, must be included in the statistical evaluation. How should the results from the first two drill holes, DDH 1 and DDH 2, be treated so that their support can be compared with the data from the other drill holes?

Solution: Since DDH 1 and DDH 2 were drilled down-dip and DDH 3 and DDH 4 were drilled perpendicular to dip, the analysed 50-cm core sections actually represent different true thicknesses. The true thickness, M_w, is calculated from the length, L_B, of the core analysed, collar inclination angle, α, of the drill hole, and dip angle of the vein, β (as shown by triangles ABC or A'B'C' respectively in Fig. 2, cf. p. 24 and Fig. 4a in Economic Evaluations in Exploration):

$$M_W = L_B \cdot \sin(\alpha + \beta)$$

For for DDH 1 and DDH 2:

$$M_{W_{1/2}} = 0.5 \cdot \sin(60°+100°) = 0.5 \cdot 0.34 = 0.17 \text{ m}$$

and for DDH 3 and DDH 4:

$$M_{W_{3/4}} = 0.5 \cdot \sin(60°+80°) = 0.5 \cdot 0.64 = 0.32 \text{ m}$$

therefore

$$\frac{M_{W_{1/2}}}{M_{W_{3/4}}} = \frac{0.17}{0.32} = 0.53 \ .$$

The true thickness of the analysed 50-cm section of core from drill holes DDH 1 and DDH 2 is about half as much as it is for drill holes 3 and 4. Therefore the results from two adjacent analysed sections in drill holes 1 and 2 should be combined into one analytical result, and this figure is then comparable with the results from drill holes 3 and 4.

If such simple corrections are not possible, then the results must be weighted accordingly, and this is discussed in Chapter 2.3.

Fig. 3. Section through an old gold vein mine

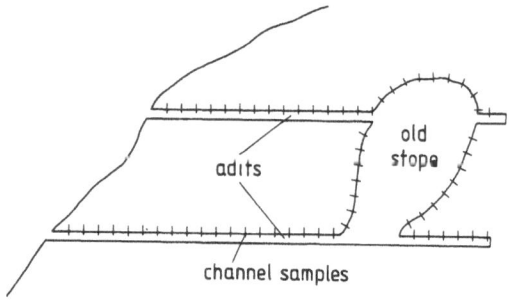

The quality and character of the samples is closely related to the sample support.

Problems related to differences in the quality of samples, such as those caused by variable recovery during core drilling, are commonly encountered in exploration and they are discussed specifically in Chapter 11.5.

The *character of the samples* varies if, for example, the systematic or random drilling pattern used to evaluate a deposit is changed subsequently to a pattern that follows a specific trend. This problem also includes questions related to samples being statistically independent or not, and an appropriate test for such cases is discussed in Chapter 13.2. This is a very common problem encountered in resampling old mines.

Example: An abandoned adit mine that exploited a gold-quartz vein is resampled. There are two adits as well as one old stope, which is also accessible. Depending on accessibility, the quartz vein is sampled at 1-m intervals (Fig. 3). Do all the channel samples have the same influence on the evaluation of the mine?

Answer: No. Both of the adits can be treated as examples of random sampling (comparable to two drill holes). However, this is clearly not the case for the stope where the previous miners advanced up to a cut-off boundary (Economic Evaluations in Exploration, p. 70). An attempt must be made to acquire the old production data from the stope in order to realistically appraise the whole vein. An average value can then be calculated from the sampling of the adits and the old production data.

1.2
Frequency Distribution

Once it has been decided, after a critical evaluation of the sampled material, that the sample specifications (character, quality and support of the samples) are the same or similar, then even substantial data packages can be treated as a population within the universe (Sect. 1.1). If the frequency of occurrences of a value x (e.g. analytical values), the so-called variate, within a very extensive data package (theoretically infinitely large) is plotted as $y=f(x)$, then the frequency distribution for this particular population is obtained. For example, assume that there are thousands of samples, all with the same specifications, from a tungsten deposit, and that they were all analysed with an absolute precision of $\pm 0.1\%$ WO_3. The analytical values comprise the value x and the number of samples with a particular analytical

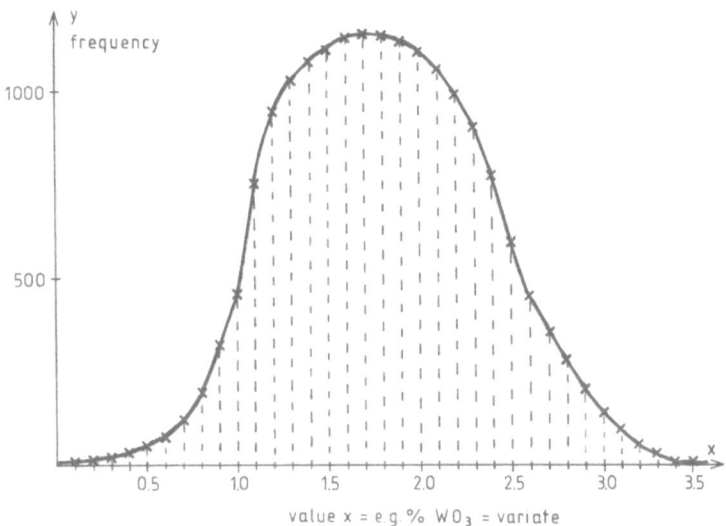

Fig. 4. Frequency distribution for the samples of a wolframite deposit

value comprise the frequency y (Fig. 4). The points plotted from the coordinates y = frequency and x = value can be interpreted now as describing a continuous frequency distribution, and discussion of this distribution will comprise a significant component of this book. The frequency distribution permits the substitution of data sets by simpler mathematical descriptions.

Frequency distributions can be classified as symmetric or skewed distributions (Fig. 5a, b). A distribution can be skewed to the right or to the left. Distributions can also be described as having either a single peak or several peaks (Fig. 5c).

The two most common cases of frequency distribution are discussed in this book. The normal distribution is a symmetric distribution with a single peak. The lognormal distribution is a skewed and single peak distribution of nonlogarithmic values that can be transformed into a symmetric distribution by the logarithmic transformation of these values (Chap. 9.2.1).

The distribution of analytical data is commonly skewed to the right (positively skewed) because of distortion to the higher values (Fig. 5b). Similarly, minor peaks commonly occur on the flanks of the frequency distributions of analytical data (Fig. 5c), and therefore in reality these distributions have several peaks and cannot be described by a single peak. These subsidiary peaks can be real, although they can also be caused by the sample population being too small or by division into class intervals (Chap. 2.2). Mineral deposits are evaluated on the basis of average values, and these small subsidiary maxima are treated as irregularities or random variations that can always result from random sampling and/or an insufficient number of samples. The frequency distribution is therefore usually treated as having a single peak. This always assumes that the subsidiary peaks are low as compared to the main maximum, and that there is no geological reason for them to be treated separately. However, if the higher values marked by a subsidiary

Fig.5.a-c Examples for frequency distributions. a Symmetric frequency distribution. b Frequency distribution skewed to the right. c Multipeak frequency distribution.

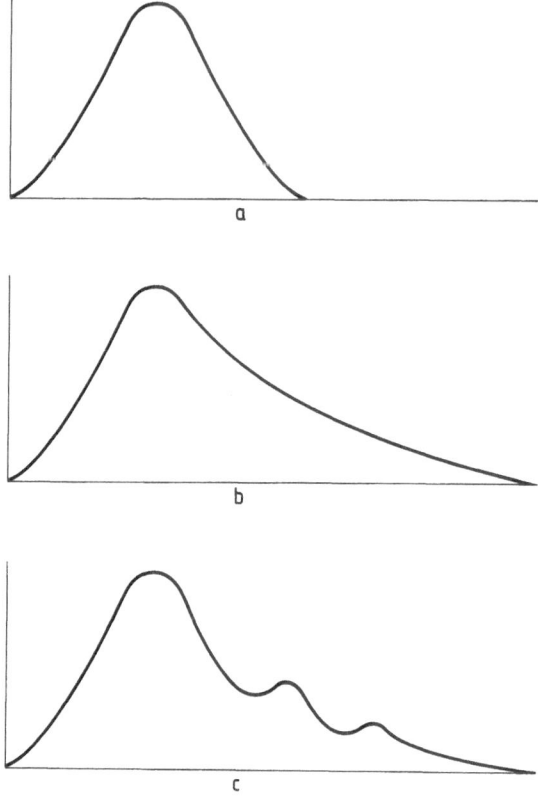

maximum correlate with a distinctly high-grade zone of mineralisation, then that would constitute a geological reason for separating the data, and another frequency distribution is calculated for the high-grade zone. Subsidiary maxima are particularly important in the evaluation of geochemical data because they can indicate the presence of separate anomalous populations, which comprise the geochemical anomalies and possible drill targets (Chap. 18.2.2).

The parameters by which the various distributions can be described are discussed in Chapters 4 and 9.2.3.

Part A Mineral Deposit Statistics

2

Treatment of the Data Set

2.1
A Simple Case of Calculating a Frequency Distribution

The *data set* is the basic component for calculating a frequency distribution. It might be a certificate on which all of the measurements or analyses are listed in the order in which they were made. With respect to analytical data, the data set is usually the certificate of analysis issued by the laboratory. Table 1 illustrates a data set of SiO_2 analyses from 20 barite samples (1-kg rock samples).

In the next step, the various analytical values in the data set are listed incrementally from the lowest to highest value. Each value on the certificate of analysis is then ticked off against the respective value in the list. This is known as the *ranking list* (Table 2).

The number of ticks for each value indicates the *absolute frequency* of that value within the whole of the sampled population. The *relative frequency* is obtained by dividing the absolute frequency by the number of samples (in this case n = 20). If this figure is multiplied by 100, the resultant figure is the *percent frequency* (relative frequency in percent). The frequency distribution of the sampled material is derived by listing the absolute and relative frequencies for each of the analytical values together in one table (Table 3).

Table 1. Basic data set – analyses sheet

Sample no.	SiO_2 grade in %	Sample no.	SiO_2 grade in %
AL 1	2.5	AL 11	2.5
AL 2	2.4	AL 12	2.5
AL 3	2.3	AL 13	2.4
AL 4	2.5	AL 14	2.6
AL 5	2.0	AL 15	2.9
AL 6	2.7	AL 16	2.7
AL 7	2.2	AL 17	2.5
AL 8	2.4	AL 18	2.8
AL 9	2.6	AL 19	2.4
AL 10	2.5	AL 20	2.6

Table 2. Ranking list for Table 1

SiO$_2$ (%)	
2.0	|
2.1	
2.2	|
2.3	|
2.4	||||
2.5	|||| |
2.6	|||
2.7	||
2.8	|
2.9	|

This frequency distribution can also be graphically displayed by plotting columns for the relative frequency above the respective analytical values (Fig. 6). This type of diagram is known as a *histogram*.

2.2
Using Class Intervals for Calculating Frequency Distributions

Analytical values usually have a wide range, and it is rare for a value to be repeated. The values are therefore grouped together in order to obtain a meaningful frequency distribution. Thirty analytical values (BX core of 1 m length) from a fluorite mine serve as an example (Table 4).

In order to evaluate the data in Table 4, the analytical values are grouped into *class intervals*.

Table 3. Frequency distribution for Table 1

SiO$_2$ (%)	Absolute frequency	Relative frequency (%)
2.0	1	5.0
2.1	0	0.0
2.2	1	5.0
2.3	1	5.0
2.4	4	20.0
2.5	6	30.0
2.6	3	15.0
2.7	2	10.0
2.8	1	5.0
2.9	1	5.0
Sum	20	100.0

Fig.6. Histogram of the SiO_2 analyses of the 20 baryte samples of Table 3

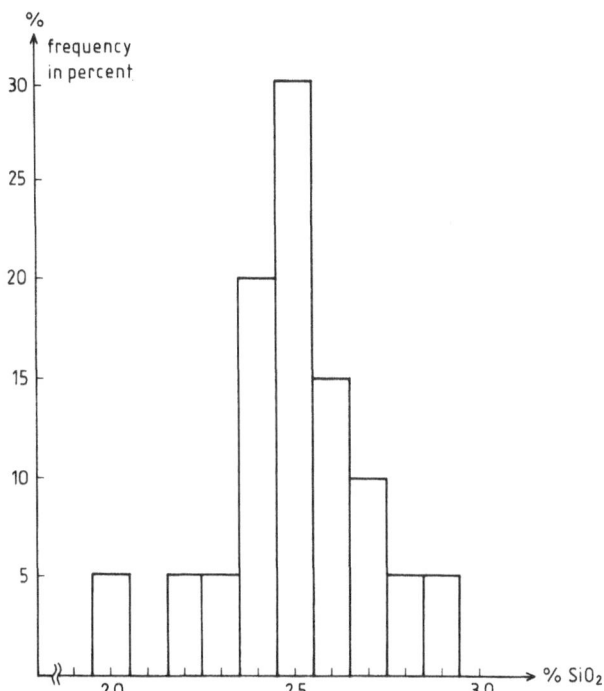

The range between the lowest and highest values (2.4% and 35.8% CaF_2 in the data shown in Table 4) is calculated and is subdivided into equal class intervals. If a fewer number of class intervals are selected, then the overview of the analytical

Table 4. Analyses sheet from a fluorspar deposit

Sample no.	CaF_2 (%)	Sample no.	CaF_2 (%)
1	34.8	16	2.4
2	16.7	17	19.8
3	18.9	18	21.5
4	5.2	19	22.6
5	28.5	20	28.5
6	11.3	21	26.5
7	7.5	22	26.9
8	18.4	23	14.4
9	21.2	24	17.1
10	23.4	25	33.5
11	35.8	26	21.0
12	17.9	27	28.8
13	12.8	28	30.7
14	13.8	29	35.4
15	15.4	30	3.8

values will be that much better, but this also results in a greater loss of information that can be derived from the values in the basic data set. It is important to understand that different class intervals from the same data set can result in different empirical frequency distributions of that data set.

In practice, two different situations can be distinguished:

a) As shown in the last example, the average grade and the grade distribution are of particular interest in the evaluation of mineral deposits and occurrences. The main emphasis is therefore focussed on the central section of the frequency distribution. In practice, about ten class intervals are usually used, and more class intervals are used only in those cases where there is a very extensive data set or if some values fall well outside the range containing most of the analytical values.

The Sturges Rule can be applied as a guide to the size of the class intervals (see below).

b) Exploration data sets acquired from geochemical surveys are usually much more comprehensive than those derived from mineral deposits or occurrences. Based on the assumption that the anomalous data, or anomalies might provide information about the presence of mineral deposits, the evaluation of geochemical data essentially involves the separation of anomalous from normal populations. The central section of the frequency distribution of the analytical values is therefore not as important as the "tail" of the higher values. Chapter 18 describes how these anomalous values are determined. Since in many cases it is necessary to identify and separate two or even more populations (one normal and one or more anomalous), even more detailed information is required as compared to case a) above. The values are therefore subdivided into between 12 and 20 class intervals. A rule of thumb in exploration statistics is that the number of class intervals should not be more than $\sqrt{n}$, where n is the number of individual values.

The example of the analytical values from a fluorite deposit (Table 4) corresponds to case (a) above, and the Sturges Rule can be applied as a guide for the range of each class interval. Sturges Rule is:

$$\text{class interval}\,(\Delta) = \frac{\text{range R}}{1+3.322\log n} \,,$$

where the range is again the difference between the highest and lowest analytical value and n is the number of values.

As an example, the values from Table 4 therefore yield:

$$\text{class interval}\,(\Delta) = \frac{35.8-2.4}{1+3.322\log 30} = \frac{33.4}{5.91} = 5.65 \;.$$

Table 5. Ranking list for the fluorspar values of Table 4

Class %	Ranking list	Absolute frequency	Relative frequency (%)
0.1 - 5 CaF$_2$	II	2	6.7
5.1 - 10 CaF$_2$	II	2	6.7
10.1 - 15 CaF$_2$	JHT	4	13.3
15.1 - 20 CaF$_2$	JHT II	7	23.3
20.1 - 25 CaF$_2$	JHT	5	16.7
25.1 - 30 CaF$_2$	JHT	5	16.7
30.1 - 35 CaF$_2$	III	3	10.0
35.1 - 40 CaF$_2$	II	2	6.7
		$\sum = 30$	$\sum = 100.1\,\%$

Whole numbers, or at least numerically simple limits to the class intervals, are normally used in order to simplify the calculations. In this case the limits to the class intervals are defined by 5, 10, 15 etc., so that there is a total of 8 class intervals[1]. A ranking list is compiled by allocating each value to its respective class interval as described in Sect. 2.1 (Table 5).

All values within a class interval together form an *interval of values,* and their total number reflects the absolute frequency of that class interval, or the absolute *class interval frequency* of the sample values. The relative class interval frequency can be calculated relative to the whole of the data set. The absolute class interval frequency is given in column 3 of Table 5, and the relative class interval frequency in column 4.

The results in the class interval frequency tables can be graphically shown by plotting the class intervals for the index values on the x-axis, in this case the class intervals of the CaF$_2$ values, and the absolute or relative frequency on the y-axis. The absolute frequency, n, or the relative frequency in percent for each class interval is displayed as a column (Fig. 7), and the result is a histogram.

The centre of each class interval is known as the *class mean* (Fig. 7). In any further treatment of the data set, it is assumed that all values within a class interval plot on the respective class mean value, and thus the individual values of the data set are no longer taken into consideration. A continuous and smooth curve could be drawn now through the histogram in Fig. 7. If the data set was sufficiently comprehensive, it could be assumed that the frequency distribution curve for the samples in the whole population corresponds to that for these specific samples in the universe (see Chap. 1.1 and 1.2).

---

[1]Perillo and Marone (1986) conclude that the optimum number of class intervals is 8 or 19.

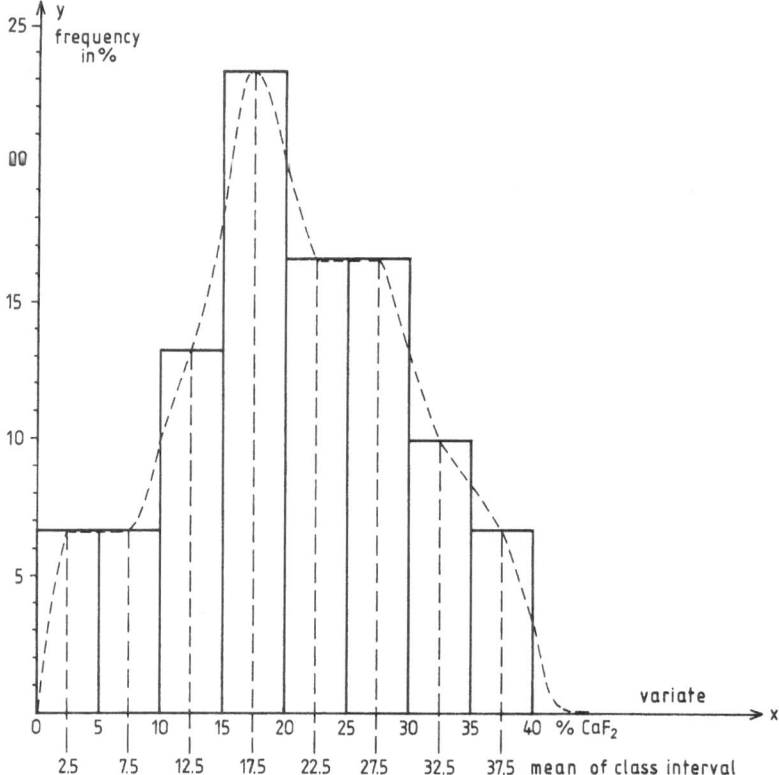

Fig. 7. Histogram for the analyses of the fluorspar deposit of Tables 4 and 5

There are occasionally examples of histograms with class intervals of varying width, and in such cases it is important to determine if the column areas (class interval x column height) remain the same. If the histogram is considered to be the preliminary stage to developing a continuous frequency distribution (as above), then the total area under the curve (or the sum of the areas of all the columns) reflects the total frequency. Hence the total area should not be changed by arbitrarily varying the class intervals. For example, the SiO_2 contents of a fluorite deposit are shown in Fig. 8. Isolated "maxima" occur in the higher values and are interpreted to be due to random variations. These "maxima" can be smoothed out by expanding the class intervals.

2.3
Frequency Distribution of Samples with Dissimilar Specifications

Samples with comparable specifications (i.e. the same sample support) were the subject of the examples in Sections 2.1 and 2.2, 1-kg chip samples in the first example and 1-m lengths of drill core in the second. However, core is often split into sections for analysis according to the geological units, and thus the core lengths are variable in length. Hence the individual analytical values have a

Table 6. Cu analyses from a Kuroko-type Cu deposit

Sample no.	Cu (%)	Length (m)	Sample no.	Cu (%)	Length (m)
1	2.8	0.8	16	6.9	1.0
2	2.5	0.3	17	5.8	1.0
3	4.6	1.0	18	8.8	0.7
4	6.8	0.5	19	7.3	0.4
5	5.6	0.5	20	10.4	0.5
6	7.2	0.9	21	15.2	0.4
7	6.8	0.9	22	8.3	0.8
8	5.5	0.4	23	3.1	1.1
9	9.5	1.0	24	2.0	1.0
10	14.9	1.3	25	1.3	1.0
11	20.1	0.5	26	1.5	0.9
12	11.2	0.3	27	3.8	1.2
13	16.5	0.7	28	1.1	1.0
14	4.0	0.4	29	1.4	1.0
15	7.8	0.6	30	0.8	1.0
					$\sum = 23.1$ m

dissimilar sample support (see Chap. 1.1). An example of this is given in Table 6, which lists analytical data for copper from a Kuroko-type volcanogenic Cu-Pb-Zn deposit.

These values cannot be simply evaluated by using a ranking list such as that described in Sections 2.1 and 2.2. It would be wrong, for example, if sample 2 with a 0.3-m length was treated the same as sample 9 with a 1-m length. The samples must therefore be weighted according to their length.

Table 7. Ranking list for the Cu analyses of Table 6

Class % Cu	Lengths of drill core in each class (m)	Sum of lengths (m)	Relative frequency (%)
0.1 - 3	0.8; 0.3; 1.0; 1.0; 0.9; 1.0; 1.0; 1.0	7.0	30.3
3.1 - 6	1.0; 0.5; 0.4; 0.4; 1.0; 1.1; 1.2	5.6	24.2
6.1 - 9	0.5; 0.9; 0.9; 0.6; 1.0; 0.7; 0.4; 0.8	5.8	25.1
9.1 - 12	1.0; 0.3; 0.5	1.8	7.8
12.1 - 15	1.3	1.3	5.6
15.1 - 18	0.7; 0.4	1.1	4.8
18.1 - 21	0.5	0.5	2.2
		$\sum = 23.1$ m	$\sum = 100.0$ %

As described in Section 2.2, the Sturges Rule is used as a guide to the width of the class intervals for the analytical values, and the samples are divided into 7 class intervals with a width of 3.

The absolute and relative frequency distribution is determined on the basis of the core lengths. The sum of the core lengths is 23.1 m, and from this the relative frequency for each class interval can be calculated (columns 3 and 4 in Table 7). The histogram is constructed as described in Section 2.2, and it can be drawn as in Fig. 7.

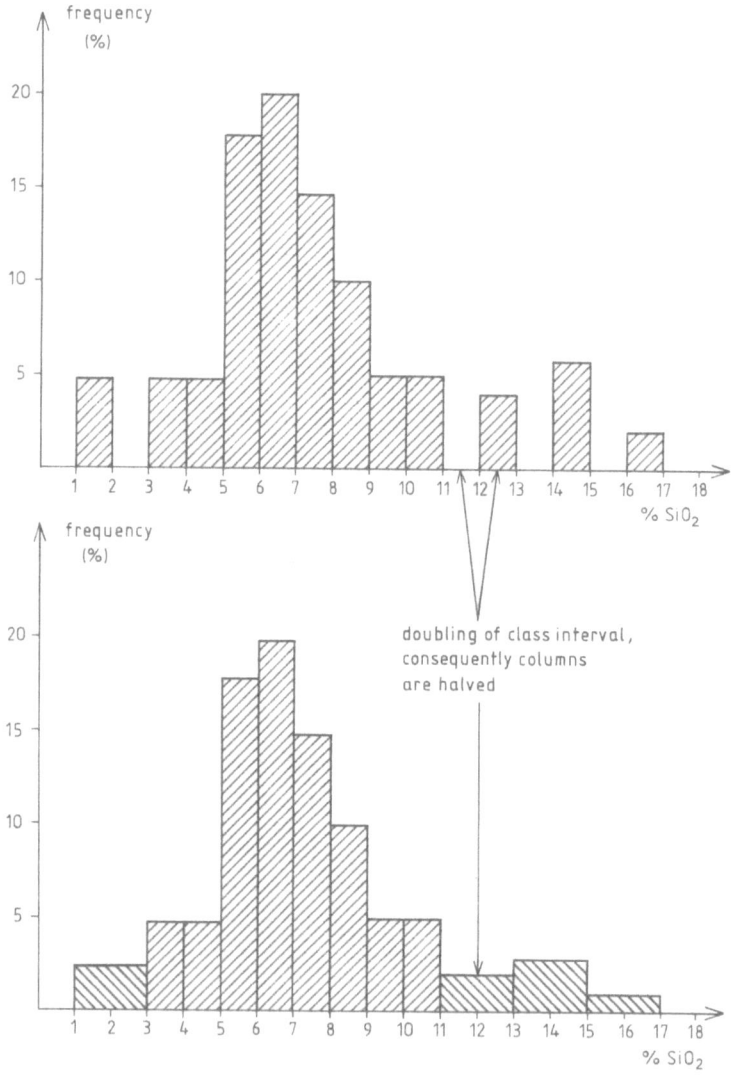

Fig. 8. Histogram for the SiO$_2$ values of a fluorspar deposit. In the lower histogram, the class intervals are doubled at the tail ends

3
Mean, Variance and Standard Deviation

Frequency distributions that describe an approximately symmetrical bell-shaped curve (Fig. 9) with a single peak can be described by two parameters: the *mean*, or more precisely the arithmetic mean, which is the point on the x-axis intersected by the axis of symmetry of the "bell curve"; and the so-called *standard deviation*, which is a measure for the width of the "bell".

3.1
The Mean

3.1.1
The Mean of Equally Weighted Values

The mean, $\bar{x}$, is defined as the arithmetic mean:

$$\bar{x} = \frac{x_1 + x_2 + x_3 + \ldots + x_n}{n},$$

whereby x_1, x_2 etc. are equivalent individual samples or measurements (i.e. they have the same sample support) and n is the total number of samples or measurements. If a sum symbol is used to denote the summation of the individual values, then the formula can be shortened to:

$$\bar{x} = \frac{1}{n} \sum_{i=1}^{n} x_i .$$

Fig.9. A single peak and symmetric frequency distribution with mean $\bar{x}$ and standard deviation s

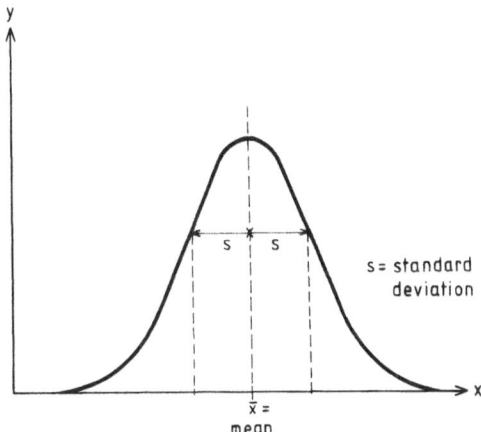

Table 8. The first five Cu
analyses of Table 6

Sample no.	Cu (%)	Length (m)
1	2.8	0.8
2	2.5	0.3
3	4.6	1.0
4	6.8	0.5
5	5.6	0.5

Example: The following five values derive from density measurements: 3.2, 3.3, 3.4, 3.5, 3.6 (g/cm³). The mean is then

$$\overline{x} = \frac{3.2+3.3+3.4+3.5+3.6}{5} = 3.4(g/cm^3).$$

3.1.2
The Mean of Unequally Weighted Values

If the samples are not equivalent, and therefore each of the samples has a different support, then they must be weighted. One method has already been described in Economic Evaluations in Exploration (Chap. 4.1 under weighting).

If x_i is again the individual sample value, and ai is the weighting factor, for example core lengths of the analysed sections, then the mean is:

$$\overline{x} = \frac{x_1 \cdot a_1 + x_2 \cdot a_2 + x_3 \cdot a_3 + ... + x_n \cdot a_n}{a_1 + a_2 + ... + a_n},$$

or shortened to:

$$\overline{x} = \frac{\sum_{i=1}^{n}(x_i \cdot a_i)}{\sum_{i=1}^{n} a_i}.$$

Example: As an example, the first five values are selected from Table 6 in Chapter 2.3:
The mean of these five values is then:

$$\overline{x} = \frac{2.8 \cdot 0.8 + 2.5 \cdot 0.3 + 4.6 \cdot 1.0 + 6.8 \cdot 0.5 + 5.6 \cdot 0.5}{0.8 + 0.3 + 1.0 + 0.5 + 0.5}.$$

$$\overline{x} = \frac{13.79}{3.1} = 4.45\%Cu.$$

3.1.3
The Mean of Data Within Class Intervals

If the individual values are already grouped into categories or there are numerous non-equivalent individual values, then the mean can be best determined with sufficient accuracy by using class intervals. In the case of numerous non-equivalent individual samples, then the weighting calculations can be simplified.

The mean is then:

$$\overline{x} = \frac{\sum\limits_{i=1}^{n} (x_i \cdot f_i)}{\sum\limits_{i=1}^{n} f_i} .$$

x_i is again an individual value, which in this case is the mean of the class interval, and f_i is the frequency of the respective class interval. For this calculation it is assumed that the deviation of the values from the mean of the class interval within each class interval will be more or less balanced. For this example, the data in Table 7 in Chapter 2.3 are repeated in Table 9.

The mean value is then:

$$\overline{x} = \frac{1.5 \cdot 7.0 + 4.5 \cdot 5.6 + 7.5 \cdot 5.8 + ... + 19.5 \cdot 0.5}{7.0 + 5.6 + 5.8 + 1.8 + 1.3 + 1.1 + 0.5} ,$$

$$\overline{x} = \frac{143.55}{23.1} = 6.21\% \text{ Cu} .$$

Table 9. Summary of Table 7

Class % Cu	Class mean x_i % Cu	Sum of lengths = length frequency f_i (m)
0.1 - 3	1.5	7.0
3.1 - 6	4.5	5.6
6.1 - 9	7.5	5.8
9.1 - 12	10.5	1.8
12.1 - 15	13.5	1.3
15.1 - 18	16.5	1.1
18.1 - 21	19.5	0.5
		$\sum$ = 23.1 m

The mean calculated from the individually weighted values ($\bar{x} = 6.25$) is less than 1% different as compared to the above value, and this difference is usually of no practical importance.

3.2 Variance and Standard Deviation of Sample Size and Population

3.2.1
Calculation for Equivalent Samples

It is clear from the bell-shaped curve of a symmetric frequency distribution as shown in Fig. 9 that different broad and narrow bell-shaped curves can have the same mean value. Consider a numerical example: four values 1, 2, 3 and 4 have a mean of 2.5; but the four values 2.3, 2.4, 2.6 and 2.7 also have the same mean. However, both groups of values are very different, since the values of the second group are much closer to the mean as compared to those of the first group. A value that provides a measure of the average difference of the individual values from the mean is obviously required to define this variation in the distribution numerically. This difference from the mean, however, is either a positive or a negative figure, depending on whether the individual value is less than or greater than the mean, and a mathematical solution must be selected that eliminates the positive and negative prefix. One method of achieving this is by squaring the deviation, summing the obtained values, and then determining the average. The result of this calculation is the variance, which is an important concept that will be encountered frequently later in this book. The positive square root of the variance is the standard deviation, s.

The variance of the sample size is denoted by s^2 and is calculated by the formula[2]:

$$s^2 = \frac{1}{n-1} \cdot \sum_{i=1}^{n}(x_i - \bar{x})^2 \text{ für } n > 1. \tag{1}$$

In the above example of the samples $x_i = 1$, 2, 3 and 4, the variance would be calculated as:

$$s^2 = \frac{1}{4-1} \cdot \left\{ (1-2.5)^2 + (2-2.5)^2 + (3-2.5)^2 + (4-2.5)^2 \right\}$$

$$s^2 = \frac{1}{3} \cdot (2.25 + 0.25 + 0.25 + 2.25)$$

$$s^2 = \frac{5}{3} = 1.67.$$

---

[2] Why is the expression (n-1) rather than the number of samples n used in this equation? The number of degrees of freedom is shown in the denominator. With a large number of samples the difference between n and n-1 in the denominator is scarcely significant. The interested reader is referred to Weber (1956) or Koch and Link (1970, p. 73) for further discussion of degrees of freedom.

The standard deviation s is therefore

$$s = \sqrt{1.67} = \pm 1.29 \ .$$

For the second example, in which the samples are $x_i = 2.3$, 2.4, 2.6, 2.7 with the same mean of $\bar{x} = 2.5$, the variance is calculated as:

$$s^2 = \frac{1}{4-1}\left[(2.3-2.5)^2 + (2.4-2.5)^2 + (2.6-2.5)^2 + (2.7-2.5)^2\right] \cdot$$

$$s^2 = \frac{1}{3} \cdot (0.004 + 0.001 + 0.001 + 0.004)$$

$$s^2 = \frac{1}{3} \cdot 0.01 = 0.0033 \ .$$

The standard deviation s is therefore

$$s = \sqrt{0.0033} = \pm 0.06 \ .$$

For the second example, in which the values are closer to the mean $\bar{x}$, both the variance and the standard deviation are significantly smaller values.

Formula (1) above has a specific disadvantage in that the differences $(x_i - \bar{x})$ have to be calculated and this involves repetitive calculation to several decimals. A more convenient formula for calculators is:

$$s^2 = \frac{1}{n-1}\left[\left(\sum_{i=1}^{n} x_i^2\right) - n\bar{x}^2\right] \text{ Fußnote 3} \ . \tag{2}$$

---

[3] This formula results from the following computation:

$$\sum_{i=1}^{n}(x_i - \bar{x})^2 = \sum_{i=1}^{n} x_i^2 - \sum_{i=1}^{n} 2x_i\bar{x} + \sum_{i=1}^{n} \bar{x}^2$$

$$\sum_{i=1}^{n} \bar{x}^2 = n \cdot \bar{x}^2$$

$$\sum_{i=1}^{n} 2x_i\bar{x} = 2\bar{x}\underbrace{\left(x_1 + x_2 ... x_n\right)}_{n \cdot x} \ .$$

Therefore the result is:

$$\sum_{i=1}^{n} x_i^2 - 2\bar{x}^2 \cdot n + n\bar{x}^2 = \sum_{i=1}^{n} x_i^2 - n\bar{x}^2 \ .$$

In the first calculated example for the samples x_i=1, 2, 3 and 4 and a mean of $\bar{x}$ =2.5, the calculation procedure would be as follows:

$$s^2 = \frac{1}{4-1}\left[\left(1^2 + 2^2 + 3^2 + 4^2\right) - 4 \cdot 2.5^2\right]$$

$$s^2 = \frac{1}{3}(30-25) = \frac{5}{3}, \text{ as above}$$

$$s^2 = 1.67$$

$$s = \pm 1.29 .$$

The statistical parameters of the sample size, such as variance and standard deviation, are not necessarily identical to those of the population. The population has a single, properly defined, probability distribution. The samples (random samples) that can be selected from this population are usually different from each other and, consequently, they will therefore have different statistical parameters. The statistical parameters will only approach being the same as those of the population if the sample size is sufficiently large and representative. The computational procedures obviously do not change but, since they are of different dimensions, the variance and the standard deviation of the population are denoted by σ^2 and σ respectively. For the geological sample size considered here or for a random sample, the notation is s^2 and s respectively. The same applies to the mean; the mean of the population is denoted by μ, and that of a sample by $\bar{x}$.

It is very improbable that one will ever know the true μ or σ^2 for a geological population, and the known values $\bar{x}$ and s^2 can be used as *estimators* for μ and σ^2. These estimators yield more or less satisfactory results for symmetrical distributions[4].

---

[4] Some authors (e.g. Moroney 1970) work with the so-called Bessel correction in order to estimate σ^2 from s^2 .

$$\hat{\sigma}^2 = \left(\frac{n}{n-1}\right) \cdot s^2$$

($\hat{\sigma}^2$ means that this is the best estimate for σ^2)

This Bessel correction is used in Chapter 11.2.2 in Eq. (1) for the statistical comparison of two analytical sequences. It is clear that the quotient $\left(\frac{n}{n-1}\right)$ approaches 1 as the number of values, n, increases, so that s^2 itself becomes a better estimator for σ^2 as n increases.

3.2.2
Determination of the Variance and Standard Deviation for Non-Equivalent or Categorised Values

3.2.2.1
Graphical Determination of the Variance and Standard Deviation

In the case of non-equivalent values, it is easiest to determine the standard deviation and variance graphically from the cumulative frequency curve and probability grid, and this will be discussed later in Chapter 5.1.3.

3.2.2.2
Calculation of Variance and Standard Deviation of Non-Equivalent or Categorised Values

The calculation of the variance of non-equivalent values is best achieved by treating the data in class intervals (Chap. 2.3).

The variance is then calculated from the following formula by modification of Eq. (2) in Chap. 3.2.1:

$$s^2 = \frac{\sum_{i=1}^{n} f_i \cdot x_i^2 - (\sum_{i=1}^{n} f_i) \bar{x}^2}{\sum_{i=1}^{n} f_i - b}. \tag{1}$$

x_i are the individual values, which in this case are the class interval means, and f_i are the frequencies. b is the average of all the weighting factors[5].

> *Example:* The figures for the copper values from a Kuroko-type deposit in Tables 6 and 7, Chapter 2.3, are used again for this example. These values were previously summarised in Table 9, Section 3.1.3, for the determination of the mean from data in class intervals. The table is repeated again.
>
> The value b, which is the mean of all the lengths, is derived from Table 7, Chapter 2.3, as b=0.77. The mean $\bar{x}$ was calculated as $\bar{x}$ =6.21 (see Sect. 3.1.3). Hence for the variance:
>
> $$s^2 = \frac{7 \cdot 1.5^2 + 5.6 \cdot 4.5^2 + \ldots + 0.5 \cdot 19.5^2 - 23.1 \cdot 6.21^2}{23.1 - 0.77}$$
>
> $$s^2 = \frac{1380.38 - 810.33}{22.33} = \frac{489.54}{22.33} = 21.92 \; .$$

---

[5] The number of degrees of freedom is in the denominator (see footnote 2) of the formulae (1) and (2) in Chap. 3.2.1 (n-1). Since the individual values are non-equivalent and the frequencies are different, and can be non-integer numbers, the denominator tends to (f_i-b). If the sample support values are of the size of 1 (for example for equivalent samples for which only the total number is considered) then the denominator will be (f_i-1).

This result of the calculation of the variance is actually somewhat too high because of the categorising of the data into class intervals. In a continuous distribution, the class interval mean does not exactly correspond to the centre of gravity of the segment. The over-estimation can therefore be corrected by means of the Sheppard correction:

$$Sh = \frac{\Delta^2}{12} \quad \text{Sheppard correction,} \tag{2}$$

in which Δ is the width of the class interval, in this case 3. The Sheppard correction is therefore:

$$Sh = \frac{3^2}{12} = 0.75 ,$$

and this figure is subtracted from the figure for the variance that was calculated above.

The corrected variance is therefore:

$$s_*^2 = s^2 - Sh = 21.92 - 0.75 = 21.17 .$$

The standard deviation s_*^2 is therefore:

$$s_* = \sqrt{21.17} = \pm 4.60 .$$

The Sheppard correction is commonly not carried out, and for geological problems the magnitude of the correction factor is often irrelevant. The standard deviation without applying the Sheppard-correction factor is:

$$s^2 = \sqrt{21.92} = \pm 4.68 .$$

The difference is therefore less than 2%.

3.3
Coefficient of Variation

The standard deviation itself is commonly of less interest than its relation to the arithmetic mean, $\bar{x}$.

The coefficient of variation C is defined therefore as the quotient $\dfrac{\sigma}{\mu}$ and is estimated by

$$\frac{s}{\bar{x}} .$$

This can be considered as a relative standard deviation. For the two numerical examples in Section 3.2, C would be

$$C_1 = \frac{1.29}{2.5} = 0.52 \quad \text{and} \quad C_2 = \frac{0.06}{2.5} = 0.02 .$$

The coefficient of variation is sometimes also expressed in percent. This value would be 52% for the case of C_1 .

The coefficient of variation is a parameter that, in the early stages of an evaluation programme, is very suitable for providing a quick indication of the

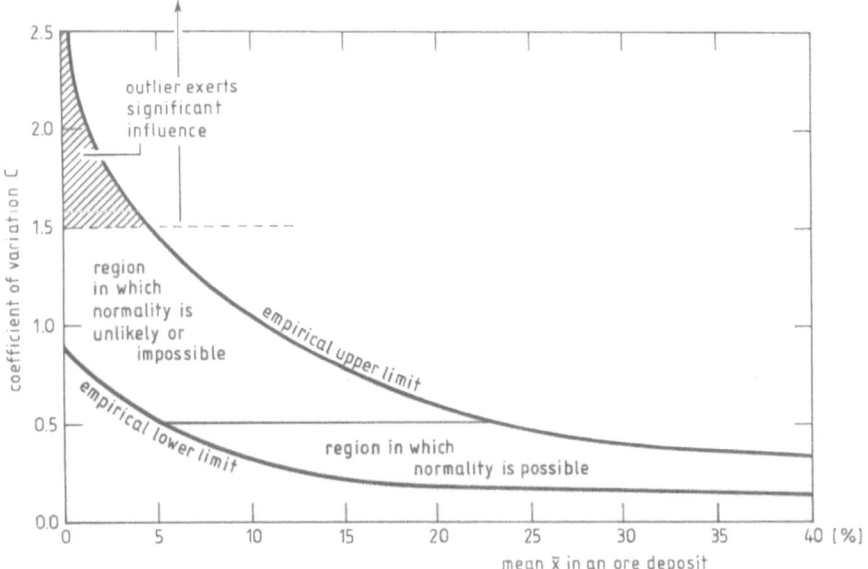

Fig. 10 Relationship between the arithmetic mean $\bar{x}$ and the coefficient of variation C. (After Koch and Link 1970; Champigny and Armstrong 1989) (with permission of the authors Koch and Link)

variability of the sample data and the grades of the proposed exploitation, for example by comparing the coefficient of variation with known values derived from other deposits. Information from other deposits of the same type can also help as a-priori information in the first order of magnitude estimation of statistical parameters. The coefficients of variation calculated from the drill hole data of various deposits are listed in Appendix Table 1.

The sample support is obviously also an important factor. Hazen (1967) derived a sample volume-variance relationship, in which the product of volume x variance should remain constant or, if the variances s_1^2 and s_2^2 from two volumes v_1 and v_2 are compared, then the following should apply:

$$s_1^2 \cdot v_1 = s_2^2 \cdot v_2 .$$

However, Koch and Link (1970) have shown from various drill hole examples that with increasing volume the variances (and therefore obviously the standard deviations[6]) decrease by a much smaller degree than is theoretically predicted[6]. The coefficients of variation from drill data on various deposits as listed in the Appendix Table 1, provide a useful initial estimate for the drill hole diameters (BQ to HQ, see Economic Evaluations in Exploration, Appendix C 2) that are usually applied today during the systematic investigation of mineral deposits.

It is commonly found that, for certain parts of a deposit, the coefficients of variation fluctuate only within a particular range. However, in the high-grade zones the variability and therefore the coefficients of variation also often increase. Isolated high-grade values, or so-called outliers, represent a special problem and will be discussed in Chapters 8 and 9.3.6. According to Champigny and Armstrong (1989), isolated high-grade values have a marked effect if the coefficient of variation exceeds 1.5. Parker (1991) uses the coefficients of variation for separating the outliers from the rest of a log- normal population. This method will be discussed in Chapter 9.3.6.

In deposits with very high grades, the coefficient of variation must decrease because grades cannot be higher than 100%. For example, C of a barite deposit with values of between 85 and 95% and a mean $\bar{x}$ of 90% can only be a maximum of:

$$C \leq \frac{0.05\sqrt{2}}{0.9} \leq 0.08 .$$

---

[6] This discrepancy can be explained geostatistically by the concept of extension variance (Royle, 1988), whereby the variance of a volume v_1 is extended to a volume v_2. This concept is discussed further in Chapter 13.3.4.3.1.

The volume-variance relationships in geostatistics have another meaning. Krige (1966) provided an empirical volume-variance relationship, which is also known as the Krige relationship. It states the following:

if $s^2_{(o/D)}$ is the variance of sample (o) in the deposit (D)

$s^2_{(o/V)}$ is the variance of sample (o) in the mining block (V) and

$s^2_{(V/D)}$ is the variance of the mining block (V) in the deposit (D),

then $s^2_{(o/D)} = s^2_{(o/V)} + s^2_{(V/D)}$

Also considering high grade zones within a deposit, the coefficient of variation must decrease for such zones, because for a data subset, the coefficient of variation is always lower than the coefficient of variation for the entire data set.

In Chapter 4, the so-called normal distribution, which is of fundamental statistical importance, is discussed.

The possibilities for testing the existence of a normal distribution are described in Chapter 5. The coefficients of variation provide a quick estimate of whether this is a reasonable possibility and, as an illustration of this, Koch and Link (1970) have developed a diagram that is reproduced in Fig. 10.

3.4
Other Parameters (Median, Mode Value)

Two other statistical parameters are introduced here, namely the *mode value*, h, (or most common value) and the *median value*, m. The median (or central value) is the midpoint value that divides a sample size or population into two equal halves, whereby one half has values less than the median and the other has values greater than the median.

On probability grids, which are based on cumulative frequencies and are described in more detail later in Chapters 5.1.2 and 9.2.2, the median can be directly read from the 50% cumulative frequency intersept (Appendices A1, A2, B1 and B2).

The median value is very important in geochemical exploration for determining the background and threshold values, and for identifying anomalous geochemical values (see Chap. 18.2.1.1).

The arithmetic mean, μ, the mode value, h, and the median, m, are obviously all the same for a single-peaked, symmetrical distribution such as the normal distribution, but they are not the same for a skewed distribution (Fig. 11).

For those cases where the distribution is skewed to the right (positively skewed), such as commonly occurs for analytical values, the arithmetic mean μ is greater than the median m, which in turn is greater than the mode value h. Figure 11 clearly shows that these three values become even more separated from each other as the skewness of the distribution increases. The arithmetic mean μ marks the point on the x-axis that is the centre of gravity for the area beneath the distribution curve.

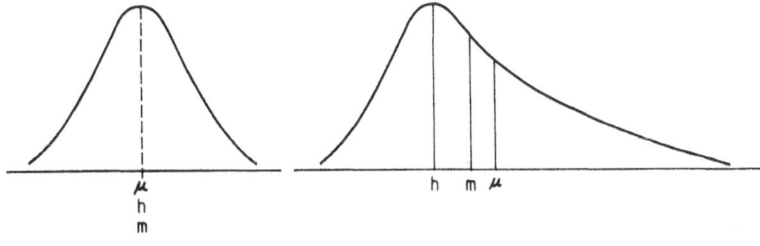

Fig. 11. Position of the arithmetic mean μ, the mode, h, and the median, m, for a single-peak symmetric distribution and a distribution skewed to the right

4
The Normal Distribution

A symmetrical bell-shaped frequency distribution is often described as a normal distribution or a Gaussian distribution. It is known as the Gaussian distribution since it was derived by C. F. Gauss and included in his work about the theory of measurement of errors. The simple assumption of a normal distribution occurs only rarely for geological data. For example, if the density of an ore sample is measured ten times, then the results are usually distributed normally around the mean, presuming that there is no systematic error. On the other hand, there is no reason why the sample values from a mineral deposit should have a normal distribution (or any other simple, mathematically definable, distribution). It can, of course, be argued that the deposit is formed by definable physical-chemical and chemical processes and that therefore the grade distribution must also be mathematically definable. However, there are so many parameters involved, all of which can vary from location to location, that it is just not possible to postulate a general conclusion about the frequency distributions of mineral deposits.

In spite of the above comments, the normal distribution is fundamentally important for the evaluation and treatment of geological data. The use of this distribution has a long history; its specific properties make it convenient to use as will be shown below, and frequently there are no other suitable distributions available. It is often possible to describe real distributions with the necessary precision by the parameters of a normal distribution, or to transform them into an approximately normal distribution, for example by the logarithmic transformation (see Chap. 9.2). The *central limit theorem* is, however, of critical importance for the treatment of geological data, and it states:

If independent random samples of the same size, n, are repeatedly collected from a population, which can have any distribution, meaning different sample series but with the same sample size n, then the distribution of the mean of the samples for a sample size n is approximately normal. The approximation to the normal distribution increases with an increase in the sample size n (see, for example, Kreyszig, 1968).

This theorem is critically important for the determination of confidence limits in calculations of the mean (see Chap. 7.1).

The central limit theorem does not define the size of the sample that is required for the distribution of the mean of the samples to be approximately normal. Kreyszig (1968) mentions a figure of about 30 for those distributions in which the sampling indicates that the population is not unusually asymmetrical. Koch and

Fig.12. Relationship between the standard deviation and the area segments below a normal distribution

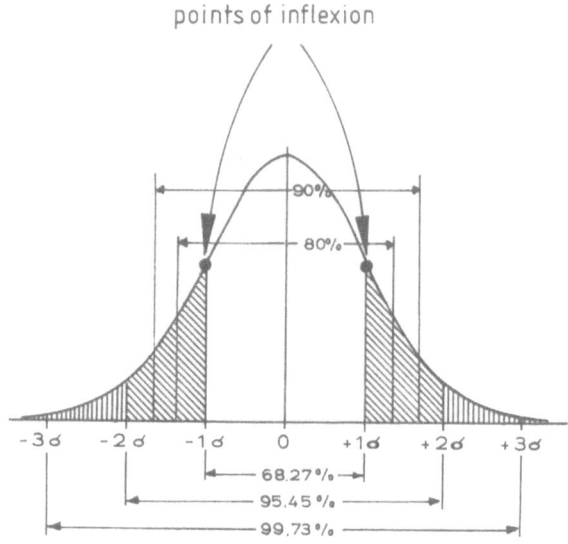

points of inflexion

Link (1970) quote the figure of 50 to 100 for most geological distributions, except for analyses for gold and various trace elements.

If the randomly distributed population has a mean μ and the standard deviation σ, then the approximately normal distribution of the means of the repeated sample series has the mean μ and the standard deviation $\dfrac{\sigma}{\sqrt{n}}$. This is described in more detail in Chapter 6.1.

Mathematically, the normal distribution is exactly defined[7]. If the curve is symmetrical about the zero point, and thus the mean = 0, then the points of inflexion are located at $\pm\sigma$, which is at an interval of one standard deviation (Fig. 12). The normal distribution is defined unequivocally by the standard deviation or the variance respectively (see the equation in footnote 7). The percent proportion of values that differ from the mean by less than or greater than σ (or multiples of σ, i.e. the size of the area under the normal distribution curve between limits defined by fractions or multiples of σ) is important for subsequent discussions (see Chap. 7). For the normal distribution (see Fig. 12):

---

[7]The normal distribution is defined by the equation (e.g. Kreyszig, 1968):

$$y = \frac{1}{\sigma\sqrt{2\pi}} \cdot e^{-\frac{1}{2}\left(\frac{x-\mu}{\sigma}\right)^2},$$

This equation will not be referred to again in this book. The important feature of the equation is the integral that describes the area under the normal distribution curve.

68.27% of the total area under the normal curve (about 2/3 of all values) lies between +σ and -σ (i.e. between the points of inflexion);
95.45% of the total area under the normal curve (excluding about 5% of all values) lie between +2σ and -2σ;
99.73% of the total area under the normal curve lie between +3σ and -3σ.

It can therefore be stated that the size of the area under the normal distribution curve between the defined limits is related to the probability with which the value of a random variable, the variate x, is located between the defined limits.

There are tables that list the area under the standardized normal distribution curve, or the definite integral value, between the defined limits. The cumulative frequency or distribution function Φ(x) (see Appendix Table 2) is used, and will be discussed in more detail later (see Chap. 5.1.2).

The standardized normal distribution has the following characteristics (see Fig. 12):

a) The total area under the normal distribution curve from - ∞ to +∞ is 1. Therefore the values for fractions of this area need only to be multiplied by 100 in order to derive a percentage figure for the number of values that lie between the upper and lower limits of an area fraction.

b) The mean value is 0 (i.e. the y axis is the axis of symmetry) and therefore 50% of the area is located between - ∞ and 0 (i.e. Φ(0)=0.5) (see Appendix Table 2).

c) The standard deviation is 1, and therefore the points of inflexion of the standardized normal distribution are located at ±1.

In the cumulative frequency table in Appendix Table 2, the area beneath the standardized normal distribution curve is increased in stages from the "left-hand tail" (i.e. increasing from - ∞), so that at 0 the value is 0.5 and at +∞ the value is 1.

1st Assignment: By using the cumulative frequency tables in the Appendix Table 2, demonstrate that 68.27% of all values of the normal distribution lie between +σ and -σ (see above).

Solution:

1. As explained above, +σ and -σ of a standardized normal distribution are located at +1 and -1.

2. From Appendix Table 2, for x=-1 the Φ(x) value is 0.1587, therefore 15.87% of the values lie outside (to the left of) the area from - ∞ to -1.

3. For the value x=+1, Φ(x)=0.8413. Since the standardized normal distribution is symmetrical about the y axis, this value must be 1-0.1587=0.8413.

4. In other words: 84.13% of all values lie between the limits - ∞ to +1.In order to determine how many values are located between the limits -1 and +1, subtract 15.87% from 84.13% [see step (2)], so that there remains 68.26%. [The difference of 0.01% (see above) is a rounding error.]

A further assignment will explain how to standardize the normal distribution.

2nd Assignment: In a quarry, relatively homogeneous and isotropic granite is crushed to a nominal diameter of 40 mm. The size of the crushed material is distributed normally. The standard deviation is 10 mm. The above-size fraction of 60 mm and over is sieved off. What is the proportion of the over-size fraction?

Solution: In the problem it is stated that the crushed material has a normal distribution. However, the table reproduced in Appendix Table 2 refers to the standardized normal distribution with a mean $\mu=0$, standard deviation $\sigma=1$ and the area under the normal distribution curve $\Phi=1$.

In this problem, however, the mean μ is 40 mm and the standard deviation $\sigma=10$ mm. The distribution must therefore be standardized for the crushed material. This is undertaken in two steps:

a) The distribution of the crushed material is subjected to a parallel transformation so that the new mean value is zero. To achieve this, the mean value μ must be subtracted from each fragment diameter measurement x_i to give the difference $x_i-\mu$.
b) The distribution of the crushed material must be "condensed" so that the standard deviation $= 1$. To achieve this, the difference calculated in (a) must be divided by the standard deviation of the crushed material $s=10$ mm.

The values of the standardized normal distribution must be found in order to use Appendix Table 2, and therefore the diameter of the fragments x_i must be recalculated into the standardized variate values of the table 2:

$$x_\mu = \frac{x_i \cdot \mu}{\sigma},$$

Since only the relative frequencies are of importance in this problem, there is no need to standardize the size of the area to 1. The whole standardization procedure, including the standardization of the area, is described once again in Chapter 5.2.

1st Step: With a mean of $\mu=40$ mm and a standard deviation σ of 10 mm, the value $x_i=60$ mm the standardized variate value, is

$$X_\mu = \frac{x_i - \mu}{\sigma} = \frac{60-40}{10} = 2,$$

which is twice the standard deviation.

2nd Step: For the value $x_\mu=+2$ in Table II in the Appendix, $\Phi(x) = 0.9772$, so that 97.72% of all values are smaller than $x_u=2$, which is 60 cm. In other words, the probability that a fragment has a diameter of greater than 60 mm is 100-97.72=2.28%. In terms of the problem, this means that the over-size fraction represents a proportion of 2.28% of the total material.

5
Testing the Normal Distribution Hypothesis

The exact mathematical check of whether the distribution is truly normal is accomplished by means of mathematical tests. Examples are the χ^2 test (or Chi-square test, see Sect. 5.2) or the Kolmogoroff-Smirnov test, although the latter will not be discussed in this book. However, an approximate graphical test is commonly used in practice. An even simpler estimate, which is described in Chapter 18.2.1.1.2, is sometimes used to test geochemical exploration data.

5.1
Graphical Test

The graphical test is undertaken with the probability grid. The upper value limit is marked on the x-axis, and the percent values of the cumulative frequency are on the y-axis. A probability grid (probability paper) is enclosed as enclosure A1 at the back of this book. This will be the standard probability graph used in this textbook. In some English-speaking countries, x- and y-areas are often reversed. For readers already more used to the reversed mode of the probability graph also such a version is enclosed as enclosure A2 at the back of this book. For a reader who has already some experience using the reversed mode (the A2-version), Fig. 14 (and later Fig. 24 for the lognormal distribution) shows an example in both versions; thus the reader can more easily become familiar with the convention used in this textbook. In order to understand the configuration of the probability grid, the terms upper value limit and cumulative frequency are explained by an example in the next section.

5.1.1
Calculation of the Cumulative Frequency

For the cumulative or accumulative frequency, the relative frequencies of the individual class intervals are simply summed from the lowest value upwards. The procedure can be explained on the basis of the example in Table 5 (in Chap. 2.2) as follows (Table 10):

The relative cumulative frequency is the relative frequency of all values that are *less than* or *equal to* an upper value limit x, which is the upper value in each of the class intervals in Table 10. For the first class interval, for example, this is 5%, and for the second class interval it is 10%, etc.

Table 10. Calculation of the relative cumulative frequency with the fluorspar values of Table 5

Class	Relative frequency (%)	Relative cumulative frequency (%)
0.1 - 5 % CaF_2	6.7	6.7
5.1 - 10 % CaF_2	6.7	13.3
10.1 - 15 % CaF_2	13.3	26.6
15.1 - 20 % CaF_2	23.3	49.9
20.1 - 25 % CaF_2	16.7	66.6
25.1 - 30 % CaF_2	16.7	83.3
30.1 - 35 % CaF_2	10.0	93.3
35.1 - 40 % CaF_2	6.7	100.0

The function of the relative cumulative frequency is known as the *cumulative frequency function* or the *cumulative distribution function*, or simply as the *distribution function*.

5.1.2
Cumulative Fequency Function of the Normal Distribution and the Derivation of the Probability Grid

If the subdivisions on the y-axis are linear, then the cumulative frequency of the normal distribution is a distorted S-shaped curve, as shown below in Fig. 13. It is clearly demonstrated by Fig. 13 that the S-shape of the cumulative frequency curve becomes even more elongate as the standard deviation σ increases.

For the graphical test, the cumulative frequency curve of a real distribution is compared to that of a normal distribution. However, it is clearly difficult to compare different, distorted curves with each other. In order to overcome this problem, the y-axis is distorted so that the S-shape of the cumulative frequency curve becomes a straight line for a normal distribution. As an example, it is shown in Fig. 13, for the +σ and -σ values (i.e. 16 and 84% cumulative frequency, see the problem in Chap. 4), how the y-axis is symmetrically distorted about the 50% mark in order to obtain a straight line. On this transformed y-axis, the values for o and 100% tend towards infinity because the normal distribution is asymptotic with respect to the y-axis, and the cumulative frequency asymptotically approaches the limiting values - ∞ and + ∞. In practice, the determination of such extreme values near o or 100% is only of any importance for very special problems such as the recognition of so-called outliers (see Chap. 8[8]).

---

[8] If the highest values are important to the problem, then it is better to plot the so-called plotting percentage rather than the cumulative frequency, particularly if the total number of values is less than 100 (cf. Chap. 8.3.1 and 8.3.3.3).

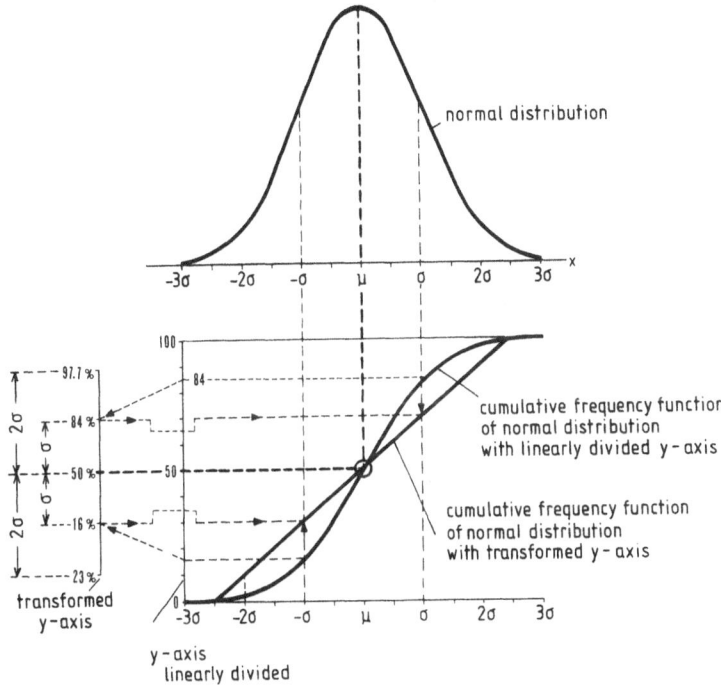

Fig. 13. Derivation of the graphical display of the cumulative frequency function of the normal distribution and of the Hazen line on the cumulative probability paper

A y-axis transformed in this way is the basis of the probability grid (see enclosure A). It can be seen from Fig. 13 or the enclosure A1 that, as a result of this transformation, the y-axis is now subdivided linearly about the median line (50% line) by multiples or fractions of the standard deviation σ or s. This s-value scale (or SW-scale; see enclosure A1, left hand side) shows the standardized distance of a value from the mean of the standardized normal distribution. The s-value will be required later in Chapter 8.3.3.3.

Since this probability grid was generated by transforming the S-shaped cumulative curve of the normal distribution into a straight line, the converse is also valid in that the cumulative frequency of any normal distribution will plot as a straight line on this grid. The straight line relationship is also known as the Hazen straight line (Hazen, 1914). Figure 13 also demonstrates that the gradient of this straight line decreases as, respectively, the standard deviation σ or s increases.

The Hazen straight line is the standard with which real distributions can be graphically compared and tested for the normality of their distribution.

5.1.3
Plotting the Cumulative Frequency Values of a Real Distribution on the Probability
Grid

The cumulative frequency percent values are plotted on the probability grid as
functions of the upper value limits. If class intervals are used, then the upper limit
of the interval is taken (see Sect. 5.1.1). Sometimes the error is made of plotting the
cumulative frequency value against the class interval mean - this is incorrect! The
graph illustrates the percentage of the population (or the sample size) beneath or
equal to the upper class interval limit or the upper value limit.

In an ideal normal distribution, all the points must plot on a straight line (see
above). In practice, however, points outside the 16 and the 84% limits (i.e. outside
the +σ and -σ or the +s and -s limits) generally do not plot on this straight line. The
reason for this is caused mainly by the distortion of the y-axis. In most cases it can
be assumed, as a rule of thumb, that the distribution is close to normality if those
points, for which the y-values are between the 16 and 84% limits, plot on a straight
line (for example see Fig. 14a[9]). Geological problems can then be solved
satisfactorily on the basis of the parameters estimated from this distribution.

Presuming that the values have an approximately symmetrical distribution, this
graphical presentation allows the mean value μ or $\bar{x}$ to be derived from the
intersection with y=50%, and the standard deviation σ or s to be read off
immediately from the difference between the respective intersection points at
y=15.6% and y=50%, or y=50% and y=84.4%. Or in more precise terms: the
interpolated Hazen straight line illustrates the distribution of the whole population
as estimated from the cumulative frequency curve for the samples. μ and σ can be
read off the grid and serve as estimates for $\bar{x}$ and s. The square of the standard
deviation σ or s then gives the variance σ² or s². The graphical method is the
quickest way of determining the variance, in particular for unequally weighted
values (Chap. 3.2.2.1).

The example of the fluorite analyses in Table 10 is shown plotted on the
probability grid on Fig. 14.

---

[9] As pointed out in Section 5.1, Fig. 14a is the normal version of a probability grid used in this text
book, Fig. 14b the inversed mode for readers familiar with the reversed mode to become familiar with
the convention applied in this text book.

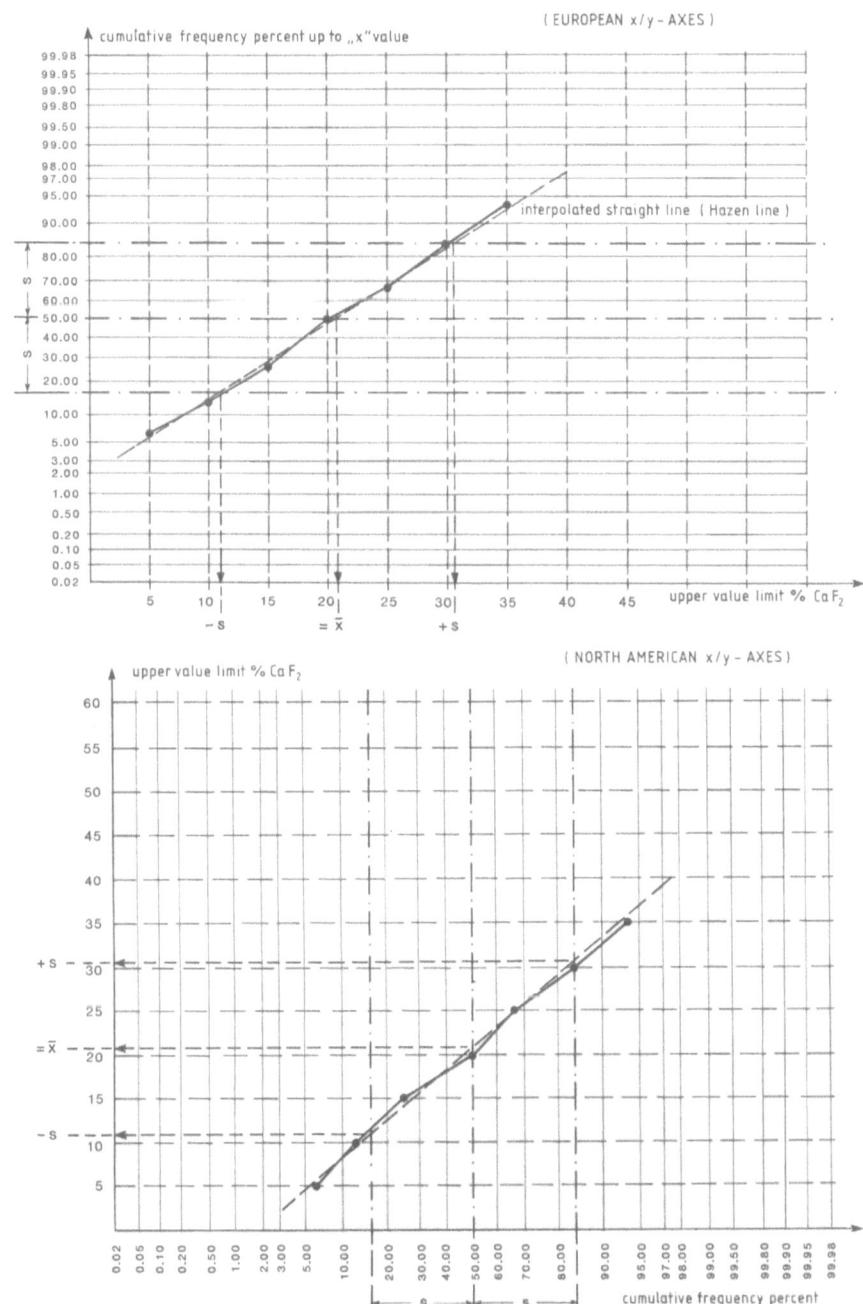

Fig. 14.a Cumulative frequency line of analyses of a fluorspar deposit from Table 10, plotted on cumulative probability paper (convention of x- and y-axes used in this textbook). b As a, only inversed x- and y-axes (North American standard)

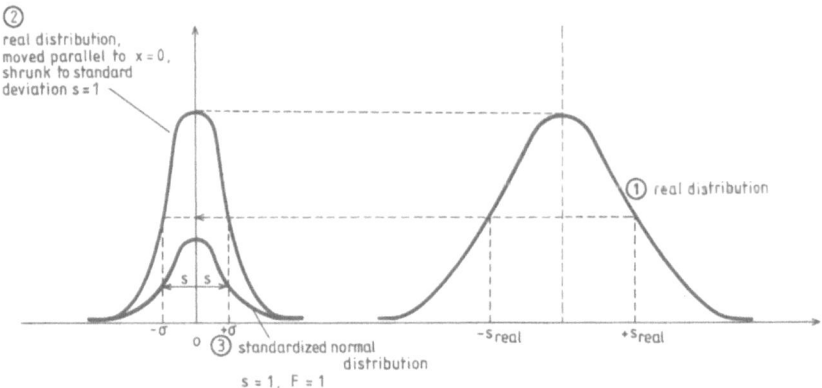

Fig. 15. Schematic display of the standardization of a frequency distribution, which is normally distributed

5.2
Chi-Square Test

The chi-square test (or χ^2 test) can be used to determine mathematically how closely the natural distribution can be compared to a normal distribution. Thus the "closeness" of the approximation is tested. The basic idea is simple (see Fig. 15).

The normal distribution is standardized (i.e. shifted by parallel displacement and condensing of the distribution so that the mean = 0 and the standard deviation = 1), and then compared with the standard normal distribution. In order to do this, the x-axis is subdivided into intervals, each of which should include at least five samples. If this is not the case, then the intervals are combined. If a value is the same as a limiting value between class intervals, then each of the class intervals is credited with one half. The test can be most easily explained by means of an example:

Table 11. 10 m composite samples from a Cu-Zn deposit

Interval (% Zn-equivalent)	No. of 10 m composites
≤ 18	13
18 - 19	5
19 - 20	12
20 - 21	8
21 - 22	13
22 - 23	12
23 - 24	7
24 - 25	11
≥ 25	6
	Sum = 87

Example: A volcanogenic Cu-Zn deposit was drilled, and the Zn equivalents have been calculated for mine planning (see Economic Evaluations in Exploration, Sect. 5.3) by means of the equation $Znequ = 1 \cdot Zn + 3.9 \cdot Cu$. This equation effectively homogenises the range of the zinc and copper grade values. For mine planning, the analysed drill core sections were combined into 10-m composites in order to simulate the interval between the sub-levels. This yielded 87 sections of 10 m, which could be subdivided into the following class intervals:
The mean was determined to be $\bar{x} = 21.4\%$ Zn equivalent, and the standard deviation $s = 2.9$.

1st Step: In the first step the natural distribution is standardized by the procedure already illustrated by the 2nd problem in Chapter 4 (i.e. shifted by parallel displacement and condensing of the distribution so that the mean $= 0$ and the standard deviation $= 1$; Fig. 15). As before, this is done by subtracting the mean value $\bar{x} = 21.4$ from the class interval limits and dividing the difference by the standard deviation $s = 2.9$ (Table 12, column 2).

2nd Step: The area under the standard normal distribution curve for both the selected and then the recalculated intervals is calculated. These figures can then be compared with the actual class interval frequency in step 5 (see below), for which the values of the cumulative frequency function or distribution function of the normal distribution are required. The cumulative frequency is denoted by $\varnothing(u)$. The values are provided in Appendix Table 2 (there as $\varnothing(x)$ equivalent to $\varnothing(u)$ in this example) , and the respective values for the class interval limits are obtained from this table (see Table 12, column 3).

3rd Step: Since the cumulative frequency values $\varnothing(u)$ are the sum of all values that are smaller than or equal to the class interval limiting value u , the differences must then be calculated (column 4) in order to find the theoretical relative frequency for each of the class intervals. (Since these are actually the areas of sections under the normal distribution curve, the sign is of no significance.)

For each interval the difference $p_i = \varnothing$ [upper limit] $- \varnothing$ [lower limit] states the probability that a value falls in this interval, presuming that the distribution is normal. The relative frequencies of these intervals must therefore be the same as these probabilities.

4th Step: If the probability p_i is multiplied by the absolute frequency 87, then an expectancy value $e_i = p_i \cdot 87$ (column 5 in Table 12) is obtained for each class interval.

5th Step: This value e_i can now be compared with the actual frequency values in Table 11, which are denoted by b_i (column 6 in Table 12).

6th Step: A reference number can now be calculated for the measure of the deviation of the actual frequency b_i from the theoretical frequency e_i. Since the differences can be positive or negative, the square of the differences is used as it was for the calculation of the variance (see Chap. 3.2.1). This squared difference must now be standardized by dividing it by the theoretical value e_i. The value so obtained is denoted as A_i (Table 12 column 7).

$$A_i = \frac{(b_i - e_i)^2}{e_i}.$$

The sum of all A_i is χ_o^2.

In this example $\chi_o^2 = 7.43$.

Table 12. The data from Table 11 used in a chi-square test calculation

Column 1	Column 2	Column 3	Column 4	Column 5	Column 6	Column 7
Interval boundaries	Transformed interval boundaries $u = \frac{x - 21.4}{2.9}$	$\varnothing\left(\frac{x - 21.4}{2.9}\right) = \varnothing(u)$	p_i	$e_i = 87\, p_i$	b_i from Column 2, Table 11	$A_i = \frac{(b_i - e_i)^2}{e_i}$
< 18	$-\infty^{a)}$ to -1.17	0 - 0.1210	0.1210	10.53	13	0.58
18 - 19	-1.17 to -0.83	0.1210 - 0.2033	0.0823	7.16	5	0.65
19 - 20	-0.83 to -0.48	0.2033 - 0.3156	0.1123	9.77	12	0.51
20 - 21	-0.48 to -0.14	0.3156 - 0.4443	0.1287	11.20	8	0.91
21 - 22	-0.14 to +0.21	0.4443 - 0.5832	0.1389	12.08	13	0.07
22 - 23	+0.21 to +0.55	0.5832 - 0.7088	0.1256	10.93	12	0.11
23 - 24	+0.55 to +0.90	0.7088 - 0.8159	0.1071	9.32	7	0.58
24 - 25	+0.90 to +1.24	0.8159 - 0.8925	0.0766	6.66	11	2.82
> 25	+1.24 to $+\infty^{a)}$	0.8925 - 1	0.1075	9.35	6	1.20
						$A_i = 7.43$

[a] Because the normal distribution approaches zero asymptotically < 18 means the interval stretches to $-\infty$ and > 25 to $+\infty$.

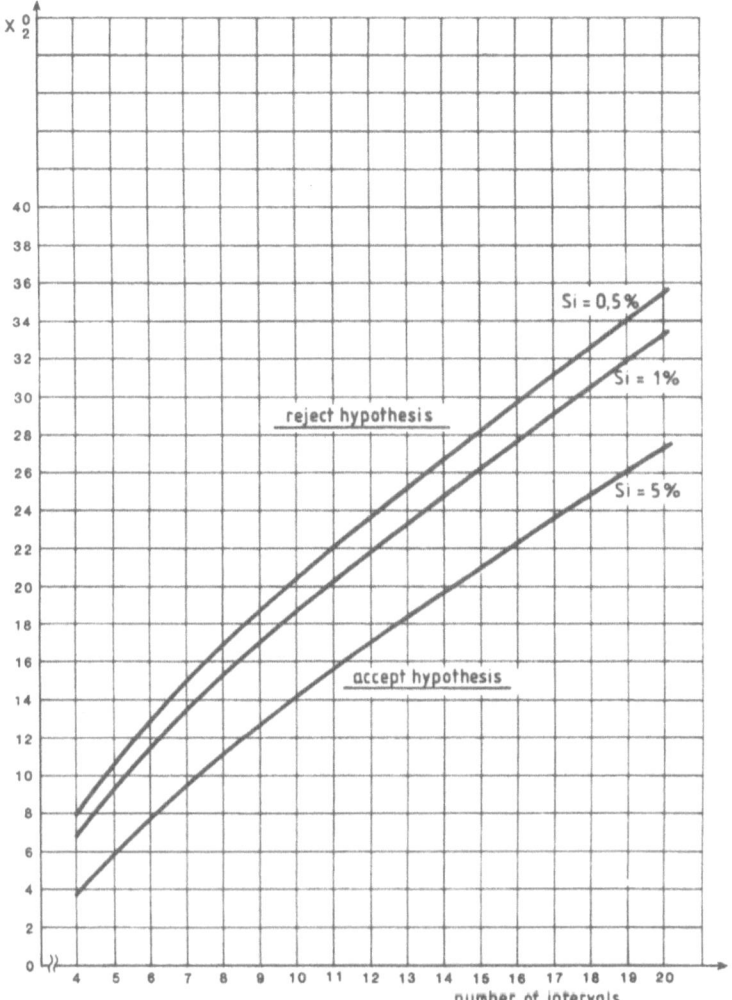

Fig. 16. Diagram for testing the hypothesis of a normal distribution

7th Step: A significance level is now selected in this step. The significance level denotes the probability of the error that the hypothesis the actual distribution complies with the normal distribution is rejected, although it is in fact true. The significance level is denoted by Si, and is selected to be 5 %. The diagram in Fig. 16 shows the number of the class intervals plotted against the value

There are nine intervals, and $\chi_o^2 = 7.43$ plots in the field "hypothesis accepted". The data does not provide any ("certain") indications that the distribution is anything other than normal, and it can therefore be assumed that the 87 composites are normally distributed. (However, this is not in itself an indication that the distribution is normal in fact.)

If the χ^2 test is used regularly for geological data, it is found that the hypothesis of a normal distribution of the data is often not true. However, if the distribution has a single peak and is approximately symmetrical, then the assumption of a normal distribution generally leads to acceptable results for geological and geochemical problems.

The same is also valid if a skewed distribution is approximated with the lognormal distribution, and this procedure will be described in Chapter 9. The χ^2 tests also show that for this case the hypothesis frequently is not true at a significance level of Si=5%. However, in spite of this, under certain conditions that are discussed in Chapter 9, this type of approximation can provide satisfactory results for geological and geochemical problems.

6
Standard Deviation and Variance of the Mean

6.1
Calculation of the Standard Case

If the mean of a population is calculated from a random selection of samples, this mean value for the sample will not correlate exactly with the true mean value for the population. If several samples are taken from the population, then the mean values of these samples will vary, and a standard deviation can also be calculated for this variation. It is known as the standard deviation of the mean (or the standard error of the mean) or the variance of the mean, respectively. The standard deviation or variance of the mean can be calculated from the standard deviation or variance of the samples. It is easy to see that the range of the different mean values must decrease in proportion to the increase in the number of individual samples in the random samples. The variance of the mean s_x^2 is:

$$s_x^2 = \frac{\sigma^2}{n} \ .$$

If σ^2 is estimated from the variance of the sample s^2, and by applying the Eq. (1) in Chapter 3.2.1:

$$s^2 = \frac{1}{n-1} \sum_{i=1}^{n} (x_i - \bar{x})^2 \ ,$$

then the variance of the mean is:

$$s_x^2 = \frac{1}{(n-1)\cdot n} \sum_{i=1}^{n} (x_i - \bar{x})^2 \ .$$

The standard deviation of the mean s_x is again the square root of s_x^2:

$$s_x = \frac{s}{\sqrt{n}} = \sqrt{\frac{\sum_{i=1}^{n} (x_i - \bar{x})^2}{n \cdot (n-1)}} \ .$$

Example: The five values derived from a density measurement in Chapter 3.1.1 are used:

$3.1, 3.3, 3.4, 3.5, 3.6 \ (g/cm^3)$.

The mean value was $3.4 \ (g/cm^3)$.

The standard deviation of the mean is now:

$$s_{\bar{x}} = \sqrt{\frac{(3.2-3.4)^2 + (3.3-3.4)^2 + (3.4-3.4)^2 + (3.5-3.4)^2 + (3.6-3.4)^2}{5 \cdot 4}}$$

$$s_{\bar{x}} = \sqrt{\frac{0.04 + 0.01 + 0 + 0.01 + 0.04}{20}}$$

$$s_{\bar{x}} = \sqrt{\frac{0.1}{20}} = \pm 0.07.$$

It is important to realise that this formula is only valid if the individual values are independent of each other, i.e. they are *statistically independent* individual values. This is the case for density measurements such as those in the above example. However, the extent to which this is the case for the sampling of mineral deposits will be investigated in Chapter 13.2. It is definitely not the case for the individual analyses from a drilling programme of a mineral deposit. The number of analysed sections from a drill hole is an artificial, man-made, quantity.

The values from the individual drill holes, for example, can be independent quantities. Sometimes the standard deviation of the mean grade of a mineral deposit is calculated from all of the individual analyses. The variance of the individual analyses is obviously different from that of the values of the drill holes (if the drill holes are subdivided into numerous individually analysed sections) since their "support" is different (see Chap. 1.1). If there are n individual analyses, the standard deviation of the mean can be reduced by subdividing the samples and thereby artificially increasing the figure n. Thus an accuracy can be simulated, although it does not represent the true situation.

6.2
Weighting Different Variances of the Mean

During the calculation of the variances for the reserve figures of a mineral deposit, it is commonly found that the variances from different parts of the deposit must be combined (see Chap 13.3, for example).

Assuming that the overall mean of the deposit is calculated on the basis of weighting the volume (or tonnage),

$$\overline{x}_{tot} = \frac{\sum\limits_{i=1}^{n} \overline{x}_i \cdot v_i}{\sum\limits_{i=1}^{n} v_1},$$

the variances can also be weighted with the volume (or tonnage) of the deposit:

$$s^2_{tot} = \frac{\sum\limits_{i=1}^{n} s_i^2 \cdot v_i^2}{(\sum\limits_{i=1}^{n} v_i)^2}.$$

Example: The average grades of a carbonate-hosted Pb-Zn deposit are determined for three homogeneous zones. The variances for zinc in each of these zones are:

Table 13. Variances for three zones of a Zn deposit

Zone	Variance s^2 $[(\% \, Zn)^2]$	Ore deposit volume (m^3)
I	64.5	$1.8 \cdot 10^6$
II	83.2	$2.0 \cdot 10^6$
III	105.8	$1.6 \cdot 10^6$

The total variance is therefore:

$$s^2_{tot} = \frac{64.5 \cdot 1.8^2 + 83.2 \cdot 2.0^2 + 105.8 \cdot 1.6^2}{(1.8 + 2.0 + 1.6)^2}$$

$$s^2_{tot} = \frac{812.63}{(5.4)^2} = \frac{812.63}{29.16} = 27.87 (\%Zn)^2$$

$$s_{tot} = \pm 5.27\% \, Zn \, .$$

The standard deviation of the mean for the whole of the deposit is therefore ±5.27% Zn.

7
Estimation of the Error

7.1
Confidence Intervals of a Mean Value

In the conceptual idea described in Chapter 6.1, it was assumed that the mean of the sample would deviate from that of the population from which it was collected. This therefore raises the question of how "precisely" does the mean value of the sample reflect that of the population. In other words, how large is the uncertainty of the mean value, or what is the magnitude of "the error"?

In the discussion of the central limit theorem in Chapter 4, it was demonstrated that the distribution of the mean values derived from a repeated selection of samples is approximately normal for large sample series of comparable size. In a new example it is assumed that several lance samples are collected repeatedly from a consignment of zinc concentrates. The mean is 50.5% Zn, the variance of the mean is $s_x^2 = 1.56$ (% Zn)^2 and the standard deviation of the mean is therefore $s_x = \sqrt{1.56} = \pm 1.25 \%Zn$, i.e. the square root of the variance. All the mean values now describe a normal distribution with $\bar{x} = 50.5$ and $s = \pm 1.25$, as shown in Fig 17. The true mean value of the zinc concentrates lies somewhere within this normal distribution. It has been demonstrated in Chapter 4 that for a normal distribution practically 2/3 of all values lie between +s and -s, and nearly 95% of all values lie between +2s and -2s. If numerous series of lance samples are collected from a consignment of zinc concentrates, this implies that 2/3 of all the sample series would lie between 50.5 ±1.25 (i.e. between 49.25 and 51./5% Zn) and only every twentieth series would lie outside 50.5 ±2 x 1.25 (i.e. either smaller than 48% or greater than 53% Zn).

This shows that an indication of error is completely meaningless unless it is accompanied by a degree of certainty (or level of confidence), in which case the following statements are comparable:-

at a 95% level of confidence:
50.5% ±2·1.25 = 50.5% ±2.5% Zn,

at a 68% level of confidence
50.5% ±1·1.25 = 50.5% ±1.25% Zn.

The confidence level therefore corresponds to the area under the normal distribution curve between the confidence limits, and the confidence limits defined by the level of confidence is the "error".

Fig.17. Errors of the average grade of a zinc concentrate as a function of the level of confidence

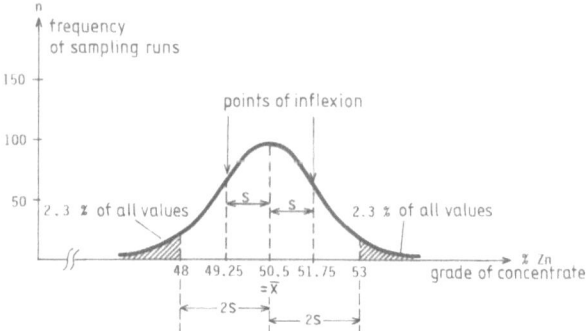

Alternatively the "tails", which are the areas outside those areas defined by ±s or ±2s (see Fig. 17), could be used for the definition of the level of confidence. The areas under the "tails", which lie outside the defined limits, are then indicative of the probability of error.

Very high confidence levels (99, 99.5%) are required for biological and medical research work, but this is not realistic for geological situations and confidence levels of 90 or 95% are usually applied. A GDMB working group (GDMB = German Association for Metallurgists and Mining Engineers, which is a German equivalent to the Institution of Mining and Metallurgy in the UK or the Society for Mining, Metallurgy, and Exploration of the USA) has been studying this question with respect to the problem of ore reserve classification, and it has recommended that a confidence level of 90% be applied to all categories of reserves (Wellmer 1983a, b). Other proposals for levels of confidence for reserve calculations range from 84 to 95%. They are summarized in Appendix Table 3.

Usually, only the standard deviation, s, of the sample is known, while that of the population, σ, and therefore its normal distribution, remains unknown. This results in a serious practical problem, and another distribution, the so-called Student's-t distribution, must be applied for smaller samples. In the Student's-t distribution the cumulative frequency distribution approaches that of a normal distribution as long as there are a sufficiently large number of values (samples in this case). The values of the Student's-t distribution, to which frequent reference will be made later in this book, are listed in Appendix Table 4.

The calculation procedure is best explained by an example:

Example: A consignment of copper concentrates was sampled with eight lance samples. The analytical values of these samples are as follows:

1st step: The mean and the variance, and thus standard deviation, are calculated (see Chap. 3.1.1 and 3.2.1).
The mean is $\bar{x}$ =25.6% Cu.
The variance is s^2=0.29.
The standard deviation is s=±0.54.

Table 14. Cu analyses from a copper concentrate shipment

Sample no.	Cu analysis (%)	Sample no.	Cu analysis (%)
1	25.8	5	26.3
2	25.0	6	25.2
3	24.8	7	26.1
4	25.9	8	25.7

2nd step: The confidence interval is denoted by ki . The limits are

$$ki = \pm \frac{s \cdot t}{\sqrt{n}}, \tag{1}$$

where t is the corresponding factor from the Student's-t distribution.

The t factor is dependent on the number of samples and the level of confidence, and 90% is selected. From Appendix Table 4, t=1.90 for 8 samples (columns 1 and 2).

The confidence interval is therefore:

$$ki = \pm \frac{s \cdot t}{n} = \pm \frac{0.54 \cdot 1.9}{8} = \pm 0.36 \approx \pm 0.4.$$

It can now be stated: the copper concentrate has a grade of 25.6 ±0.4% Cu at a 90% confidence level.

The relative confidence interval can also be calculated by dividing ki by the mean $\bar{x}$. In this case, therefore:

$$\frac{ki}{\bar{x}} = \pm \frac{0.36}{25.6} = \pm 0.014,$$

i.e. at a confidence level of 90%, the true mean value varies by a maximum of 1.4% (relative) from the calculated mean value of $\bar{x} = 25.6$.

Under the GDMB 1983 convention, the errors, which are the basis of the reserve classification system, are calculated as above from the confidence interval. The standard deviations should, when possible, be calculated geostatistically from extension and dispersion variances, which are discussed in Chapter 13.3. The GDMB recommendations are summarized in Appendix Table 3. More in detail the aspect of reserve classification is discussed in Chapter 13.3.

7.2
The Average Error

The average error for a series of measurements is quoted as a single value,

e.g. $\bar{x} \pm s$,

where $\bar{x}$ is again the arithmetic mean and s is the standard deviation, and therefore $\pm s$ is the average error. It is quoted without confidence levels, i.e. the above formula implies a confidence level of 68% (see Chap. 4).

Example: Once again, as in Chapter 3.1.1, the five density measurements are used, 3.2, 3.3, 3.4, 3.5, 3.6 (g/cm^3). The mean was $\bar{x}$ =3.4 g/cm^3. The standard deviation is therefore (Chap. 3.2.1):
s=±0.16, rounded up to ±0.2, i.e. the average error is ±0.2 g/cm^3.

It must be emphasized again that the average error is applied to a series of measurements that can be determined as accurately as possible, such as physical determinations similar to the example of the density of a sample. It is thus the precision of numerous measurements.

The example of the consignment of concentrates in Section 7.1 or the determination of the average grade of an ore deposit represent another problem. The analytical data is distributed over a wide range (in comparison to which the analytical error is assumed to be negligibly small). A mean value of the concentrate consignment or of the ore deposit shall be calculated from this range of values, in which case the standard error of the mean ($\bar{x} \pm \dfrac{s \cdot t}{\sqrt{n}}$) is applicable (Sect. 7.1).

7.3
The Law of Error Propagation

In Sections 7.1 and 7.2 the confidence levels or the errors of direct observations and measurements were discussed in terms of the grades of concentrate consignments or ore deposits as well as the density of a rock sample. In this chapter the error derived from indirect observations will be discussed. A typical example is the calculation of the true thickness of an ore intersection. The appropriate formula is taken from Economic Evaluations in Exploration, Chapter 3.1.1:

$$M_W = L_B \sin(\alpha + \beta) \quad . \tag{1}$$

M_W is the true thickness, L_B is the length of the intersection, α is the angle of inclination of the drill hole at the ore intersection, and β is the dip of the ore body.

The true thickness is therefore an indirect observation, i.e. it is derived from several direct observations – L_B, α and β. The length of the intersection L_B can usually be measured very accurately. The angle of inclination of the drill hole at the ore intersection is usually determined by surveying the hole, but this dimension is often not measured, especially in short holes. The dip angle of the ore deposit is commonly the most uncertain of these dimensions. The dip is measured either at the outcrop or by reconstructions from three or more drill hole intersections, i.e. it is impossible to make a direct measurement of the dip at the site of the intersection. Dip commonly can vary over short distances in, for example, ore veins. The question is therefore posed, how does this problem of measurement affect the level of accuracy of the true thickness?

Simple case: It is assumed that a stratiform Pb-Zn horizon, dipping at $\beta=80°$, is being drilled. The drill hole is inclined at $\alpha=60°$ normal to the dip. It is estimated that the sum of the possible deviations from the angles of drill inclination and dip is $5°$. With a $5°$ deviation, how large is the relative error in the true thickness?

Answer: The estimated error of $5°$ is divided into a deviation of $2.5°$ upwards and downwards. The relative error is then:

$$\frac{\Delta M_W}{M_W} = \frac{L_B \cdot \sin(60°+80°-2.5°) - L_B \cdot \sin(60°+80°+2.5°)}{L_B \cdot \sin(60°+80°)}$$

$$\frac{\Delta M_W}{M_W} = \frac{\sin(137.5°) - \sin(142.5°)}{\sin(140°)} = \frac{0.676-0.609}{0.643}$$

$$\frac{\Delta M_W}{M_W} = \frac{0.067}{0.643} = 0.104 \text{ , i.e. approximately } \pm10\%.$$

If the effects of a $5°$ error are calculated for different angles, it can be shown that for holes drilled at very oblique angles (i.e. $\alpha+\beta$ approaches $180°$ or $180°-(\alpha+\beta)$ approaches zero), the relative error for the true thickness increases sharply (see Fig. 18)[10]

Complex case: The error is known for an angle of inclination of a drill hole at the ore intersection ($\Delta\alpha=5°$) and for the dip of the ore body ($\Delta\beta=10°$). In this case the Gaussian law of the error propagation, which uses partial derivatives, must be applied.

With respect to this problem, the following is valid:

$$\Delta M_W \approx \sqrt{(\frac{\partial M_W}{\delta\alpha})^2 \cdot \Delta\alpha^2 + (\frac{\delta M_W}{\delta\beta})^2 \cdot \Delta\beta^2} \; , \tag{2}$$

where ∂ is the partial derivative. Thus for example:

$$\frac{\partial[\sin(\alpha+\beta)]}{\delta\alpha} = \cos(\alpha+\beta),$$

---

[10] With this procedure, the error attains the value 0 at $90°$, since $\sin 87.5° = \sin 92.5°$. This is, of course, meaningless. If a maximum or a minimum lies within the range of angles under consideration, then it must be questioned which maximum and minimum values can be attained. The smallest value in the range $87.5°$ to $92.5°$ is $\sin 87.5°$ or $\sin 92.5°=0.999$, the largest value in this range is $\sin 90°=1$, so that in this range the relative error is $\frac{1-0.999}{1}=0.001$, i.e. 0.1%.

Fig.18. Error in percent of thickness as function of angle between orebody and drill hole $[180° -(\alpha+\beta)]$ for a 5°) deviation

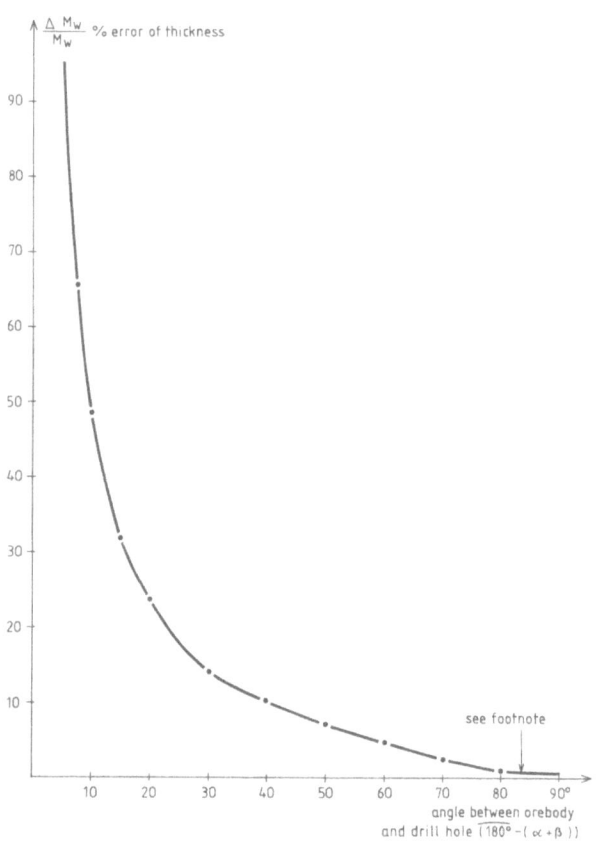

i.e. the function $\sin(\alpha+\beta)$ should be partially deviated to α, and the result is then $\cos(\alpha+\beta)$ (see common mathematical tables).

$\Delta\alpha$ and $\Delta\beta$ are the errors for the angles α and β. The relationship (1) above is substituted for M_W on the right-hand side of Eq. (2) above, which results in:

$$\Delta M_W = \sqrt{\left[\frac{\delta[L_B \cdot \sin(\alpha+\beta)]}{\delta\alpha}\right]^2 \cdot \Delta\alpha^2 + \left[\frac{\delta[L_B \cdot \sin(\alpha+\beta)]}{\delta\beta}\right]^2 \cdot \Delta\beta^2}.$$

If the partial derivatives are carried out for α and β, the result is (L_B is considered to be constant, since it has only a minimal error):

$$\Delta M_W \approx \sqrt{L_B^2 \cdot \cos^2(\alpha+\beta) \cdot \Delta\alpha^2 + L_B \cdot \cos^2(\alpha+\beta) \cdot \Delta\beta^2}$$

$$\Delta M_W = L_B \cdot \cos(\alpha+\beta) \cdot \sqrt{\Delta\alpha^2 + \Delta\beta^2}.$$

The relationship (1) above is applied again for M_W, and this results in the relative error:

$$\frac{\Delta M_W}{M_W} \approx \pm \frac{L_B \cdot \cos(\alpha+\beta) \cdot \sqrt{\Delta\alpha^2 + \Delta\beta^2}}{L_B \cdot \sin(\alpha+\beta)} = ctg(\alpha+\beta) \cdot \sqrt{\Delta\alpha^2 + \Delta\beta^2} \ .$$

It is again assumed that the angle of dip of the strata is $\beta=80°$ with a drill inclination of $\alpha=60°$. The error of the angle is therefore determined by $\Delta\alpha=\pm5°$ and $\Delta\beta$ with $\pm10°$. This is then recalculated as an arc measurement:

$180°$ represents the arc π, so that $5°$ is:

$$5° = \frac{\pi \cdot 5}{180} = 0.087$$

and $10°$ is naturally double:

$$10° = \frac{\pi \cdot 10}{180} = 0.174 \ .$$

The relative error is therefore:

$$\frac{\Delta M_W}{M_W} \approx \pm ctg(80°+60°) \cdot \sqrt{0.087^2 + 0.174^2} = 1.19 \cdot \sqrt{0.038}$$

$$\frac{\Delta M_W}{M_W} \approx \pm 1.19 \cdot 0.195 = \pm 0.23,$$

i.e. $\pm23\%$.

8
Skewed distributions

8.1
Introduction

Most of the natural distributions encountered in geology are not symmetric, but are usually more or less skewed to the right (i.e. positively skewed). Thus, higher grades occur in addition to the average grade tenor, and they extend beyond the range considered as the normal distribution. These irregularly high values or enrichments frequently convert mineralisation into orebodies.

Experience shows that, even after taking the usual dilution factors into account (see Economic Evaluations in Exploration, Chap. 6.1), a simple arithmetic mean often yields an average grade which is too high in comparison with the grade of the ore that is actually mined. On the other hand, these exceptionally high grades cannot be ignored because they indicate the location of high-grade zones that are often essential for an economic exploitation of the deposit (cf. Economic Evaluations in Exploration, Chap. 11.2). The treatment and assessment of the high values are therefore of decisive importance during the economic evaluation of a mineralized occurrence.

The problems associated with the evaluation of high grades are extremely complex. Intricate geostatistical techniques, such as kriging (Chap. 13.4.5), have been developed in order to avoid the overestimation of the high grades, but they are not really applicable in the early phases of evaluating a mineral deposit. It is also important to take the mining methods into consideration since, for example, the selective mining techniques used in gold mines in Western Australia (e.g. Carras, 1986) are more suitable for yielding the high grades indicated by drilling, as compared to an open pit bulk mining method.

The influence of the highest grade values (i.e. richest samples) is more significant, when cut-off grades are applied. This is discussed later in Chapter 12.

The lognormal distribution, in which the logarithms of the individual values can be described by a normal distribution, has become very important for the treatment of skewed distributions in exploration geology. This aspect is discussed in Chapter 9.

In principle, two different methodologies are distinguished for the estimation of the actual average grades that might be attained during mining:

a) The high values are examined as to whether they are acceptable, or if they must be treated as erratically high values that are then either rejected or reduced. The mean is calculated as a normal arithmetic average. This is discussed in the following sections of this chapter.

b) The values are transformed and examined as a lognormal distribution. The geometric average is increased by a multiplication factor in order to estimate the average grade. The geometric average, which is the same as the median, is always less and also more stable than the arithmetic average (Chap. 3.4). This method is described in Chapter 9.

Some of the methods described in the following sections of this chapter have no theoretical basis. They are rules of thumb that have been established and proved effective by experienced geologists and mining engineers.

Whenever possible, the high values are evaluated by various methods, which result in a range of estimates for the mean values from which an average mean can be selected as the best estimate.

If there are only a few high values, then the main techniques applied would fall within methods (a) (Sects. 8.3 and 8.4). If there are so many high values that they appear to be a component of the population, then the lognormal distribution techniques are preferred (Chap. 9). If there are any nearby mines, which exploit the same or similar style of mineralisation, then the procedures used for handling the high values occurring should be investigated.

8.2
Measurement of Skewness (Asymmetry)

The coefficient of skewness Sf can be calculated from the following formula (measurement of skewness after Johannsen):

$$Sf = \frac{\sum\limits_{i=1}^{n}(x_i - \bar{x})^{3}}{n \cdot s^{3}}.$$

x_i are again individual values, $\bar{x}$ is the arithmetic mean, s is the standard deviation, and n is the number of individual values. Since $\sum(x_i - \bar{x})$ is ζ for symmetric distributions, then Sf must be also zero for the normal distribution. If the distribution is skewed to the right, then Sf>0 (therefore the expression is positively skewed), and if it is skewed to the left, then Sf<0 (negatively skewed).

Table 15. Analyses from six intersections of a Cu-Ni deposit

Drill hole	Ni grade (%)	$x_i - \bar{x}$	$\left(x_i - \bar{x}\right)^3$
1	0.9	-0.85	-0.61
2	1.2	-0.55	-0.17
3	1.5	-0.25	-0.02
4	1.9	+0.15	0.00
5	2.4	+0.65	+0.27
6	2.6	+0.85	+0.61
			$\Sigma = +0.10$

Example: There are six intersections in a Cu-Ni mineral deposit with the following nickel grades, as shown in Table 15:

$$\overline{x} = 1.75$$

$$s = 0.67$$

$$\overset{3}{s} = 0.30$$

$$Sf = \frac{+0.10}{6 \cdot 0.30} = +0.06.$$

Experience shows that the simple arithmetic average of the grade values is acceptable within the range $Sf = \pm 1$.

8.3
Assessing Isolated or Only a Few High Values

In the early phases of an evaluation, it is assumed that the values are statistically independent of each other. There are usually insufficient values for the recognition of spatial interdependence by means of a so-called variogram. This will be discussed in Chapter 13.2. In most cases, only weakly defined trends can be identified, and for high-grade zones of mineral deposits these trends are based on a single or very few values.

8.3.1
Corrections Using the Graphical Cumulative Frequency Distribution (Hazen Line)

If the grade values are plotted as a cumulative frequency curve on a probability grid, it can often be observed that the final 5 - 10% deviate significantly as a "tail" from a straight line (the Hazen line, Chap 5.1.2). This can be corrected by "bending back" this tail so that it lies on the straight line.

Since the final value, that with a cumulative frequency of 100%, does not appear on the probability grid used for the usual cumulative frequency plots, the so-called plotting percentage (or plot frequency) is preferred for this technique. (The cumulative frequency at 100 % is infinity because the normal distribution has no limit, but only approaches zero asymptotically, cf. Chap. 5.1.2).

According to Koch and Link (1970), the plotting percentage or plot frequency is:

$$\text{Plotting percentage } Hp = 100 \cdot \frac{3 \cdot (\text{absolute cumulative frequency}) - 1}{3 \cdot n + 1}, \tag{1}$$

where n is again the number of values.

Example: The first two intervals are used from the example of the 30 fluorite analyses (i.e. n=30) in Chapter 2.2, Table 5:

Table 16. The first two intervals of Table 5 with their plotting percentages

Class	Absolute frequency	Absolute cumulative frequency	Plotting percentage
0.1 - 5% CaF_2	2	2	5.5
5.1 - 10% CaF_2	2	4	12.1

According to the above Eq. (1), the plotting percentage Hp for the first interval is:

$$H_p = 100 \cdot \frac{3 \cdot 2 - 1}{3 \cdot 30 + 1} = 100 \cdot \frac{5}{91} = 5.49\%,$$

and for the second interval:

$$H_p = 100 \cdot \frac{3 \cdot 4 - 1}{3 \cdot 30 + 1} = 100 \cdot \frac{11}{91} = 12.09\%.$$

Figure 19 shows an example, based on a Pb-Zn deposit, of the correction of higher values by the graphical cumulative frequency distribution. The Pb contents

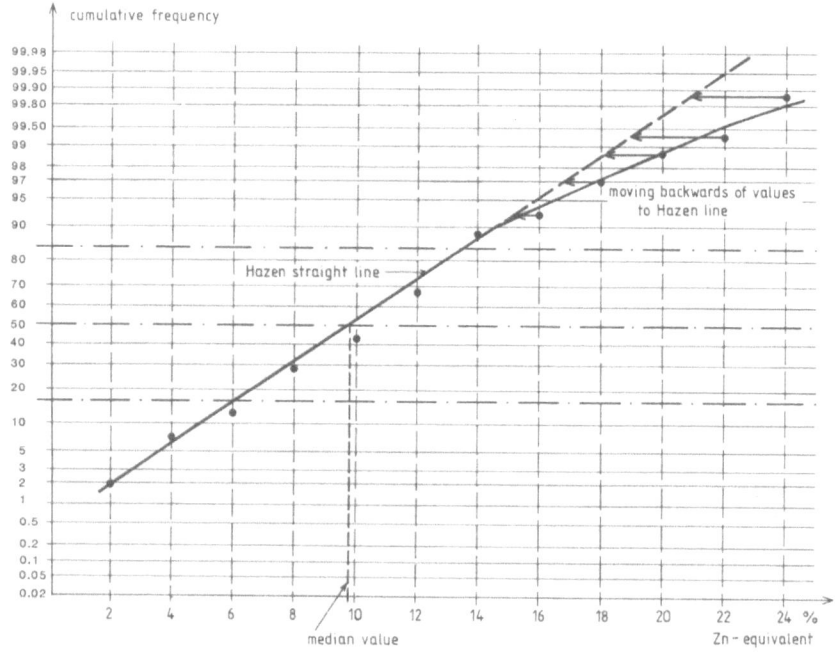

Fig. 19. Correction of high values with cumulative probability paper

are recalculated as Zn equivalents (see Economic Evaluations in Exploration, Chap. 5.3). In the cumulative frequency plot the points above about 16% Zn equivalent deviate from the Hazen straight line and are "bent back".

In practice, this means that the mean has been approximated by the median (Chap. 3.4), so that in Fig. 19 m = 9.8% Zn equivalent, which is a relatively conservative estimate. The median and the mean are the same only if the distribution is symmetric, and the median is always less if the distribution is skewed to the right (cf. Fig. 11).

8.3.2
Reducing the Highest Values to the Next Highest

If there is only one single value in a sample data set that is significantly higher than all the others, then the highest value is reduced to the next-lowest value. This avoids complicated calculations which, even so, would often not result in an average value substantially different from the value obtained by applying this rule of thumb.

Example: There are 45 drill intersections of a gold deposit that were classified according to the gold values (Table 17).

Table 17. Forty five inter-sections from an Au deposit	Drill hole	Thickness (m)	Grade (g Au/t)
	1-39	1.7-4 m	1.5-8.3 g/t
	40	8.0	9.1
	41	3.0	10.1
	42	1.0	12.0
	43	2.0	13.4
	44	3.0	21.7
	45	2.0	44.0

The final and highest value is reduced to the next highest, in this case from 44.0 g/t to 21.7 g/t. The mean thus obtained was $\bar{x}$ = 4.3 g/t, in contrast to 4.8 g/t. The difference of 0.5 g/t results from:

$$\Delta = \frac{44.0 \text{g/t} - 21.7 \text{g/t}}{45} = 0.5 \text{g/t}.$$

8.3.3
Statistical Outlier Tests

In statistics, an analytical value is denoted an outlier if it deviates so strongly from the other values that it must be assumed to be caused by an analytical error. The case of a high analytical value from within a mineral deposit presents another problem, since it reflects a real analytical value and not an error (always supposing that sampling or analytical error can be rejected; see Chap. 11.2.1).

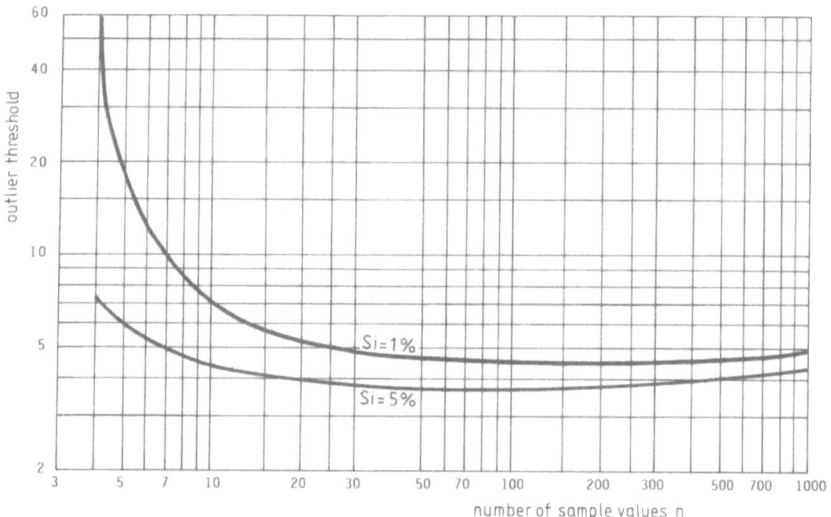

Fig. 20. Outlier threshold, g, as a function of the number of sample values, n. (Doerffel 1962) Si = 5% for outlier test for high values (Chap. 8.3.3.1); Si = 1% for analytical problems (Chap. 11.2.2)

It is clear that in such cases the distribution is not normal, although this is assumed in this outlier test.

However, experience has shown that this method is suitable for determining the cut level as a first approximation method (Chap. 8.4.1). The selected level of significance for the probability of error (Chap. 5.2, 7th step) is Si=5%[11]

There are numerous articles and even text books (Hawkins 1980; Barnett and Lewis 1984) on the recognition of outliers. Three simple tests are described in the following sections.

8.3.3.1
Test for an Extensive Data Set

Doerffel (1962) provided a diagram for the determination of the outlier threshold (Fig. 20, see also footnote 11).

The mean $\bar{x}$ and the standard deviation s are calculated without the highest value. A value x_A is then an outlier if it lies outside a defined range:

$$x_A \geq \bar{x} + s \cdot g \,, \tag{1}$$

where g is the outlier threshold and can be determined from the diagram in Fig. 20.

---

[11] There are two levels of significance in Fig. 20: Si=5% and Si=1%. The Si=1% level is selected, for example, for the comparison of two series of analyses (Chap. 11.2.2).

Example: A gold-bearing quartz vein was investigated by an adit. The vein was systematically sampled every 3 m by chip sampling along a profile. The samples, classified according to their grades, have the following values (Table 18):

Sample no.	Grade (g Au/t)	Sample no.	Grade (g Au/t)
2	1.5	12	9.0
11	2.3	15	12.0
9	2.4	5	12.1
4	2.6	7	13.0
3	3.2	16	13.7
1	3.5	14	17.2
6	4.8	8	17.8
10	5.5	13	24.5
17	7.7	18	38.9
		19	68.5

Table 18. Chip samples from an adit into an Au bearing quartz vein

The final values are conspicuously high. The test will show if they should be reduced.

1st Step: The mean and standard deviation are calculated without the highest value, 68.5 g/t. The mean is calculated to be $\bar{x}$ = 10.65 g/t, the standard deviation to be 9.60. The cut level g for a level of significance Si=5 % for 19 values is derived from Fig. 20: g=4.

The following is then calculated

$$x_A \geq \bar{x} + s \cdot g = x_A \geq 10.65 + 9.60 \cdot 4 = 49.05 \, ,$$

i.e. the value 68.5 must be reduced.

2nd Step: Consider the next highest value 38.9, and the mean and standard deviation are calculated again but without the value 38.9 (and of course without the value 68.5 that has just been eliminated). In this case:

$$\bar{x} = 8.99 \text{ and } s = 6.72 \, .$$

g is again g=4, so that the following is derived

$$x_A \geq \bar{x} + s \cdot g = 8.99 + 6.72 \cdot 4 = 35.87 \, ,$$

i.e. the value 38.9 must be reduced.

3rd Step: The next highest value 24.5 is now examined, and the mean and standard deviation are calculated as in Steps 1 and 2, but in this case the three highest values 24.5, 38.9 and 68.5 g/t are omitted.

$\bar{x} = 8.02$ and $s = 5.58$,

and therefore

$x_A \geq \bar{x} + s \cdot g = 8.02 + 5.58 \cdot 4 = 30.34$,

and thus the value 24.5 is acceptable.

4th Step: The value to which the two higher values 38.9 and 68.5 should be reduced (the cut level) is determined in this step. The outlier threshold, which was the last of the values to be reduced, then was calculated in Step 2 as:
$x_A \geq 35.8$.
Whole numbers are always selected for the cut level, and thus in this case it is 36 g/t.

5th Step: After reducing the two highest values 38.9 and 68.5 to 36 g/t, the mean is calculated. This is:

$\bar{\bar{x}} = 11.8$,

If the rule of thumb outlined in Section 8.3.2 was applied, and the highest value 68.5 was reduced to the next lowest value 38.9, then the mean would be 12.1 g/t. A difference of this amount is of no significance in the early phases of an investigation.

8.3.3.2
Test for a Restricted Data Set

One is often required to evaluate mineralisation in the early stages of exploration when there are only a few drill intersections. Sometimes, one or two of these intersections yield very high analytical values as compared to the others. Initially, of course, the significance of these high grades is investigated with respect to the geology, and a geological overview of the area is obtained. The geological arguments should always predominate. However, a small statistical test, which was also developed by Doerffel (1967), has been adapted to the geological context and can be very helpful. It is the Q test:

The Q value is defined as:

$Q = \dfrac{x_A - x_r}{R}$.

Table 19. Q values (Doerffel 1967; Dean and Dixon 1981)

No. for value n	Q value for Si = 5% (according to Doerffel 1967)	Q value for Si = 10% (according to Dean and Dixon 1981)
3	0.97	0.94
4	0.84	0.76
5	0.73	0.64
6	0.64	0.56
7	0.59	0.51
8	0.54	0.47
9	0.51	0.44
10	0.49	0.41

x_A is the high value that is to be checked, x_r is the next adjacent value, and R is the range between the lowest and highest values. The high value is acceptable if Q is smaller than the corresponding Q value in Table 19.

Example: A tungsten deposit is tested by seven drill holes (Table 20):

Table 20. Results of seven drill holes from a tungsten deposit

Drill hole	WO$_3$ (%)	Drill hole	WO$_3$ (%)
1	0.8	5	4.6
2	1.4	6	2.1
3	0.7	7	1.5
4	2.4		

The value of 4.6% WO$_3$ in drill hole 5 should be checked for its statistical acceptability. The next adjacent drill hole is No. 4, which yielded a value of 2.4% WO$_3$. From this equation

$$x_A = 4.6\% \quad x_r = 2.4\% \quad R = 4.6 - 0.7\% ,$$

i.e.

$$Q = \frac{x_A - x_r}{R} = \frac{4.6 - 2.4}{4.6 - 0.7} = \frac{2.2}{3.9} = 0.56 .$$

The significance level Si=5% is selected. Q is smaller than the corresponding value in Table 19 for Si=5%, and therefore the value 4.6 is acceptable in statistical terms.

In this situation, a spatial interdependence can be important and this will be explained together with geostatistical considerations later in Chapter 13. If the next adjacent drill hole had been drill hole 1, where the intersection was only 0.8% WO_3, then Q=1 and therefore clearly above the value determined from Table 19, and thus this value would be regarded as statistically exceptional.

8.3.3.3
The FUNOP Method

The FUNOP (Full Normal Plot) method was developed by Tukey (1962), and some of the nomenclature is derived from Koch and Link (1970). As in Chapter 8.3.1, the theoretical cumulative frequency distribution is used as a standard. In the FUNOP method a new value, known as the Judd, is derived. The Judd is defined as:

$$JUDD = \frac{x_i - m}{SW(x_i)} . \tag{1}$$

x_i are again the individual values, m is the median (cf. Chap. 3.4), SW (x_i) is the s-value for the value x_i. On the normalized cumulative frequency scale (see Appendix A1, left-hand scale), the s-value is the distance of the plot frequency (or the plotting percentage) point for the corresponding value x_i from the mean $\bar{x}$, for which the s-value is SW = 0 (Chap. 5.1.2).

Example: If the plotting percentage value for a value x_i =84.4%, then the corresponding s-value SW =1.

As explained in Chapter 3.4, the median m and the mean $\bar{x}$ are identical for a normal distribution. Thus if the individual values x_i are normally distributed, then the Judd value is the same as the standard deviation s.

If a Judd for a value suspected of being an outlier is greater than twice the interpolated Judd value of the median, then the value should either be ignored or reduced by the procedures described in the above sections.

As in Chapter 8.3.1, the plotting percentage is used rather than the cumulative frequency [Eq. (1), Chap. 8.3.1].

Example: There are 15 values for a phosphate deposit that have been listed with increasing grade (Table 21). Must the highest value be reduced?

1st Step: The plotting percentage Hp is calculated from the absolute cumulative frequency (Table 21, column 3) by applying Eq. (1), Section 8.3.1:

$$Hp = 100 \cdot \frac{3 \cdot (\text{absolute cumulative frequency}) - 1}{3 \cdot n + 1}$$

n is the total number of values, and therefore n=15.
Therefore for the first value Hp is:

$$Hp = \frac{100 \cdot (3 \cdot 1 - 1)}{3 \cdot 15 + 1} = \frac{100 \cdot 2}{46} = 4.3\% .$$

Table 21. Calculation of Judd values for a phosphate deposit

Column 1	Column 2	Column 3	Column 4	Column 5	Column 6	Column 7	Column 8
P_2O_5 analysis	Absolute frequency	Absolute cumulative frequency	Plotting percentage (%) (relative)	s-value SW	$x_i - m$	$Judd = \dfrac{x_i - m}{SW}$	Judd, ordered according to size
18.5	1	1	4.3	- 1.77	- 3.50	1.98	0.75
19.6	1	2	10.9	- 1.25	- 2.40	1.92	0.88
20.3	1	3	17.4	- 0.95	- 1.70	1.79	1.78
21.3	1	4	23.9	- 0.80	- 0.70	0.88	1.92
21.4	1	5	30.4	- 0.80	- 0.60	0.75	1.98
21.7	2	7	43.5				
22.0	1	8	50.0				2.13
22.5	1	9	60.9				
23.2	1	10	63.0				
23.8	1	11	69.6	+ 0.50	+ 1.80	3.60	2.27
23.9	1	12	76.1	+ 0.70	+ 1.90	2.71	2.71
24.4	1	13	82.6	+ 1.07	+ 2.40	2.27	2.96
25.7	1	14	89.1	+ 1.25	+ 3.70	2.96	3.60
33.8	1	15	95.7	+ 1.73	+ 11.80	6.82	6.82

2nd Step: The s-value SW is determined graphically for the plotting percentage from Appendix A1, left-hand scale, for the upper and lower third of the values (column 5 in Table 21).

3rd Step: The numerator, $x_i - m$, in the above Eq. (1) is determined. m is the median, which is the 8th value in Table 21, or 22.0% P_2O_5. Therefore, the value in the first row:

$$18.5 - 22.0 = -3.5 \ .$$

4th Step: The Judds for the upper and lower third of the values are calculated according to the above Eq. (1). Therefore for the first row:

$$Judd = \frac{x_i - m}{SW} = \frac{-3.5}{-1.77} = 1.98 \ .$$

5th Step: The Judds are classified according to size and the Judd for the median is interpolated. The interpolated Judd value for the median is therefore:

$$Judd(median) = \frac{2.27 + 1.98}{2} = 2.13 \ .$$

6th Step: This interpolated Judd value for the median is compared to the Judd, 6.82, for the value suspected of being an outlier. The latter is clearly more than twice as large.

Answer: The final analytical value 33.8% P_2O_5 is therefore suspected of being an outlier, and it must be either ignored or else reduced (see Sects. 8.3.3.1 and 8.3.3.2).

8.4
Practical Experience with the Cut Levels

8.4.1
Experience from the Gold Sector

Gold mining is specially affected by the problem of the high values, and there are several case histories that provide practical solutions for this problem. Numerous gold mines have developed practical rules for reducing values that are based on a long-term comparison of the arithmetic average of all relevant analytical values with the average grade of the run of mine ore (ROM). These rules are not based on an evaluation of the statistical distribution of the gold mineralization, but are developed as a practical tool to satisfactorily reconcile predicted grades with achieved ROM grades. The cut level naturally increases with the average grade of the ore. Figure 21 illustrates the trend of the correlation between the cut level and the average grade for several mines, most of which are Australian gold mines. Although the cut levels are influenced by many factors specific to a mine, like the kind of distribution of the gold mineralization, the kind of reserve calculation (e.g. calculation sectionwise based on the polygonal method) or mining methods, the trend line in Fig. 21 can be used as a general guideline in the first stages of exploration.

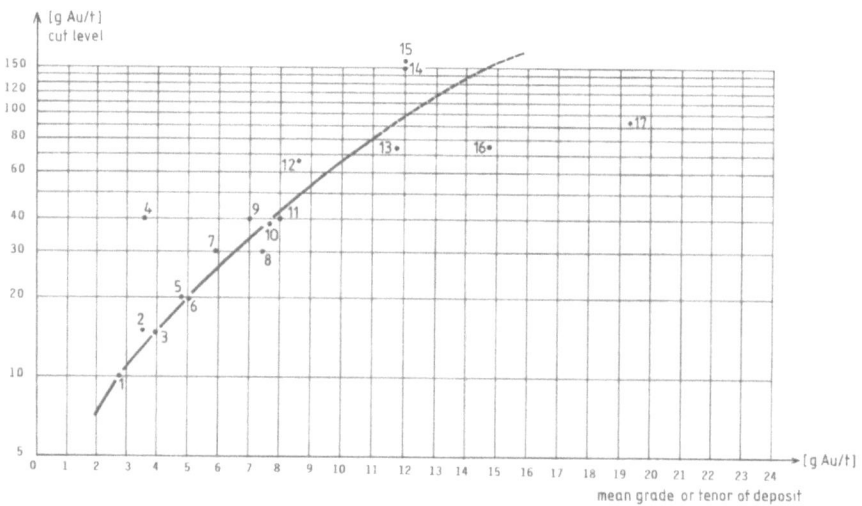

Fig. 21. Cut levels in gold mining as a function of mean grade. Data basis for Fig. 21. Number and mine/deposit (if not indicated otherwise, Australia). *1* Broad Arrow; *2* Ora Banda/BHP; *3* Dome (Canada); *4* Paddington; *5* Mt. Charlotte; *6* Victory/Kambalda; *7* Paringa; *8* Kia Ora/Marvel Loch; *9* Perseverance; *10* Great Boulder (before 1951); *11* Fimiston; *12* Great Boulder (after 1951); *13* Mararoa; *14* Lancefield; *15* GMK 1940; *16* Crown/Norseman; *17* Princess Royal

It is usually possible to determine the tenor, or order of magnitude of the grade, for gold mineralization, i.e. whether it is 3, 6 or 19 g/t gold. The cut level can then be determined for a first approximation from the diagram in Fig. 21.

8.4.2
Derivation of the Cut Level from the Lognormal Distribution

Several mining companies derive the cut level from the lognormal distribution into which skewed distributions can be transformed, (i.e. the logarithms of the values can be satisfactorily described by a normal distribution). The lognormal distribution will be discussed in Chapter 9. For these distributions the upper 2σ limit of the population is taken. It was shown in Chapter 4 and Fig. 12 that 95.45% of all values lie between the -2σ and $+2\sigma$ limits, and thus the upper tail contains half of the remaining values, therefore:

$$\frac{100 - 95.45}{2} = 2.28\% .$$

The cut level is therefore 100 - 2.28 = 97.7%.

This cut level is identical to the threshold value that distinguishes normal from possibly anomalous values in geochemical exploration (Chap. 18.2.1.1.1).

Another method in dealing with outliers using the lognormal distribution is discussed in Chapter 9.3.6.

9
The Use of the Lognormal Distribution

9.1
Introduction

Individual values are logarithmically transformed for the use of the lognormal distribution. The type of logarithm is of no importance, and either the natural logarithm, which is based on the natural number e=2.7183 (thus x is transformed to ln x), or the decimal logarithm to the base 10 (thus x is transformed to log x) can be used[12].

Natural logarithms are used in most of the following calculations, and for ease of use a table of natural logarithms is reproduced as Appendix Table 5.

Logarithmic values are used for the derivation of the mean and calculation of the variance and standard deviation, in the same way as has already been described for normal untransformed values. All values to be considered in logarithmic distribution have to be > 0, otherwise statistical parameters like the mean and the variance cannot be calculated.

Example: A gold deposit was drilled by five widely spaced drill holes that yielded the following results

| 5.2 | 1.5 | 35.9 | 9.8 | 17.7 g Au/t. |

In the first step, these figures are converted to the natural logarithms:

Table 22. Results of 5 drill holes from an Au deposit

Drill hole no.	x_i (g Au/t)	ln x_i
1	5.2	1.65
2	1.5	0.41
3	35.9	3.58
4	9.8	2.28
5	17.7	2.87

---

[12] The following relationship enables conversion between the natural logarithm and the decimal logarithm:

log a ln 10 = ln a (ln 10 = 2.3026)

ln a log e = log a (log e = 0.4343).

The arithmetic mean is now (Chap. 3.1.1):

$$\bar{x} = \frac{1}{n}\sum_{i=1}^{n} x_i \text{ , and therefore in this example } \bar{x} = 14.0 \text{ .}$$

Accordingly the mean of the logarithmic values, which is denoted by γ is:

$$\ln\gamma = \frac{1}{n}\sum_{i=1}^{n} \ln x_i \text{ , i.e. } \frac{1.65 + 0.41 + 3.58 + 2.28 + 2.87}{5} = \frac{10.79}{5} = 2.16 \text{ .}$$

Since ln $\gamma = 2.16$, the antilog for γ;

$$\gamma = 8.65 \text{ .}$$

Since the logarithms have been added, clearly a *geometric mean* has been generated because

$$\frac{\ln x_1 + \ln x_2 + \ln x_3 + \ln x_4 + \ln x_5}{5}$$

is the same as

$$\ln(\sqrt[5]{x_1 \cdot x_2 \cdot x_3 \cdot x_4 \cdot x_5}) \text{ .}$$

One fundamental point about the use of the arithmetic or geometric mean must be discussed.

The grade distribution of some deposits is skewed to the right because additional high values occur above the normal range of values. Sometimes the geometric mean is calculated under the assumption of a lognormal distribution of these grades, or the data is transformed into another distribution from which the mean value is calculated, and then this value is considered to be the average grade of the deposit.

The mean of the lognormal distribution, which is the geometric mean, is always *less* than the arithmetic mean (see the above example). Thus it is very conservative to use such a procedure, and it is also relatively easy to miss the chance of recognising an economic occurrence.

The *arithmetic* mean should always indicate the true average grade of a deposit, assuming that it has been correctly sampled. This does not depend on the mathematical distribution used to describe the sample population of the deposit, as is easily demonstrated by the following consideration. Consider that the whole of an orebody, the universe (Chap. 1.1), is subdivided into individual samples. Clearly the arithmetic mean of all the individual samples (which is the population, Chap. 1.1) corresponds to the average of the orebody. If the individual samples are then mined, then the metal contents are *added*, and not multiplied!

If the sample size is now reduced by taking a representative sample, which precisely correlates with the original sample size or population, then the arithmetic mean of this sample must also correspond to the average of the orebody.

It can be argued that for other populations other methods of deriving the mean might provide a better indication of the mean, such as the age distribution of the populace or a school class. For the case of the age distribution, all means are fictional figures, be it the arithmetic, geometric or another; but this is not so for the derivation of the mean for the evaluation of an orebody. The metal content, which is the final amount that will be produced from the orebody, is the product of the mean value of the grade and the tonnage.

In practice, the arithmetic mean is often too high in comparison to the grades that are actually obtained during mining, as has been discussed in Chapter. Methods to reduce the arithmetic mean in order to derive a better agreement between estimates and actuality were mentioned in Chapter 8. However, the geometric mean of a lognormal distribution can also be used by multiplying it by the factor of the Sichel's t-estimator, and this will be discussed further in Section 9.3.3. Experience has shown that in many deposits, especially gold deposits, the lognormal frequency curve can be satisfactorily fitted to the assay frequency histograms (De Wijs 1951; Krige 1951, 1960, 1994; Sichel 1952, 1966).

9.2
The Lognormal Distribution (for Numerous High Values)

9.2.1
Derivation of the Lognormal Distribution

The values of a population that describe a lognormal distribution are transformed logarithmically to a normal distribution, as has been explained in Section 9.1. This is illustrated schematically in Fig. 22. The skewed distribution is shown in Fig. 22a with the most common or mode value h, median m and mean $\bar{x}$ (see also Chap. 3.4). Figure 22b shows the frequency distribution of the same values after they have been transformed logarithmically. The peak of the frequency curve now marks the median value. If the distribution is exactly lognormal, then the median m is identical to the logarithm of the geometric mean, γ.

Initially it might seem surprising that a value such as the median m is shifted from the flank of one distribution curve to the peak of the other, while the mode value h is shifted from the peak to the flank. This can be best understood by considering the frequency curves to be the outline of histograms (Chap. 2.2). If the class intervals are assumed to be the same in Fig. 22a, then the logarithmic transformation will make them unequal. The class intervals are therefore converted to equal widths for the log-transformed version in Fig. 22b, and the values are reclassified into new intervals, which results in the shift of the maximum. This can be explained by an example:

Example: A gold deposit has been investigated in the first phase of an exploration programme by 20 regularly spaced trenches. These yielded the following grades that were recalculated for a minimum mining thickness of 1 m for

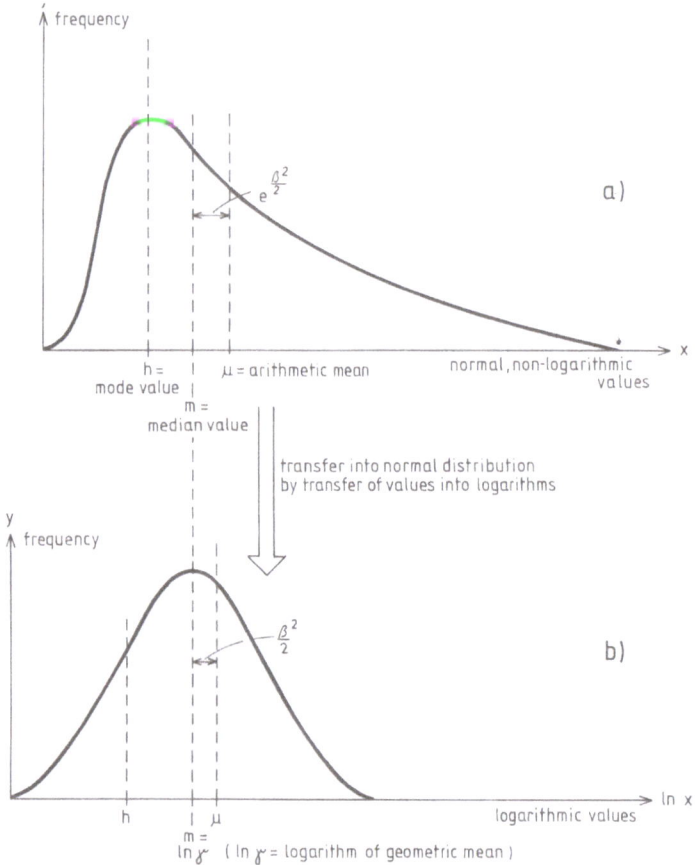

Fig. 22. Transfer of a distribution skewed to the right into a normal distribution by transfer of the single values into logarithms

selective open pit mining. These values have been classified according to their magnitude in column 2, Table 23.

The data are subdivided into eight class intervals of equal 10 g/t Au width. The class intervals and class frequencies are shown in column 3 of Table 23.

The resultant histogram is depicted in Fig. 23a. Additionally, the mode value h, the geometric mean or median value m and the arithmetic mean x have also been plotted. The x-axis is then transformed to a logarithmic scale for Fig. 23b. The arrows show how the various class intervals as well as h, m and x have shifted. Column 4 in Table 23 lists the logarithms of the gold grades of the 20 trenches in this example, and these are classified into eight equal-sized class intervals in column 5 of the same table. The result is shown in Fig. 23c, together with the mode

Table 23. Results from 20 trenches in an Au deposit

Column 1 Trench no.	Column 2 Value x_i in g Au/t	Column 3 Class frequency	Column 4 Logarithmic value $x_i = \ln x_i$	Column 5 Absolute frequency within classes
6	1.8	(0-10):	0.59	(0.5-1.9):
3	2.7		0.99	2
10	3.2		1.16	(1.0-1.5):
11	4.1	10	1.41	2
4	4.9		1.59	(1.5-2.0):
7	6.0		1.79	
12	6.5		1.87	3
2	7.5		2.01	(2.0-2.5):
8	8.9		2.19	
13	9.9		2.29	4
9	10.5	(10-20):	2.35	
1	13.0		2.56	(2.5-3.0):
14	15.3	5	2.73	
15	16.2		2.79	4
20	19.4		2.97	
3	23.8	(20-30):	3.17	(3.0-3.5):
17	27.1	2	3.30	2
19	35.2	(30-40) : 1	3.56	(3.5-4.0):
		(40-50) : 0		2
18	50.7	(50-60) : 1		
		(60-70) : 0	3.93	
16	79.7	(70-80) : 1	4.38	(4.0-4.5) : 1
	$\bar{x} = 17.3$		mean $= 2.38 = \alpha$	

value h, the median or geometric mean value m and the arithmetic average $\bar{x}$. It can be clearly seen how the median value has shifted from the flank of the histogram of the non-logarithmic values to the peak of the logarithmic values.

This example emphasises the importance, when constructing histograms, of selecting the class intervals for a lognormal distribution from the log-transformed values, and not from the original non-transformed values.

9.2.2
The Logarithmic Probability Grid

In Chapter 5.1.2 the probability grid for the normal distribution was derived. It is also possible to generate a similar grid for the lognormal distribution. The cumulative frequency curve of the lognormal distribution will also be represented by a straight line (Hazen straight line). Thus, the y-axis with the cumulative frequency percent values is the same, while the x-axis is logarithmically subdivided

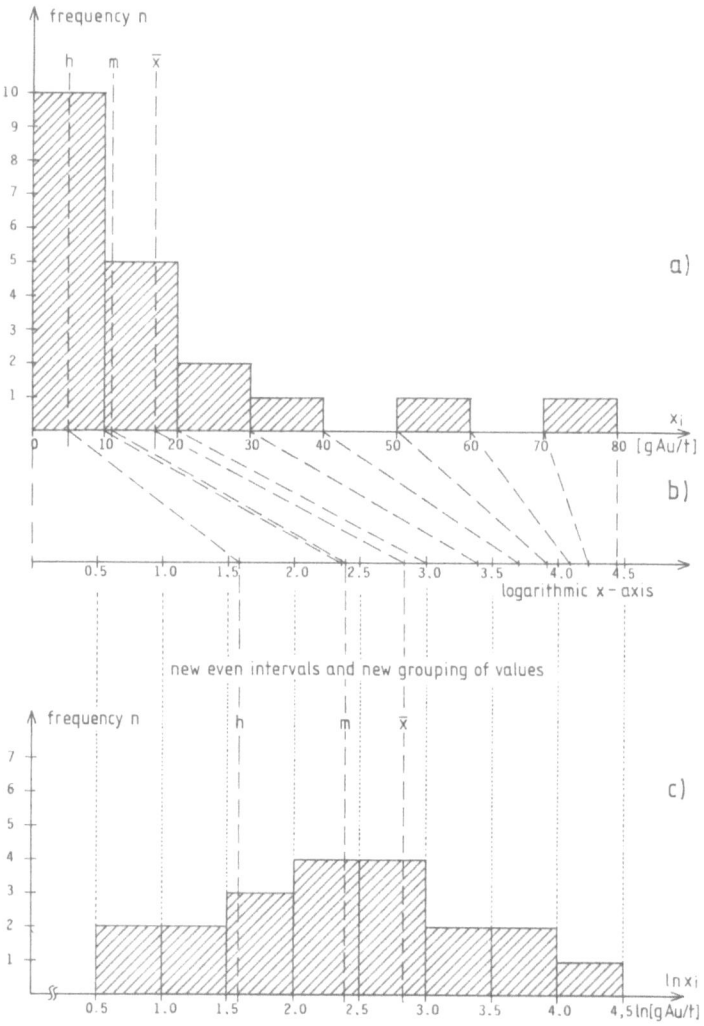

Fig. 23a-c Histogram of gold values of Table 23. a Histogram with linearly divided x-axis. b Transfer of the linearly divided x-axis into a logarithmically divided x-axis. c Histogram of gold values in Table 23 with logarithmically divided x-axis. The parameters h, m, $\bar{x}$ are the parameters of histogram a which are transferred into histogram c with logarithmic values

(Fig. 24). Krige (1978) developed a grid from which the arithmetic mean $\bar{x}$ (as well as the mode value h, however it is of no interest for this discussion) can be derived (Fig. 24a)[13] . A further example of the grid is provided as Appendix B1 at the back of

---

[13] Again, following the example of Fig. 14a and b in Chapter 5.1.3, Fig. 24a is the normal version of a probability grid used in this text book, Fig. 24b shows the inversed mode for readers already familiar with the reversed mode to more easily become familiar with the convention applied in this text book.

this book. (Again, as mentioned in Chap. 5.1 for the probability grid of the normal distribution, in some English speaking countries the x- and y- axes are often reversed. For readers already more used to the reversed version of the probability grid also such a version is enclosed as Appendix B2 at the back of this book).

The use of this grid is demonstrated with the example of the trench samples collected from a gold deposit that was introduced in Chapter 9.2.1. The data are repeated, but have been reclassified, in Table 24. Simple geometric class limits have been selected. These limits increase by approximately double the previous limit: 1; 2.5; 5; 10; 20: 40: 80. The cumulative frequency percent is then calculated for each class interval (Table 24, column 4).

The cumulative frequency percent is then plotted on the logarithmic probability paper (Fig. 24) at the upper limit of each class interval, as has already been explained in Chapter 5.1.3. The Hazen line is then extrapolated through these points, and if they plot very close to the line, then it can be presumed that the data set can be described as lognormal. This Hazen line is now shifted parallel until it

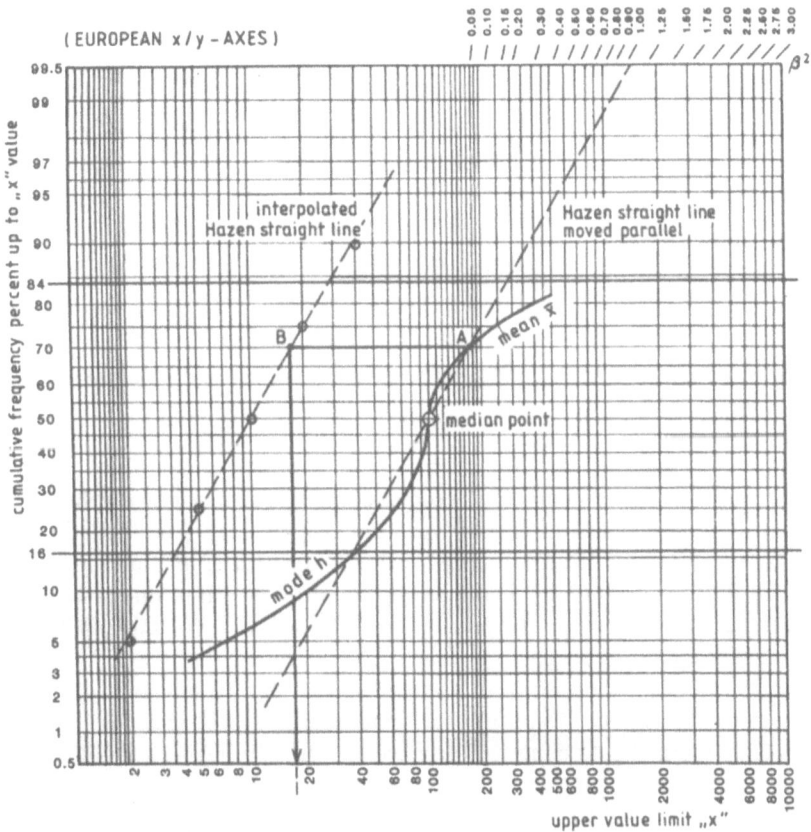

Fig. 24.a. Logarithmic cumulative probability paper for the graphic determination of the arithmetic mean and the mode (convention of x- and y-axes used in this textbook).

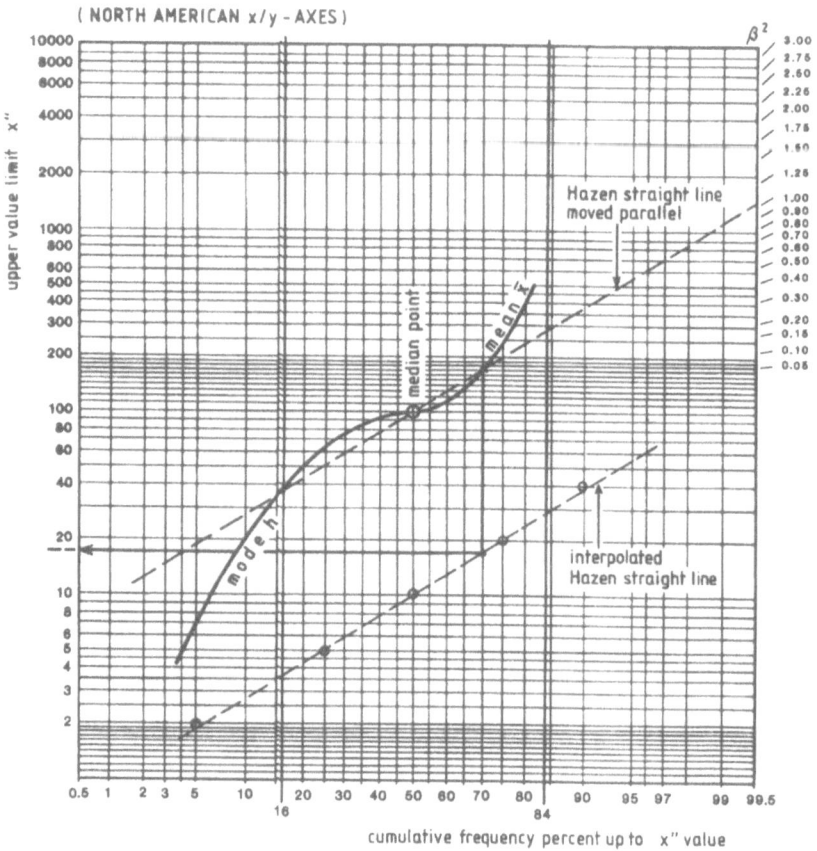

Fig. 24.b. As a, only inversed x- and y-axes (North American standard). (Krige 1978 with permission of the author)

runs through the median point in the centre of the grid. The parallel-shifted Hazen line cuts the mean value line at point A. A line is now drawn parallel to the x-axis through point A, and this line intersects the original Hazen line at point B. The value on the x-axis at point B is the arithmetic mean, which in this case is 16.6 g Au/t. If the arithmetic mean is calculated from the values in column 2 in Table 24, then it yields the value 17.3 g Au/t.

The probability paper can also be used to separate two or more lognormally distributed populations. Deposits with multiple pulse mineralization like for example epithermal gold deposits, can have two or more lognormally distributed assay populations creating irregular lognormal probability plots (see e.g. Parker 1991). In the statistical treatment of geochemical exploration data it is a standard procedure to separate two or more populations using the probability paper. This is shown is Chapter 18.2.2. In Section 9.3.6 the probability grid will be used to separate two lognormal populations for the evaluation of a mineralization with outlier data.

Table 24. Summary of the values of Table 23

Column 1 Trench no.	Column 2 Value x_i (g Au/t)	Column 3 Class	Column 4 Relative cumulative frequency (%)
6	1.8	1.0 - 2.5	5
5	2.7		
10	3.2	2.5 - 5	25
11	4.1		
4	4.9		
7	6.0		
12	6.5	5.1 - 10	50
2	7.5		
8	8.9		
13	9.9		
9	10.5		
1	13.0		
14	15.3	10.1 - 20	75
15	16.2		
20	19.4		
3	23.8		
17	27.1	20.1 - 40	90
19	35.2		
18	50.7	40.1 - 80	100
16	79.7		
	$\bar{x} = 17.3$		

9.2.3
Determination of Parameters for the Lognormal Distribution

9.2.3.1
Mean and Variance

The random variable x_i was the basis of the normal distribution. For the lognormal distribution there is a new random variable, ln (x_i). The lognormal distribution is described by the mean and variance parameters, just as it is for the normal distribution. Since these parameters are based on a new random variable, other symbols are used when working with the lognormal distribution in order to avoid any confusion. The mean value of the lognormal distribution is denoted by α, and the variance by β^2[14]:

---

[14] The lognormal distribution is defined by the equation (see for examples Koch and Link, 1970):

$$y = \frac{1}{x \cdot \beta\sqrt{2\pi}} \cdot e^{-\frac{1}{2}\left(\frac{\ln y - \ln x}{\beta}\right)^2}.$$

$\mu(\ln x_i) = \alpha$

$\sigma^2(\ln x_i) = \beta^2.$

The mean α is the logarithm of the median value m or the geometric mean y respectively. Therefore:

$\alpha = \ln y$ oder $y = e^\alpha.$

Returning to the example in Section 9.2.1, the value y can be derived from the 50% cumulative frequency line on the probability diagram in Fig. 24, and is 10 g/t Au. The fundamental concept of the Hazen line for lognormal distributions is the same as that already discussed in Chapter 5.1.3 for the probability grid of a normal distribution: the interpolated Hazen line represents the distribution of the whole population as estimated from the cumulative frequency curve of the sample size. α and β can be derived from the cumulative frequency diagram and can be used as estimates for the parameters of the sample size.

If the estimated value y of 10 g Au/t is compared with that calculated from the sample set, then using the log-transformed values in Table 23, column 4:

$$\alpha = \ln y = \frac{1}{n} \cdot \sum_{i=1}^{n} \ln x_i .$$

This yields $\alpha = \ln y = 2.38$, thus $y = 10.8$ g Au/t.

By means of the logarithms of the values, the variance β^2 of the logarithmic distribution can be calculated from the sample set in the same way as the variance is calculated for non-logarithmic values (Chap. 3.2.1).

If the variance is estimated graphically from the logarithmic standard deviation on the probability diagram, then the x-values corresponding to cumulative frequencies 16, 50 and 84% are read:

$$\beta = \frac{1}{2} \cdot \left[\ln(\frac{x_{84}}{x_{50}}) + \ln(\frac{x_{50}}{x_{16}}) \right] .$$

x_{84}, x_{50}, x_{16} are the non-logarithmic values, which constitute the quotients, and they are normally shown on the x-axis of the probability grid (Fig. 24).

If the logarithms themselves are read off the probability grid, then clearly the differences must be calculated.

The following are read off Fig. 24:

$x_{84} = 28,$

$x_{50} = 10,$

$x_{16} = 3.7$.

This yields:

$$\beta = \frac{1}{2}\left[\ln(\frac{28}{10}) + \ln(\frac{10}{3.7}) \right] = \frac{1}{2}\left[\ln 2.8 + \ln 2.7 \right] ,$$

$$\beta = \frac{1}{2}(1.03 + 0.99) = 1.01 ,$$

thus the variance $\beta^2 = 1.02$.

It is clear from this numerical example that the logarithmic standard deviation β is apparently an average *ratio* of the original data to the geometric mean and is *not*, as for the normal standard deviation, the average of the *differences* from the mean value [Chap. 3.2.1, Eq. (1)].

As a third possibility, the variance β^2 can be directly read from the probability grid in Fig. 24. The Hazen line that was shifted parallel to run through the median point intersects the variance scale on the upper margin at a value slightly more than 1.

The book by Aitchison and Brown (1969) is recommended for those wishing to study the lognormal distribution in more detail.

9.2.3.2
The Correction Factor θ

Geological data, particularly mineral deposit ore grade data with a distribution skewed to the right, can be plotted on the logarithmic probability net (Appendices B1 and B2 at the back of the book). It is commonly found that the cumulative frequency distribution plots as a curve, and not straight as is required for a lognormal distribution. Krige (1962) demonstrated that, by adding the constant θ to the values, it is possible to satisfactorily transform distributions of this type into lognormal distributions. As described in Section 9.2.3.1, normal and lognormal distributions are usually described by the two parameters mean and variance. Lognormal distributions which are created by adding the correction factor θ, a third parameter, are therefore called the *three- parameter lognormal distribution*.

The transformation into a three-parameter lognormal distribution is explained by an example.

9.2.3.2.1
Graphical Determination of the Correction Factor θ

Example: A gold deposit was exploited in the 1930s and must now be evaluated. There are numerous sample results from this previous period of exploitation, all of which are reported as pennyweights (dwt, see Economic Evaluations in Exploration, Chap. 1.1.4.5). The thickness of the veins is very variable, and this

Table 25. Accumulation values in dwt · inch from an Au deposit

Column 1 Class interval (dwt · inch)	Column 2 No. of samples	Column 3 Relative cumulative frequency (%)	Column 4 New class intervals using correction parameter θ (dwt · inch)
0.0- 4	94	11.9	57.0- 61
4.1- 8	17	14.0	61.1- 65
8.1- 16	30	17.8	65.1- 73
16.1- 32	33	21.9	73.1- 89
32.1- 64	47	27.9	89.1- 121
64.1- 128	95	39.9	121.1- 185
128.1- 256	135	56.9	185.1- 313
256.1- 512	131	73.4	313.1- 569
512.1- 1024	114	87.8	569.1- 1081
1024.1- 2048	62	95.7	1081.1- 2105
2048.1- 4096	27	99.0	2105.1- 4153
4096.1- 8192	6	99.7	4153.1- 8249
8192.1-16384	0	99.7	8249.1-16441
16384.1-32768	2	100.0	16441.1-32825
	$\Sigma = 793$		

dimension is reported in inches. The thickness x grade product, the accumulation values or intensity factors, are calculated in order to take the varying thickness of the vein into account, and are reported as the unit dwt·inch. In metric units 1 dwt·inch = 3.95 g ·cm (see Economic Evaluations in Exploration, Chap. 1.2.4). In order to avoid unnecessary calculations, the evaluation is carried out in the original units and the result is then converted into metric units.

A statistical evaluation of all the accumulation values in dwt·inch units yielded the following (Table 25, columns 1 to 3):

These cumulative frequency values are now plotted on the logarithmic probability grid (Fig. 25). This shows that the lower sections of the cumulative frequency curve are strongly bent, and that the curve only becomes linear for the higher class intervals.

The constant θ is derived by the following method of approximation: the straight section of the cumulative frequency curve is extrapolated to the lower values (extrapolation line in Fig. 25). The difference is then calculated between the upper class limit of the lowest cumulative frequency values and the corresponding value on the extrapolation line. This is the point on the probability grid where the absolute difference is at greatest.

The first three cumulative frequency values are selected, and the following differences are determined:

θ_1 (between A and B):   51
θ_2 (between C and D):   57
θ_3 (between E and F):   64

The average is then θ   = 57.3, rounded to 57.

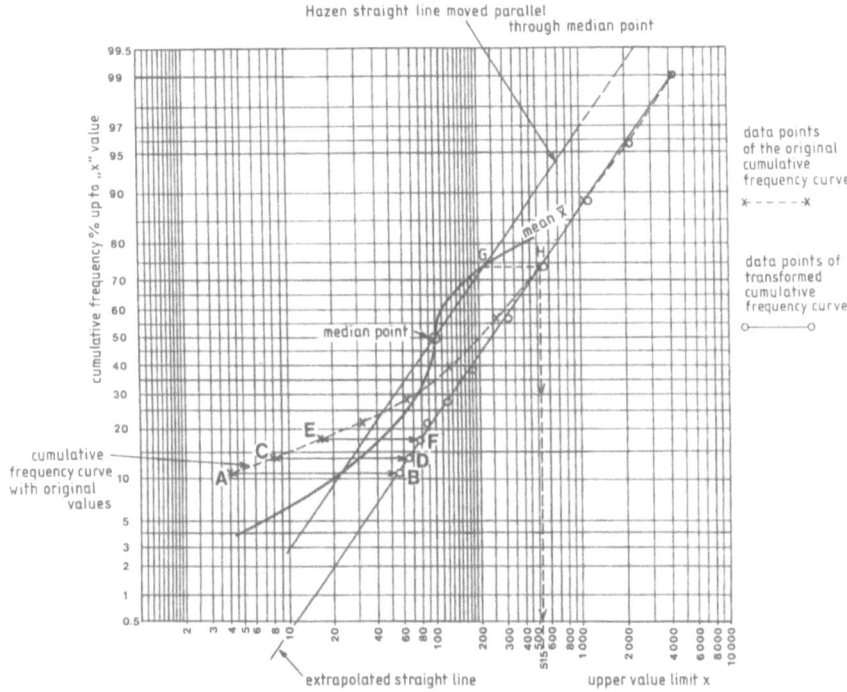

Fig. 25. Example for the derivation of the correction factor θ on the logarithmic cumulative frequency paper

This value is then added to each of the class limits of the individual intervals (column 4 in Table 25).

The cumulative frequency curve is thus transformed and can be seen to be a satisfactory approximation to a Hazen line.

This means that many of the skewed distributions that are found in geology can be approximated to lognormal distributions, which are described by three parameters:

the mean of the distribution α,
the variance β² and
the correction factor θ .

If the arithmetic mean is now determined graphically, as was demonstrated in Section 9.2.2, then the constant θ must be subtracted from this graphically derived value in order to find the arithmetic mean of the original data set.

The calculation procedures for the example based on the data in Table 25 and Fig. 25 would be as follows: the Hazen line for the transformed data is shifted parallel again so that it runs through the median point on the grid. It intersects the mean value line at point G. A line parallel to the x-axis is drawn from point G and intersects the original Hazen line at point H. The x-value at point H is the arithmetic mean of the transformed data: x_t. It can be seen to be the value 515. The correction factor θ must now be subtracted from this value in order to derive the arithmetic mean $\bar{x}$ for the original data:

$$\bar{x} = \bar{x}_t - \theta = 515 - 57 = 458,$$

which is rounded to 460 dwt·inch, or converted to the metric system 460 x 3.95 = 1817 g·cm.

9.2.3.2.2
Mathematical Determination of the Correction Factor θ

Rendu (1978) provides a simple formula for the determination of the correction factor θ:

$$\theta = \frac{\gamma^2 - f_1 \cdot f_2}{f_1 + f_2 - 2\gamma} \; .$$

The value γ is again the median value that can be read from the 50% line on the probability grid. f_1 and f_2 are characteristic limiting values that correspond to the probabilities p and 1-p. For the sake of simplicity, the values 16 and 84% on the y-axis are selected for p and 1-p.

The example from Section 9.2.3.2.1 and Fig. 25 is used once again, and the following values are derived from the diagram in Fig. 25 (using the cumulative frequency curve for the primary data, which is the dashed line):

γ = 195 dwt · inch,

f_1 (value at the 16 % cumulative line) = 11 dwt · inch,

f_2 (value at the 84 % cumulative line) = 850 dwt · inch.

Therefore substituting in the formula for q:

$$\theta = \frac{195^2 - 11 \cdot 850}{11 + 550 - 390} = \frac{28675}{471} = 60.9 \text{ dwt · inch},$$

which is a somewhat higher value as compared to that obtained graphically in the previous Section 9.2.3.2.1.

9.2.3.2.3
Using a-Priori Information for the Estimation of the Correction Factor θ

Sichel (1968, as referenced by Krige et al., 1990) showed that the correction factor θ is extremely badly estimated from a small number of samples. Krige et al., (1990) therefore recommended using a priori information of similar mineralisations of the same type in the same geological region for the estimation of the correction factor θ and thereby finally improving the estimate of the means of the grades.

9.3
Determination of the Arithmetic Mean Value for Skewed Sample Distributions Described by Lognormal Distributions

9.3.1
Introduction

The problems of obtaining a satisfactory arithmetic mean value by applying the normal formula

$$\bar{x} = \frac{1}{n}\sum_{i=1}^{n} x_i$$

have been discussed in Chapter 8. The influence of high values was emphasized and the question of which rules of thumb should be used to correct erratically high values was addressed. In the following, the lognormal distribution is used for the correction, since as outlined above, experience shows that the lognormal frequency curve can be satisfactorily fitted to assay histograms of many deposits.

9.3.2
Determination of the Arithmetic Mean from the Logarithmic Mean and Logarithmic Variance

In Sections 9.2.2 and 9.2.3.2 the arithmetic mean was determined for skewed distributions by using a lognormal distribution as an approximation of the true sample distribution. The arithmetic mean $\bar{x}$ of the lognormal distribution satisfies the relationship:

$$\bar{x} = \gamma \cdot e^{\frac{\beta^2}{2}}, \tag{1}$$

where γ is the geometric mean and β^2 is the variance (Sect. 9.2.3.1). Furthermore:

$$\gamma = e^{\alpha}, \tag{2}$$

where α is the mean of the lognormal distribution, then the arithmetic mean[15] is:

---

[15] Thus $\ln\bar{x} = \alpha + \frac{\beta^2}{2}$. This explains in Fig. 22b that the difference between the logarithm of the

geometric mean and that of the arithmetic mean is $\frac{\beta^2}{2}$, or $e^{\frac{\beta^2}{2}}$ in the frequency distribution of the

non-logarithmic values in Fig. 22a.

$$\bar{x}=e^{\alpha}\cdot e^{\frac{\beta^2}{2}}=e^{(\alpha+\frac{\beta^2}{2})}.$$
(3)

For a three parameter lognormal distribution the arithmetic mean would be

$$\bar{x}=e^{\left(\alpha+\frac{\beta}{2}\right)}-\theta$$
(4)

where θ is the correction factor for the three parameter lognormal distribution (Sect. 9.2.3.2)

It must be emphasized once more that the relationships in Eq. (1) or (4) above represent only one of several possible estimators. The usual relationship for the arithmetic mean (Chap. 3.1.1) is

$$\bar{x}=\frac{1}{n}\sum_i^n x_i,$$

which is also a possible estimator, although it is has very marked weaknesses if there are high values, as has been explained in Chap. 8.1.

The relationship given in Eq. (3) above is the arithmetic mean of the distribution of the population, and it presumes that α and β of the population are already known, or have been correctly calculated from the limited data size only. The estimator is known, however, only on the basis of this sample material. This means that the arithmetic mean calculated with these estimators can significantly deviate from that of the population. Link and Koch(1975) have shown, for example, that the high values can strongly influence the determination of the estimated parameters.

9.3.3
Sichel's t-Estimator

The South African, Sichel (1966), developed a factor, Sichel's t-estimator, to solve the problem of obtaining the best estimation of the arithmetic mean for skewed sample sets that have an approximately lognormal distribution. It is a value by which the geometric mean, γ, is multiplied. The concept is simple:

assume there are the following analytical values for gold:

1, 4, 5, 8, 10, 14 g Au/t .

Then the arithmetic mean is according to the relationship:

$$\bar{x}=\frac{1}{n}\sum_{i=1}^n x_i, \quad \text{therefore } \bar{x} =7\text{ g Au/t}.$$

The geometric mean is:

$$\gamma=\sqrt[n]{x_1\cdot x_2\cdots x_n}, \quad \text{i.e. } \gamma=5.3\text{ g Au/t}.$$

If the lowest of the above values is now changed to 0.1 g Au/t, which, in absolute terms, is a relatively small difference to the value, and the highest of the above values is increased by tenfold to 140 g Au/t, then the arithmetic mean changes to

$\bar{x} = 27.9$ g Au/t ,

however, the geometric mean remains constant since

$1 \cdot 4 \cdot 5 \cdot 8 \cdot 10 \cdot 14 = 0.1 \cdot 4 \cdot 5 \cdot 8 \cdot 10 \cdot 140$.

It can be seen that, with respect to high values, the geometric mean is considerably more stable than the arithmetic mean.

The determination and application of Sichel's t-estimator is best explained by examples.

1st Example: The results from the 20 trenches listed in Table 23 are used. The mean of the logarithmic values in column 4 is determined to be $\alpha = 2.38$, so that the geometric mean γ is:

$$\gamma = e^{\alpha} = e^{2.38} = 10.8 .$$

The logarithmic variance was graphically determined (Fig. 24) in Chap. 9.2.3.1 to be $\beta^2 = 1.02$. Appendix Table 6 reproduces the tables for Sichel's t-estimator. From this table, the t-estimator t_{si} can be derived as a function of the logarithmic variance β^2 and the number of samples n. Thus for $\beta^2 = 1.02$ and the number of samples n = 20, and by interpolating between $\beta^2 = 1.0$ and $\beta^2 = 1.1$, we obtain:

$t_{si} = 1.646.$

The arithmetic mean is now:

$$\bar{x}_{si} = \gamma \cdot t_{si} = e^{\alpha} \cdot t_{si} ,$$

$\bar{x}_{si} = 10.8 \cdot 1.646 = 17.8$ g Au/t .

This value of 17.8 can be compared as follows:

a) the arithmetic mean calculated from formula (1) in Section 9.3.2, where $\gamma = 10.8$ and $\beta^2 = 1.02$ (Sect. 9.2.3.1):

$$\bar{x}_1 = \gamma \cdot e^{\frac{\beta^2}{2}} = 10.8 \cdot e^{0.501} = 10.8 \cdot 1.65 = 17.8 \text{ g Au/t};$$

b) $\bar{x}_2 = 17.3$ g Au/t from the normal mathematical mean given in Table 23, and

c) $\bar{x}_3 = 16.6$ g Au/t on the basis of the graphical determination explained in Section 9.2.2 and Fig. 24.

2nd Example: For the gold deposit described in Section 9.2.3.2, it was demonstrated that a lognormal distribution can describe the accumulation value (grade x thickness) so long as the correction factor $\theta = 57$ is added. In this example there are eight values from a part of the ore deposit listed in Table 26 (and, if it is assumed that the lognormal distribution of the whole deposit is the same, then this applies also):

The logarithmic variance can be determined from the logarithms of the values by the standard formula, which is [Chap. 3.2.1, Eq. (1)]:

$$s^2 = \frac{1}{n-1} \cdot \sum_{i=1}^{n}(x_i - \bar{x})^2 \ .$$

If this relationship is now used for the example, then the following substitutions are made:

$$s^2 \rightarrow \beta^2 \ ,$$

$$x_i \rightarrow \ln(x_i + \theta) \ ,$$

$$\bar{x} \rightarrow \alpha = \frac{1}{8} \sum_{i=1}^{8} \ln(x_i + \theta) \ .$$

β^2 is now calculated, and the result is:

$$\beta^2 = 2.36.$$

Table 26. Eight accumulation values from an Au deposit

Accumulation value: x_i (dwt · inch)	Transformed value: $x_i + \theta$	Logarithmic value: $\ln(x_i + \theta)$
356	413	6.02
137	194	5.27
67	124	4.82
128	185	5.22
481	538	6.29
377	434	6.07
511	568	6.34
17515	17572	9.77
$\bar{x}_i = 2446.5$		$\alpha = \frac{1}{n}\sum_{i=1}^{8}\ln(x_i + \theta)$ $\alpha = 6.23$

Sichel's t-estimator is now determined from Appendix Table 6 for $\beta^2 = 2.36$ (interpolated between 2.3 and 2.4) and the number of samples n = 8:

$t_{si} = 2.881.$

The arithmetic mean is than calculated as:

$\bar{x}_{si} = \rightarrow e^{\alpha} \cdot t_{si} = e^{6.23} \cdot 2.881 = 507.76 \cdot 2.881$,

$x_{si} = 1462.8.$

The correction factor θ must be subtracted from this value:

$\bar{x}_{si}$ (transformed back) $= 1462.8 - 57 = 1405.8$,

or rounded to 1406 dwt $\cdot$ inch.

If this value is now compared to the arithmetic mean,

a) calculated from the normal averaging of the values in the 1st column in Table 26, then:

$\bar{x}_1 = 2446.5$ dwt $\cdot$ inch,

b) calculated from the formula (3) in Section 9.3.2, where $\alpha = 6.23$ (Table 26) and $\beta^2 = 2.36$ (see above), then

$$\bar{x}_2 = e^{(\alpha + \frac{\beta^2}{2})} = e^{(6.32 + 1.18)} = e^{7.41} = 1652.4 .$$

from which the correction factor $\theta = 57$ must be subtracted, so that:

$\bar{x}_2 = 1652.4 - 57 = 1595.4$ dwt $\cdot$ inch.

This example clearly demonstrates the extreme influence of the highest value on the normal arithmetic mean $\bar{x}_1$. It also has an effect on the arithmetic mean $\bar{x}_2$ calculated by the relationships of the logarithmic distribution. The example shows how this effect on the arithmetic means can be reduced by using Sichel's t-estimator.

A comparison of Example 1, in which the mean calculated with Sichel's t-estimator was higher than the normal arithmetic mean, with Example 2, in which the converse was the case, shows the following:

If there are only a few isolated very high values, then Sichel's t-estimator generally yields a lower estimate of the arithmetic means; if there are numerous high values, which result in a relatively continuous but definitely skewed distribution, then the estimate of the normal arithmetic mean is generally lower. An experienced geologist will always select the lower value, even though this has absolutely no statistical basis. It has already been discussed in Chapter 8.2 that an experienced geologist will always use various methods for the calculation of the mean value, and will often select a final value on the basis of geological considerations. The following is an example of a relevant geological argument: the accumulation values are the product of thickness and grade. High values can therefore be caused by both extreme thicknesses and/or extreme grades. If the reasons for high accumulation factors are primarily major variations in thickness and locally extreme thicknesses, then the higher estimate of the mean value would be more reliable since extreme thicknesses are more continuous than extremely high grades, which can occur very locally. The latter is the so-called nugget effect, which will be discussed in more detail in Chapter 13.2. If there is a major nugget effect in the grades, then the lower estimate of the mean value is always preferred.

9.3.4
Finney's Diagram

The previous two chapters have demonstrated the problems encountered in the estimation of the arithmetic mean of data with an approximately lognormal distribution. Finney (1941) published a diagram that is particularly helpful for the selection of the various estimators (Sect. 9.3.2). This diagram shows the effectiveness of the estimate of the normal arithmetic mean from the sample size as compared to the true value of the arithmetic mean for a lognormally distributed population (Fig. 26). So long as the coefficient of variation is less than 1.2, then the simple arithmetic averaging is 90% effective in the estimation of the true arithmetic mean. Therefore, for coefficients of variation that are less than 1.2 (the corresponding logarithmic variance must then be less than about 0.9), it is recommended to choose the simple arithmetic averaging method. This applies of course also to distributions that can be satisfactorily interpolated as a straight line, the Hazen line, on the logarithmic probability paper in Appendices B1 and B2, and whose gradient is steeper that the marginal marking for $\beta^2 = 0.9$.

The coefficient of variation for approximately logarithmic distributions of data C_{log} is calculated from the following equation:

Fig.26. Efficiency of estimation of arithmetic mean $\overline{x}$ as a function of the logarithmic coefficient of variation. (After Finney 1941; Koch and Link 1970) (with permission of the authors)

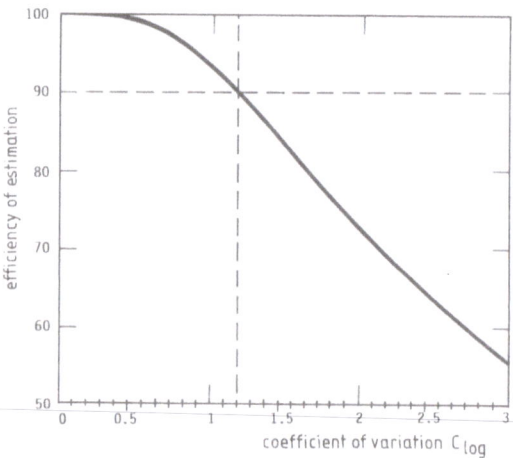

coefficient of variation C_{log}

$$c_{\log.} = \sqrt{e^{\beta^2} - 1} \ ,$$

where β^2 is again the logarithmic variance (see Koch and Link, 1970, for details).

Example: The logarithmic variance was $\beta^2 = 2.36$ in the 2nd example in Section 9.3.3.
Therefore the coefficient of variation $C_{\log}$ is:

$$C_{\log.} = \sqrt{e^{\beta^2} - 1} = \sqrt{e^{2.36} - 1} = \sqrt{10.59 - 1} = \sqrt{9.59} \ ,$$

$$C_{\log} = 3.10 \ ,$$

thus, according to Finney's diagram (Fig. 26), the effectiveness of the estimator by normal arithmetic averaging is only about 50 %. This result underlines that derived from the calculation with Sichel's t-estimator in Example 2, Section 9.3.3.

9.3.5
Confidence Interval of the Arithmetic Mean of Lognormally Distributed Data

Both Sichel (1966) and, in a more refined way, Wainstein (1975) have developed tables for calculating confidence intervals for the arithmetic mean of lognormal or approximately lognormal distributions of data.

First we have to assume a level of confidence as outlined in Chapter 7.1. Following various recommendations for reserve classification as discussed in Chapter 7.1 and Appendix Table 3, we again select a level of confidence of 90%. Because of the asymmetry, the upper and lower limits of the confidence level are calculated separately. A 90% confidence level means that 90% of the values (in this case those of the frequency distribution of repeated determinations of the mean) lie within the confidence limits and, as a consequence, 10% lie outside them. This 10% is subdivided with one half in each of the "tails", so that the lower limit is 5% and the upper limit is 95%. Wainstein's tables for this (Wainstein, 1975) are reproduced as Appendix Table 7a and b.

The use of this table is best explained by the two examples in Section 9.3.3:

1st Example: In Section 9.3.3, the arithmetic mean was calculated for the results from 20 trenches (Table 23) with the assistance of the Sichel's t-estimator:

$$\bar{x}_{si} = 17.8 \text{ g Au/t} \ .$$

The logarithmic variance was $\beta^2 = 1.02$. The lower confidence limit is, of course, of the most practical interest, and it is calculated first. From Appendix Table 7 a, a factor is derived for n=20 and $\beta^2 = 1.02$

$$W_u = 0.6796 \ .$$

The lower limit for a 90 % level of confidence is therefore calculated as:

$$K_u = \bar{x}_{si} \cdot W_u = 17.8 \cdot 0.6796 = 12.1 \text{ g Au/t.}$$

The upper limit is also derived in a similar way. From Appendix Table 7 b, for n=20 and β^2 =1.02, the factor

$$W_O = 1.8587.$$

The upper limit for a 90% level of confidence is therefore calculated as:

$$K_o = \bar{x}_{si} \cdot W_o = 17.8 \cdot 1.8587 = 33.08 \text{ g Au/t.}$$

2nd Example: Eight accumulation values (grade x thickness product) were considered in the 2nd example in Chap 9.3.3. The arithmetic mean was calculated using the correction factor θ = 57 and by Sichel's t-estimator as

$$\bar{x}_{si} = 1462.8 \text{ dwt} \cdot \text{inch,}$$

from which the arithmetic mean is finally calculated to be 1406 dwt · inch by subtraction of the correction factor θ=57. The calculation of the confidence limits must be made using the value $\bar{x}_{si}$ including the correction factor θ, since the approximation to a lognormal distribution occurs only as a result of the addition of the correction factor (Sect. 9.2.3.2). The Wainstein tables can only be used if a lognormal distribution is assumed.

The logarithmic variance β^2 was β^2 = 2.36. The lower confidence limit is calculated first: the factor W_u is determined, by two interpolations, from Appendix Table 7 a for n = 8 and β^2 = 2.36.

1st Step: The factor W_u is determined for n = 8:

for β^2 = 2.3 and β^2 = 2.4,

for β^2 = 2.3, W_u = 0.3937,

for β^2 = 2.4, W_u = 0.3891.

2nd Step: The value for β^2 = 2.36 is now determined by interpolation between β^2 = 2.3 and 2.4. This results in:

$$W_u = 0.3917.$$

The lower limit for a 90% level of confidence is therefore calculated as:

$$K_u = \bar{x}_{si} \cdot W_u = 1462.8 \cdot 0.3917 = 573.0 \, .$$

The correction factor $\theta = 57$ must now be subtracted, which results in:

$$K_{u(\text{transformed back})} = 573 - 57 = 516 \, \text{dwt} \cdot \text{inch.}$$

The same procedures are used for the calculation of the upper confidence limits by means of the factor K_o from Appendix Table 7b for $n = 8$ and $\beta^2 = 2.36$:

$$W_o = 25.2761 \, ,$$

then:

$$K_o = \bar{x}_{si} \cdot W_o = 1462.8 \cdot 25.2761 = 36973.9 \, .$$

The correction factor $\theta = 57$ must now be subtracted, which results in:

$$K_{o(\text{transformed back})} = 36976.4 - 57 = 36916.9 \, \text{dwt} \cdot \text{inch.}$$

This demonstrates that the upper limit for the confidence interval is only of theoretical interest. The most important values for *practical* purposes, after removing the correction factor θ, are:

the arithmetic mean $\bar{x} = 1406 \, \text{dwt} \cdot \text{inch}$
and the lower limit for a 90% level of confidence: $K_u = 516 \, \text{dwt} \cdot \text{inch}$.

9.3.6
Statistical Treatment of Outlier Data Using Two Lognormal Distributions

Parker (1991) uses the method of ordering the data in descending order and calculating the cumulative coefficient of variation (Chap. 3.3.1) from the bottom up, that is the coefficient of variation of all samples with values less than or equal to the current sample and including the current sample. He observes the point at which the cumulative coefficient of variation accelerates upward. This is where the influence of the outliers in the upper tail becomes strong. This point, the so-called breakpoint, is normally around a cumulative coefficient of variation of 1.

The procedure (slightly modified from Parker 1991) again is best explained by an example.

Example: Thirty three channel samples (accumulation values grade x thickness) are available from a first investigation of an epithermal gold mineralization (Table 27) showing a wide range of data. The task is to estimate a mean grade for this group of samples.

Table 27. Thirty three assay values from an epithermal Au mineralisation

Column 1 No. of value n	Column 2 Value x (g Au/t·m)	Column 3 Cumulative values Σ' (metal content)	Column 4 Frequency % of total sum	Column 5 Cumulative frequency % (top to bottom)	Column 6 Cumulative frequency % (bottom to top)	Column 7 Stepwise mean (bottom to top)	Column 8 Cumulative standard deviation (bottom to top)	Column 9 Cumulative coefficient of variation C
33	140.0	140.0	24.20	24.20	100.00	17.53	33.78	1.93
32	120.0	260.0	20.74	44.94	96.97	13.70	26.06	1.90
31	80.0	340.0	13.83	58.77	93.94	10.27	17.69	1.72
30	60.0	400.0	10.37	69.14	90.91	7.95	12.26	1.54
29	42.0	442.0	7.26	76.40	87.88	6.16	7.46	1.21
28	12.0	454.0	2.07	78.48	84.85	4.88	2.91	0.60
27	8.8	462.8	1.52	80.00	81.82	4.61	2.60	0.56
26	8.5	471.3	1.47	81.47	78.79	4.45	2.51	0.56
25	8.4	479.7	1.45	82.92	75.76	4.29	2.42	0.56
24	8.1	487.8	1.40	84.32	72.73	4.12	2.31	0.56
23	7.5	495.3	1.30	85.62	69.70	3.94	2.20	0.56
22	7.4	502.7	1.28	86.90	66.67	3.78	2.10	0.56
21	7.0	509.7	1.21	88.11	63.64	3.61	1.99	0.55
20	6.7	516.4	1.16	89.27	60.61	3.44	1.88	0.55
19	6.0	522.4	1.04	90.30	57.58	3.27	1.76	0.54
18	5.9	528.3	1.02	91.32	54.55	3.12	1.68	0.54
17	5.8	534.1	1.00	92.32	51.52	2.95	1.58	0.53
16	5.1	539.2	0.88	93.21	48.48	2.78	1.44	0.52
15	4.4	543.6	0.76	93.97	45.45	2.26	1.35	0.52
14	4.2	547.8	0.73	94.69	42.42	2.49	1.30	0.52
13	4.0	551.8	0.69	95.38	39.39	2.36	1.26	0.53
12	3.9	555.7	0.67	96.06	36.36	2.23	1.21	0.54
11	3.8	559.5	0.66	96.72	33.33	2.07	1.14	0.55
10	3.3	562.8	0.57	97.29	30.30	1.90	1.04	0.55
9	3.0	565.8	0.52	97.80	27.27	1.74	0.97	0.56
8	2.9	568.7	0.50	98.31	24.24	1.59	0.91	0.57
7	2.5	571.2	0.43	98.74	21.21	1.40	0.80	0.57
6	2.2	573.4	0.38	99.12	18.18	1.22	0.69	0.57
5	1.6	575.0	0.28	99.39	15.15	1.02	0.55	0.54
4	1.5	576.5	0.26	99.65	12.12	0.88	0.52	0.59
3	1.1	577.6	0.19	99.84	9.09	0.67	0.38	0.57
2	0.5	578.1	0.09	99.93	6.06	0.45	0.07	0.16
1	0.4	578.5	0.07	100.00	3.03	0.40		

1st Step: The accumulation values in Table 27 have to be ordered in descending order. This is already done in Table 27, column 2. The cumulative coefficients of variation are then calculated from the bottom up (column 9 in Table 27). Table 27 and Fig. 27 show that these coefficients of variation first fluctuate between 0.5 and 0.6. Starting with the value of 12 gAu/t·m the cumulative coefficient of variation suddenly increases. This is taken as the so-called breakpoint.

2nd Step: On lognormal probability paper a Hazen line is fitted to the data above the breakpoint (Fig. 28).

This Hazen line is extended to the 84 and 16% line (84 and 16% cumulative frequency percent). The x-values for the 84 and 16% cumulative frequency percent are read: x_{16} and x_{84}

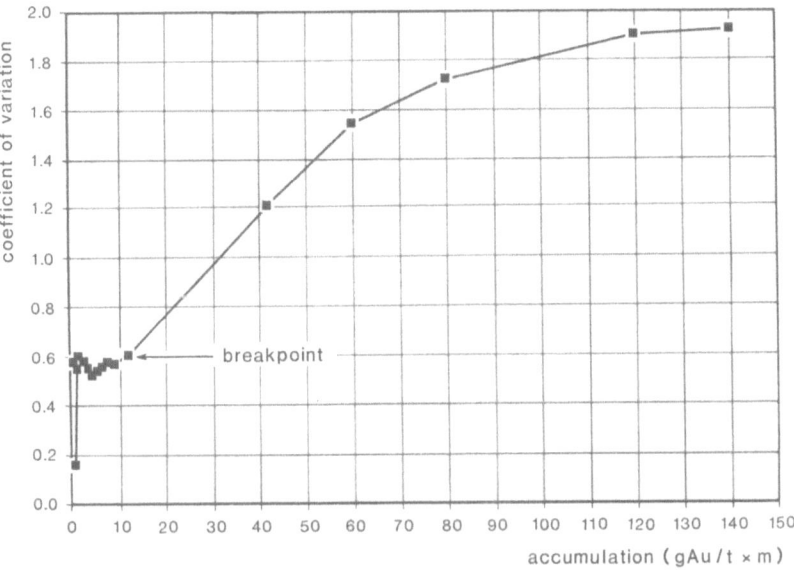

Fig. 27. Plot of cumulative coefficients of variation of data in Table 27 versus accumulation (grade x thickness) values

$x_{16} = 6$ g Au/t·m

$x_{84} = 48$ g Au/t·m.

From this fitted lognormal distribution the mean α of the log values and the standard deviation of the log values β (cf. Sect. 9.2.3.1) are calculated.

$$\alpha = \frac{\ln x_{16} + \ln x_{84}}{2}$$

$$\alpha = \frac{\ln 6 + \ln 48}{2} = \frac{1.791 + 3.871}{2} = 2.831$$

$$\beta = \frac{\ln x_{84} - \ln x_{16}}{2} =$$

$$\beta = \frac{\ln 48 - \ln 6}{2} = \frac{3.871 - 1.791}{2} = 1.04.$$

3rd Step: With the breakpoint we have partitioned the total set of data into two distributions. The distribution with the lower values has as the highest value sample the value, at which we recognized the breakpoint (12 g Au/t·m). The 2nd

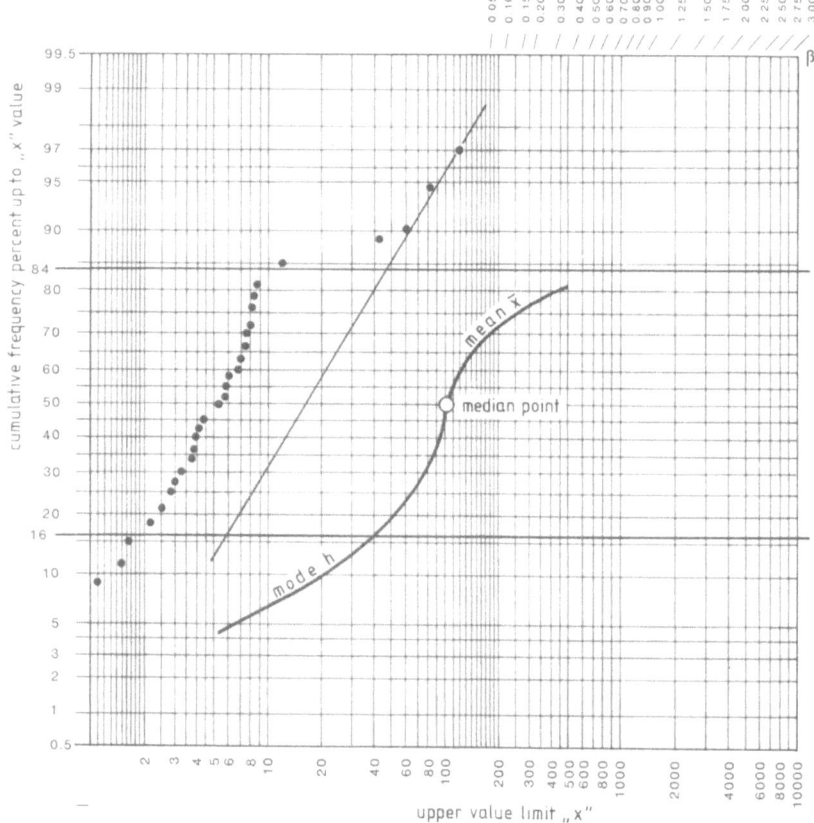

Fig. 28. Lognormal probability plot of data of Table 27 with fitted lognormal model for upper portion of distribution

population starts with the next value (42 g Au/t·m). For the population with the high values we determine the cumulative frequency distribution as 87.88 % (i.e. 0.8788) (see column 6 in Table 27). If we take the cumulative frequency distribution of the normal distribution in Appendix Table 2 the cumulative frequency is represented by the area under the normal distribution curve $\emptyset$ (see Chap. 4). We have now to determine the equivalent x-value, x_B.

$$\emptyset_{(xB)} = 0.8778$$

From Appendix Table 2 we can read and interpolate for $x_B = 1.169$.

4th Step: We have to calculate now the quantile Q_B of the fitted (not actual) frequency distribution at the break point according to the formula

$$Q_B = e^{(x_B\,\beta + \alpha)}$$

$$Q_B = e^{(1.169 \cdot 1.04 + 2.831)}$$

$$Q_B = {}^{4.047} = 57.226 \,.$$

*5th Step:*We will estimate now the mean grade above the break point of the fitted distribution

$$\bar{x}_{B+} = \frac{e^{\left(\alpha + \frac{\beta^2}{2}\right)}\left(1 - \emptyset\left[\frac{\ln Q_B - \alpha - \beta^2}{\beta}\right]\right)}{1 - \emptyset\left(\frac{\ln Q_B - \alpha}{\beta}\right)}\,.$$

First we will calculate the $\emptyset$-values. The $\emptyset$-values are read from Appendix Table 2.

a) $\emptyset\left(\dfrac{\ln Q_\beta - \alpha - \beta^2}{\beta}\right) = \emptyset\left(\dfrac{\ln 57.226 - 2.831 - 1.04^2}{1.04}\right)$

$\emptyset\left(\dfrac{4.047 - 2.831 - 1.082}{1.04}\right) = \emptyset\left(0.1288\right) = 0.5513\,.$

b) $\emptyset\left(\dfrac{\ln Q_\beta - \alpha}{\beta}\right) = \emptyset\left(\dfrac{\ln 57.266 - 2.831}{1.04}\right)$

$\emptyset\left(\dfrac{4.047 - 2.831}{1.04}\right) = \emptyset\left(1.169\right) = 0.8788\,.$

Therefore now

$$\bar{x}_{B+} = \frac{e^{\left(2.831 + \frac{1.04^2}{2}\right)} \cdot \left(1 - 0.5513\right)}{1 - 0.8788}$$

$$\bar{x}_{B+} = \frac{e^{(2.831 + 0.541) \cdot 0.4487}}{0.1212} = \frac{e^{3.372} \cdot 0.4487}{0.1212}$$

$$\bar{x}_{B+} = \frac{29.137 \cdot 0.4487}{0.1212} = 107.87 \,.$$

6th Step: We estimate now the mean grade of the total data set by combining the mean grade of the subpolulation below the break point $\bar{x}_{B-}$ and the mean grade of the subpopulation above the break point $\bar{x}_{B+}$. We will weight these mean grades by the percentage of each subpopulation below and above the break point (I_{B-} and I_{B+}), as can be seen from Table 27, column 6.

$$I_{B-} = 0.8485$$

$$I_{B+} = 1 - 0.8485 = 0.1515 \ .$$

$\bar{x}_{B+}$ we discussed in the 5th step. We still have to determine the mean grade of the subpopulation below the break point. Here we take the normal arithmetic mean, which is

$$\bar{x}_{B-} = 4.88 \text{ g Au/t·m.}$$

The estimate of the total mean $\bar{x}_{tot}$ of the data set in Table 27 therefore is

$$\bar{x}_{tot} = I_{B-} \cdot \bar{x}_{B-} + I_{B+} \cdot \bar{x}_{B+} = 0.8485 \cdot 4.88 + 0.1515 \cdot 107.87$$

$$\bar{x}_{tot} = 4.41 + 16.34 = 20.48 \approx 20.5 \text{ g Au/t·m.}$$

10
Other Distributions for the Evaluation of Mineral Deposit Data

The analytical data for mineralization or mineral deposits as well as geochemical exploration programmes have been evaluated, insofar as mathematically defined distributions were used, on the basis of the normal and the lognormal distribution. It was previously mentioned that there is no geological or geochemical reason why populations or sample data sets of analytical data should abide by a particular, mathematically defined, distribution such as, for example, the normal or the lognormal distribution. Experience shows only that, for most cases, the application of normal or lognormal distributions does yield usable results for the evaluation of data sets. However other distributions have been applied, for example the binomial distribution. This will not be discussed any further with respect to the treatment of populations of analytical data, since it is of only subordinate importance.

However, the binomial distribution, also known as the Bernoulli distribution, is useful in such cases as, for example, the selection of the correct sample size when the analysed component is only contained in a few mineral particles. The next chapter, Chapter 11, discusses the selection of the correct sample size. The binomial distribution is also of importance for the evaluation of irregular mineral deposits. Sometimes it can be seen that some deposits such as vein deposits, in spite of their irregularities, can be divided into two sections by a sort of natural cut-off level. This is comparable with the problem of different populations, which is discussed in Chap. 18 for the statistical evaluation of geochemical exploration data. There is one population with background values of the vein mineralization (see also Chap. 18.2.1.1) and another higher grade population that essentially enables the vein to be mined. If different areas of the vein are investigated, it can sometimes be shown that the ratio of the mineable to non-mineable areas varies only relatively slightly, in any case less that the average grade. This information can be used to advantage if, for example, the potential extension to depth of a vein must be evaluated, and any significant changes predicted. If there is only a minor amount of drill hole data, then it is difficult to recognize any trend in the grades or even to calculate an average value with limits of error. The number of drill holes with intersections above and below the natural cut-off are then compared, and the binomial distribution is used to determine if there is a significant deviation from those values derived from long term experience.

The binomial distribution [named after the binomial coefficient $\binom{n}{k}$, which is derived from the calculation of a "bi-nomen" power $(a+b)^n$], which is also known as the Bernoulli distribution, describes the probability distribution for the probability that, during n independent random experiments, an event occurs exactly k times. The example of throwing dice suffices to explain the binomial distribution. The probability of throwing a six is p = 1/6 for each throw, regardless of the number of times (be it several times or none) that the six has been thrown previously. If the throw of a six is considered to be a success, and the throw of other numbers to be a failure, then the

probability of success is p = 1/6 and the
probability of failure is q = 5/6,

and therefore (and this is valid for the binomial distribution and not only for throwing the die):

$$p + q = 1. \tag{1}$$

A further example is the repeated selection of marbles from a sack full of white and black marbles, during which the selected marble (sample) is always returned to the sack. There is a close similarity to the sampling of ores, in which the valuable commodity is contained in relatively few, large particles such as gold grains. This is discussed in the next chapter. An experiment in which only two different and mutually exclusive results are possible is known as a Bernoulli experiment, and therefore the use of the term Bernoulli distribution for the binomial distribution (for example, Kreyszig 1968). If the probability of event A occurring in each experiment is p and the probability of it not occurring is q (p+q=1), then the probability P_{Bi} that an event A will occur precisely x times during n repetitions of the experiment is:

$$P_{Bi} = \binom{n}{x} \cdot p^x \cdot q^{n-x}. \tag{2}$$

The binomial coefficient $\binom{n}{x}$ is referred to as "n over x".

Hence:

$$\binom{n}{x} = \frac{n \cdot (n-1) \cdot (n-2) \cdot \ldots \cdot (n-x+1)}{1 \cdot 2 \cdot 3 \cdot \ldots \cdot x} = \frac{n!}{x!(n-x)!} \,.^{16} \tag{3}$$

---

[16] n! is known as factorial n, for example $4! = 1 \cdot 2 \cdot 3 \cdot 4$.

An example:

$$\binom{4}{3} = \frac{4 \cdot 3 \cdot 2}{1 \cdot 2 \cdot 3} = 4 .$$

Its application is best explained by an example.

Example: Several parallel veins occur in a Zn-Pb mine. A statistical evaluation has demonstrated that there is a quasi-natural cut-off in the veins of 5% combined Pb+Zn. About 35% of the area of the vein is greater than this cut-off, and about 65% less than it.

The depth extension of the veins has only been tested by four drill holes, and their results are listed in Table 28.

The deposit is offered for sale. The following question that must be addressed is: do the drill results indicate a significant change – an improvement is needed in this example – with depth or not? Only if a significant improvement can be reasonably expected the potential buyer would be interested.

Answer: The natural cut-off of 5% combined Pb+Zn is used in order to answer the question of a possible significant change in the grade.

The drill holes are considered in terms of a Bernoulli experiment, insofar as an intersection with grades of more than 5% combined Pb+Zn is treated as a success, and those with grades of less than 5% combined Pb+Zn as a failure.

There is an important precondition that must be fulfilled prior to studying the intersections in terms of a Bernoulli experiment: each experiment, or in this case drill result, must be *statistically independent* of the others, in the same way as, for example, the throw of a die (if a six is thrown, the chance of throwing another six next time is 1/6, which is just the same as if a one had been thrown previously). This type of independence is only occasionally found in ore deposits. Every geologist knows that the chance of finding a high grade intersection in the vicinity of a previous high-grade intersection is better than near a low-grade intersection. There is therefore a definite spatial dependence within an area of influence, which can be determined by a variogram calculation. This is described in the geostatistical section of Chapter 13.

1st Step: The statistical independence is tested by the first step. There are four drill intersections through each of the three veins. There is no reason to suppose that the drill results from one vein are in any way contingent on the results from the two other veins.

Table 28. Intersections of four drill holes into three parallel Pb-Zn veins

Drill hole	Vein 1			Vein 2			Vein 3		
	Pb %	Zn %	Pb+Zn %	Pb %	Zn %	Pb+Zn %	Pb %	Zn %	Pb+Zn %
Ra 1	0.5	0.8	1.3	3.7	19.4	23.1	6.7	11.9	18.6
Ra 2	1.8	5.2	7.0	1.3	1.5	2.8	1.1	2.3	3.4
Ra 3	0.3	1.2	1.5*	2.1	0.8	2.9	2.1	11.8	13.9
Ra 4	4.8	10.2	15.0	0.7	3.8	4.5	0.4	3.7	4.1

The drill results on one vein are tested by a variogram (Chap. 13.2) that is generated for the known sections of the deposit. This shows that the area of influence is 200 m. It is assumed that this character does not change with depth, i.e. it is a stationary deposit. The distance between all of the drill holes Ra1 to Ra4 on all three veins is greater than 200 m, so that it is justified to assume a statistical independence.

2nd Step: The drill results in Table 28 are evaluated in the following way:
Vein 1: 2 intersections with over 5% combined Pb+Zn, 2 below
Vein 2: 1 intersection with over 5% combined Pb+Zn, 3 below
Vein 3: 2 intersections with over 5% combined Pb+Zn, 2 below

On the basis of these few intersections, it appears that 50% of the vein area in veins 1 and 3 is economically mineable, and only 25% in vein 2. Evidently, it is not really possible to make any deductions from so few intersections. All three veins are now considered together in order to increase the significance. There are therefore now 12 intersections of which five are positive intersections (i.e. 42% in comparison to the previous example of 35%).
For the binomial Eq. (2) above:

$n = 12,$
$x = 5 .$

The probability of success, p, that a value lies above 5% combined Pb+Zn in the known sections of the ore deposit is:

$p = 0.35$, and therefore
$q = 0.65$
[the probability of failure is calculated from Eq. (1) above].

The probability p_{Bi} of finding five successes from a total of 12 drill holes is:

$$p_{Bi} = \frac{12 \cdot 11 \cdot 10 \cdot 9 \cdot 8}{1 \cdot 2 \cdot 3 \cdot 4 \cdot 5} \cdot 0.35^5 \cdot 0.65^7 =$$

$p_{Bi} = 792 \cdot 0.0053 \cdot 0.0490 = 0.206.$

3rd Step: This absolute probability, p_{Bi}, actually means very little. There are 13 possibilities, from zero to 12 successes, in 12 intersections. A probability p_{Bi} can be calculated for each case. The sum of all the probabilities of the possible combinations is then 1.

A probability of error must be chosen in order to make any meaningful statement (e.g. see Chap. 5.2). If a figure of 10% is selected, then it means that there is a 10% chance that a possibly more favourable ratio than 0.35/0.65 between mineable and non-mineable areas cannot be recognized. All the probabilities, p_{Bi}, must be calculated, starting from the most improbable combination (12 positive

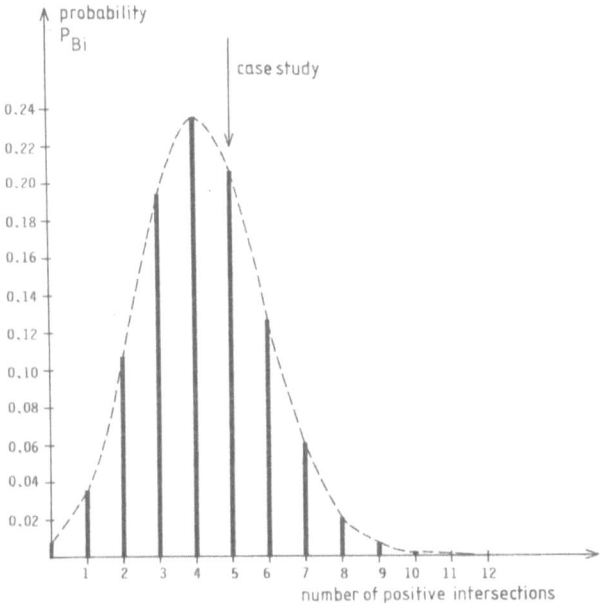

Fig. 29. Probabilities for the occurrence of positive intersections for the case study in Table 28

intersections, then 11 positive, etc.), and summed until the limit of 0.1 (i.e. 10%) is exceeded. This exercise has been carried out in Table 29.

It can be seen that at least seven positive drill holes are needed in order to be able to assume, with a 10% probability of error, that the mineable areas become relatively larger.

The five successful drill holes do not provide any indication for a significant change in the area of mineability. This is emphasised if the probabilities of the individual combinations are calculated and illustrated graphically (Fig. 29). If the most probable value is calculated from four successful drill holes, then:

Table 29. Probabilities of successful events for a probability of success p = 0.35 for twelve trials

No. of successful events X	Single probabilities P_{Bi}	Cumulative probabilities P_{bi}
12	0.0000	0.0000
11	0.0001	0.0001
10	0.0008	0.0009
9	0.0048	0.0057
8	0.0199	0.0256
7	0.0591	0.0847
6	0.1281	0.2128

$p_{Bi} = 0.237$

(in comparison with $p_{Bi} = 0.206$ for five successful drill holes).

The number of drill holes also has to be taken into consideration. If there were 100 drill holes[17] then, for the selected 10% probability of error, the critical limit would be attained with 41 successful drill holes, which means that in comparison to the most probable number of 35 drill holes for $p = 0.35$ only six, or a 17% greater number of successful drill holes, would be required in order to recognize a significant improvement in the relative proportion of the mineable area. In the previous example of 12 drill holes, the number of successful drill holes must be increased from 4 to 7, or by 75% for the same probability of error in order to attain the same result.

---

[17] The $\binom{n}{x}$ - value for so many individual values is determined from factorials, which are provided as programmes in some pocket calculators [see Eq. (3) above].

11
Statistical Problems Encountered in Sampling and the Analytical Results

11.1
Sample Collection

11.1.1
General Remarks

It is evident that, for the same confidence levels, the relevant sample quantity must be proportionately greater according to
- how much coarser the sampled material is, and
- the degree to which the economically interesting component to be analysed is concentrated as individual, discrete, mineral particles.

On the other hand, if the valuable component is very finely dispersed, as for example Ni in nickel laterites, then the only influence is the grain size on the size of the sample of the material and the geological distribution of the homogeneous zones, but not the valuable material itself. The binomial distribution was introduced in Chapter 10. If the valuable component occurs as discrete mineral particles, this distribution is fundamentally important for sample collection. The Poisson distribution, which deals with rare events encountered in the Bernoulli experiments (Chap. 10), is also critical for sample collection. A few gold particles in a sample can certainly be treated as a rare "event".

The effect that a few individual particles can have is explained by two examples.

1st Example: The dimension 1 assay tonne $\approx$ 30 g, which was previously used for gold analyses, was introduced in Economic Evaluations in Exploration (Chap. 1.1.4.5.c). It is assumed that the sample material is ground to -80 mesh, so that practically all the grains pass through an 80 mesh sieve[18], which is a sieve with a mesh width of 0.177 mm (see Economic Evaluations in Exploration, Appendix Table 4). It is further assumed there is a spherical gold particle with a diameter of d=0.177 mm which just passes through the mesh, so that it weighs $56 \cdot 10^{-6}$ g:

The volume of a sphere is:

$$V = \frac{\pi \cdot d^3}{6} \; ,$$

---

[18] After crushing and grinding, the grain size of homogeneous and isotropic material has a normal distribution. Since it ranges from $+\infty$ to $-\infty$ (Chap. 4, 2nd exercise), there is no reason to expect that all the material must pass through a specific sieve mesh size. Normally 95% does. The sample preparation technician usually choose only 80%, and refer to K 80 or p 80. K 80 = 200 mesh means that 80% of the ground material passes through a 200 mesh sieve.

and the density of gold is

$$\varrho = 19.3 \, g/cm^3.$$

Therefore the weight of the sphere is:

$$G_{Au} = \frac{\pi \cdot d^3}{6} \cdot \varrho = \frac{\pi \cdot (1.77 \cdot 10^{-2})^3 \cdot 19.3}{6} = 56 \cdot 10^{-6} g \ .$$

It is then assumed that the sample has a weight of 30 g = 30 x 10^{-6}t, and therefore the contribution of one small sphere of gold to the total gold content of the sample is

$$\frac{56 \cdot 10^{-6} g}{30 \cdot 10^{-6} t} = 1.87g \ Au/t.$$

This clearly demonstrates the extreme importance of thoroughly cleaning the crusher and mill every time after it is used. A single grain of entrained gold can result in the sample appearing to be of interest. Some laboratories therefore separate the samples before preparation into those that are expected to yield high grades and those expected to yield low grades. If a gold particle is entrained with a high-grade sample, then the relative error is much less than it would be for a low-grade or a gold-free sample.

   2nd Example: A molybdenum ore is crushed to 7.5 mm. The sample under consideration has a size of 500 g. The molybdenum occurs as molybdenite, MoS_2, which has a density of $\varrho = 4.7 \, g/ \, cm^3$. For the sake of simplicity, it is assumed that the individual MoS_2 grains are much smaller than a fragment of molybdenum ore, and that the fragment of ore itself can be approximated to a sphere. The sphere of molybdenum ore therefore weighs:

$$G_{Mo} = \frac{\pi \cdot d^3}{6} \cdot \varrho = \frac{\pi \cdot 0.75^3}{6} \cdot 4.7 = 1.04 \, g,$$

i.e. relative to a 500 g sample, that is
0.2% MoS_2.

Grades of this magnitude are economic for some ore deposits.

11.1.2
Sample Size

11.1.2.1
Introduction

There have been various studies dealing with the statistical consequences of using the binomial and Poisson distributions, the latter being the limiting distribution of the binomial distribution for rare events. These studies (e.g. Clifton et al. 1969) conclude that the sample must be of sufficient size so that it contains at least 20 particles of the component to be analysed. This can be readily checked for crushed samples in which the ore material is macroscopically visible, such as molybdenite, chalcopyrite, galena and sphalerite. However, a pan concentrate must be prepared for gold samples, and the gold flakes counted and then recalculated with respect to the original sample size.

Formulae and diagrams for the necessary sample size have been developed by various authors, the best known being by Gy (1967). Kraft (1981) provides a good overview of the different formulae. All these formulae, and the diagrams derived from them, require the initial determination of various constants or dimensions, such as the size of the gold particles, that normally necessitate detailed studies. An equation for estimating the sample size, presuming the standard deviation can be satisfactorily determined, is discussed in the following Section 11.1.2.2. The Field Geologist's Handbook, published by the Australasian Institute of Mining and Metallurgy (Monograph Series No. 9, 1982), contains a summary of the most important diagrams in Gy (1967) and Clifton et al. (1969).

Comlabs, Adelaide, which is an Australian analytical laboratory, has published, with reference to Gy (1967), a diagram that uses the grain size of the sample as a basis for determining the size of the sample. This diagram has proved to be a good rule of thumb, and is reproduced as Fig. 30. In the case of quartering samples, the diagram can also be used to assist in defining the necessary extent of sample crushing between each separation, especially if the samples are crushed by hand.

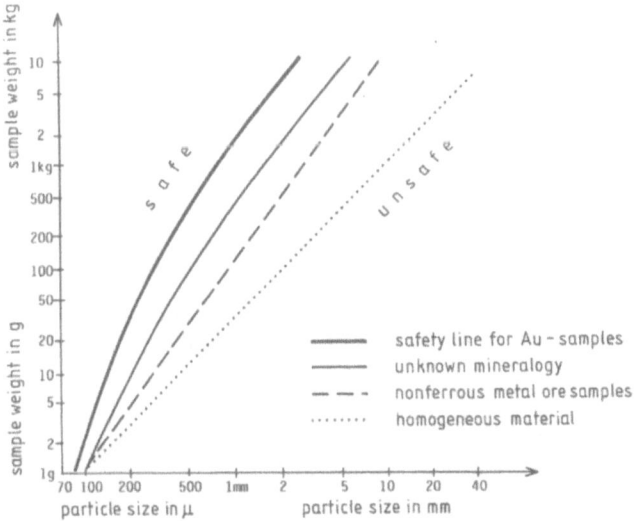

Fig. 30. Diagram for the determination of the sample size as a function of the particle size within the sample. (After Comlabs, Adelaide)

11.1.2.2
Gy's Sampling Formula

If the material is coarse-grained or first beneficiation tests have been executed (see Economic Evaluations in Exploration, Chap. 9.3.2.4) so that the necessary parameters can be estimated, then Gy's formula can be applied for the relative variance of the error s_r^2:

$$s_r^2 = \frac{C \cdot d^3}{M_E}, \tag{1}$$

d is the diameter of the material of the sample in cm (defined as the screen size which is passed of 95% of the material.

M_E is the sample weight in g.

C is a constant which is defined as:

$$C = \left(\frac{1-\alpha_L}{\alpha_L}\right)\left[(1-\alpha_L) \cdot \mu_A + \alpha_L \cdot \mu_G\right] \cdot l \cdot f \cdot g = m \cdot l \cdot f \cdot g,$$

m   is called the mineralogical parameter,
α_L is the grade of the lot considered: not the grade of the element, but of the relevant ore mineral,
μ_A is the density of the relevant ore mineral (in g/cm³),
μ_G is the density of the gangue (in g/cm³),
f   is a dimensionless form factor which for all practical purposes is equal to 0.5,
g   is a dimensionless granulometric distribution factor which for usual uncalibrated material is taken to be equal to 0.25,
l   is a liberation parameter which can be estimated from Table 30. Here d/d_o is the ratio of the largest dimension of fragments of the sample (d) over the liberation size of the relevant ore mineral (d_o). A precise estimation requires a granulometric analysis.

Gy (1979) recommends the application of the formula (1) only for s_r^2 not larger than 0.01, i.e. an error at 2x standard deviation of not more than 20%. This means

Table 30. Estimation of l from d/d_o for Gy's formula (Gy 1967; David 1977).

d/d_o	< 1	1 to 4	4 to 10	10 to 40	40 to 100	100 to 400	> 400
l	1.0	0.8	0.4	0.2	0.1	0.05	0.02

that the required precision in the normal exploration stages falls well below this safety value.

1st Example: Outcrops of a coarse-grained Zn mineralization in carbonates are chip-sampled. The maximum fragment size of the samples is 2 cm. The liberation size of the sphalerite is estimated to be 1 mm. Grades are estimated to be around 5% Zn. What is the necessary sample weight for an error of 5% at 2x standard deviation, meaning in 95 of 100 cases the error would be within ± 5% (see Chap. 7.1).

Answer:

1st Step: For 5% Zn we have to calculate the amount of sphalerite present, meaning convert from Zn grade to ZnS grade. This is done using atomic weights (see Economic Evaluations in Exploration, Chap. 1.3)

The atomic weight of Zn = 65.38 and

$$\text{of } S = 32.06$$

$$\text{sum} = 97.44.$$

Therefore the Zn share of ZnS is: $\dfrac{32.06}{97.44} = 0.67$ and

5% is the equivalent to $\dfrac{5}{0.67} = 7.46 \approx 7.5\%$ ZnS.

2nd Step: The ratio d/d_O is

$$d/d_O = 2/0.1 = 20$$

According to Table 30, $l = 0.2$

3rd Step: We calculate m:

$$m = \left(\frac{1-\alpha_L}{\alpha_L} \right) \left[(1-\alpha_L) \cdot \mu_A + \alpha_L \cdot \mu_G \right].$$

The density of the sphalerite, which has a light brown colour and is poor in Fe, is 4.0 g/cm3 (μ_A).

The density of gangue (carbonate) is 2.5 g/cm (μ_G)

α_L is 7.5% = 0.075 (see 2nd step).

$$m = \frac{(1-0.075)}{0.075} \left[(1-0.075) \cdot 4 + 0.075 \cdot 2.5 \right]$$

$$m = 12.33 \cdot [3.70 + 0.19] = 48.0 .$$

4th Step: We calculate the constant C:

$$C = m \cdot l \cdot f \cdot g$$

$$C = 48 \cdot 0.2 \cdot 0.5 \cdot 0.25$$

$$C = 1.2.$$

5th Step: We now calculate the necessary sample weight with Eq. (1)

$$s_r^2 = \frac{C \cdot d^3}{M_E}. \tag{1}$$

Therefore

$$M_E = \frac{C \cdot d^3}{s_r^2},$$

d was 2 cm.

For s_r^2 we required a 5% error at 2x standard deviation, i.e.

$$s = 2.5\% \approx 0.025 = 2.5 \cdot 10^{-2}.$$

Therefore $s^2 = 6.25 \cdot 10^{-4}$

$$M_E = \frac{1.2 \cdot 2^3}{6.25 \cdot 10^{-4}} = 1.54 \cdot 10^4 \text{ g,}$$

meaning a 15-kg sample is necessary.

François-Bongarçon (1991, 1993) further studied and modified Gy's sampling formula. In his sampling formula the exponent for d in Eq. (1) is not fixed at 3, but also considered a variable for broken gold ores. Francois-Bongarçon (1991) developed a modified formula:

$$s_r^2 = \frac{470 \cdot d^{1.5}}{M_E}, \tag{2}$$

meaning the constant C in Eq. (1) is 470 and the exponent for d is 1.5 instead of 3.

2nd Example: A gold sample is ground to 95% passing 200 mesh (= 0.0074cm). What is the necessary sampling weight again for an error of 10% at 2x standard deviation? This means one standard deviation is 5% ≈ 0.05.

Answer: $s_r = 0.05$, and therefore

$$s_r^2 = 2.5 \cdot 10^{-3},$$

$$M_E = \frac{470 \cdot 0.0074^{1.5}}{2.5 \cdot 10^{-3}} = \frac{470 \cdot 0.000637}{2.5 \cdot 10^{-3}},$$

$M_E = 126$ g of sample material is necessary.

Another useful relationship between the relative precision and the required sample weight can be deduced from Gy's sampling formula (1) but also the volume-variance relationship of Chapter 3.3.

$$s_r^2 \approx \frac{1}{M_E}.$$

If one wants, e.g. to double the precision of the sampling, meaning the standard deviation s_r has to be halved, then the mass has to be increased by the square of $2=4$.

$$s_{r_1}^2 \approx \frac{1}{M_{E_1}} \tag{1}$$

$$s_{r_2}^2 = \left(\frac{s_{r_1}}{2}\right)^2 = \frac{s_{r_1}^2}{4} \approx \frac{1}{M_E \cdot 4} \tag{2}$$

If, for example in the first example one is content with an error of 10% instead of 5%, the necessary sample volume would be then $\frac{15.4}{4} = 3.85$, i.e. around 4 kg would suffice.

11.1.2.3
Estimation of the Sample Size for a known Standard Deviation s

11.1.2.3.1
Formula for Estimating the Sample Size
Kraft (1981) proposed the use of the following formula[19] as the basis for all estimations of the sample size or the total sample quantity:

$$n = \left(\frac{t \cdot s}{\pm U}\right)^2, \tag{1}$$

---

[19] This formula is, of course, the same as that for the calculation of the confidence limits, compare Chap. 7.1, Eq. (1). The confidence limit ki in that formula is the uncertainty U in this one.

Table 31. Rules for the sampling of Sn concentrates (Kraft 1981)

	„Heterogeneous" concentrate (handcobbed) 15-25% Sn		„Homogeneous" concentrate (by machine) 40-60% Sn	
Standard deviation s of increments	1.5	2.5	0.5	1.0
Required accuracy ± (in ± % Sn)	0.1	0.1	0.1	0.1
Required number of increments n	900	2500	100	400
Percentage of lot to be taken as sample (increment 1 kg, lot 100 t)	0.9	2.5	0.1	0.4

where n is the number of sample increments of a certain, always constant, weight (e.g. a shovel full of 1 kg material) that are collected from the material to be sampled. This number n is defined as the square of the quotient of the sampling error (s = standard deviation, t = Student's t-factor, e.g. ca. 2 for a 95% confidence level, see Appendix Table 4) and the required uncertainty ±U of the expected result. The total weight of the sample G_W is then provided by

$$G_W = n \cdot g,$$

where g is the weight of each sample increment.

The estimation of the standard deviation s, which must be determined for the same sample increment, and naturally for the same material and the same sampling procedure, is the main difficulty. On the basis of the above formula (1), Kraft (1981) provides the following examples for the sampling of Sn concentrates in Table 31.

These formulae and concepts are suitable as guidelines for sampling by a smelter that regularly receives concentrates from a particular mine. It is then worth the effort of determining the standard deviation s by taking a series of samples. However, these formulae are not so appropriate for a geologist who, in a foreign country, is required to collect samples from a dump of concentrates, or hand-cobbed and possibly even untreated ores. However, it is sometimes possible to estimate the standard deviation s. It is unusual to be the very first person who is requested to make an evaluation, and it is probable that someone has already done a similar exercise. For example if concentrates, either from a plant or hand-cobbed, must be evaluated, then there is usually analytical data from previous sampling. Commonly, it is found that such concentrate dumps have been sampled several times. If at least two samples were collected from each concentrate or ore dump, and it can be assumed that the sampling was always undertaken in the same way, then the standard deviation can be estimated by methods that will be described in the following two sections.

11.1.2.3.2
Estimation of the Standard Deviation for two Samples from Each Component

If there are two samples from each concentrate or ore dump, then the standard deviation s can be estimated by a calculation that is also used for the geostatistical calculation of variograms (cf. Chap. 13.2). If it is assumed that each concentrate or ore dump has been previously sampled by two samples, x_1 and x_2, then the standard deviation s, or the variance s^2, are:

$$s2 = \frac{1}{2 \cdot n} \sum_{i=1}^{n} (x_1 - x_2)^2 \, , \tag{2}$$

summed for all concentrate or ore dumps[20]. The number of value pairs = n.
The use of this formula can best be illustrated by an example:

Example: Hand-cobbed barite concentrates are produced at a small barite mine in Turkey. The truck loads of concentrate are tipped into a central dump at the harbour, from where it can be shipped in 5000- t lots.

Previously, random grab samples were collected from the whole of the barite dump, crushed on sheet metal to fragments of about 1 cm in size, and quartered down to a 2-kg sample. After the final quartering, the halves from both sides were

---

[20] This can be explained for two samples x_1 and x_2 from one dump. The standard formula for the variance is [Chap. 3.2.1, Eq. (1)]:

$$s^2 = \frac{1}{n-1} \sum_{i=1}^{n} (x_i - \overline{x})^2$$

$\overline{x}$ is the mean value

In this case $\overline{x} = \frac{x_1 + x_2}{2}$ and $n = 2$.

Therefore the following can be stated:

$$s^2 = \frac{1}{2-1} \cdot \left[\left(x_1 - \frac{x_1 + x_2}{2} \right)^2 + \left(x_2 - \frac{x_1 + x_2}{2} \right)^2 \right]$$

$$s^2 = \frac{1}{4} \cdot \left[(2x_1 - x_1 - x_2)^2 + (2x_2 - x_1 - x_2)^2 \right]$$

$$s^2 = \frac{1}{4} \cdot \left[(x_1 - x_2)^2 + (x_2 - x_1)^2 \right]$$

Since $(x_1 - x_2)^2 = [-(x_2 - x_1)]^2 = (x_2 - x_1)^2$, then:

$$s^2 = \frac{1}{2} (x_1 - x_2)^2 \, .$$

Table 32. Sampling results from a hand-cobbed barite concentrate

Lot no.	1st sample x_1 BaSO$_4$ (%)	2nd sample x_2 BaSO$_4$ (%)	$x_1 - x_2$	$(x_1 - x_2)^2$
A	93.3	93.7	0.4	0.16
B	95.1	95.2	0.1	0.01
C	96.1	95.9	- 0.2	0.04
D	94.3	94.8	0.5	0.25
E	93.7	94.1	0.4	0.16
F	94.0	94.5	0.5	0.25
G	95.8	95.1	- 0.7	0.49
				$\Sigma = 1.36$

collected and analysed. The following results from earlier sampling were provided on certificates of analysis, and are reproduced above as Table 32. The problem to be addressed is: what must be the size of a sample so that the uncertainty U is ±0.2% BaSO$_4$?

It would be incorrect to derive the average value for all the lots, and then to calculate the variance from the standard formula [Chap. 3.2.1, Eq. (1)]. Each lot has a different average grade, and the deviation or range of the two analyses from each lot must be treated with respect to the average grade of that lot and not the mean of all the lots. The Eq. (2), which was derived above, is used:

By using Eq. (2) above, s^2 is:

$$s^2 = \frac{1}{2 \cdot n} \sum_{1}^{n} (x_1 - x_2)^2 = \frac{1}{2 \cdot 7} \cdot 1.36 = 0.097 = s^2 \, .$$

Therefore s is:

$$s = \sqrt{0.097} \ = \pm 0.31 \, .$$

This must be substituted in Eq. (1) above. For this equation, the Student's t-factor must also be obtained from column 5 of Appendix Table 4, where it is found to be 2.37 for 7 lots for a 95% confidence level[21]. Therefore, if the uncertainty is ±U=±0.2 BaSO$_4$, then the number of sample increments n is:

$$(\frac{s \cdot t}{\pm U})^2 = (\frac{0.31 \cdot 2.37}{0.2})^2 = 13.5 \, ,$$

---

[21] There are 7 lots (k = 7), and the total number of samples is 14, so that

$$F = \left(\sum_{i=1}^{k} n_i \right) - k = 14 - 7 = 7 \text{ is derived for column 5, Appendix Table 4.}$$

Table 33. d_2-Factors for estimating the standard deviation (Deming 1950)

No. of samples n per unit	d_2	No. of samples n per unit	d_2
2	1.128	9	2.970
3	1.693	10	3.078
4	2.059	11	3.173
5	2.326	12	3.258
6	2.534	13	3.336
7	2.704	14	3.407
8	2.847	15	3.472

so that the sample must consist of 14 increments of 2 kg each, which is a total weight of 28 kg, in order to fulfil the required accuracy of ±0.2% BaSO4 at a 95% confidence level.

11.1.2.3.3
Estimation of the Standard Deviation for Three or More Samples from Each Component

If there are three or more samples from each concentrate or ore dump, then the standard deviation, s, can be estimated from the range, R, using the method devised by Deming (1950):

$$s = \frac{\overline{R}}{d_2} . \tag{3}$$

$\overline{R}$ is the average range and d_2 is a factor that can be derived from Table 33 (after Deming, 1950).
The procedure can again be explained by an example.

Example: Hand-cobbed lead concentrates are produced in Morocco at a small operation exploiting a galena vein. There are about 5000 t concentrate on the

Table 34. Sampling results from a hand-cobbed Pb concentrate

Lot no.	Analyses of samples (% Pb)				Range R
A	50.4	49.5	55.7	52.8	55.7 —— 49.5 = 6.2
B	59.6	53.0	58.2	58.3	59.6 —— 53.0 = 6.6
C	60.1	55.3	58.9	55.0	60.1 —— 55.0 = 5.1
D	55.2	54.3	58,7	55.9	58.7 —— 54.3 = 4.3
E	48.5	50.2	48.2	52.8	52.8 —— 48.2 = 4.6
					$\overline{R} = 5.4$

dump. The mine owner is attempting to obtain finance for the operation, and thus the material that has already been produced must be evaluated. There are five certificates of analysis from previous sampling. In each case, four samples were collected, and *they are listed in Table 34*.

1st Step: In order to determine the range R, the highest and lowest value is identified in each_of the lots, and they are subtracted from each other. The arithmetic mean $\bar{R}$ is then calculated from five values for the range R.

2nd Step: The standard deviation s is now estimated by the following formula (3). The value for the factor $d_2=2.059$ is obtained for four samples from Table 33.

$$S = \frac{\bar{R}}{d_2} \, , \tag{3}$$

$$s = \frac{5.4}{2.059} = 2.62 \, .$$

The next stages in the procedure for calculating the size of the sample are then exactly the same as the example in Section 11.1.2.3.2.

11.1.3
The Special Case of Gold

A number of occurrences of gold mineralization are extreme cases insofar as the gold often occurs as coarse grains, and thus the sample has a very strong nugget effect. Furthermore, in most cases the gold occurs as the free metal. The nugget effect will be discussed later in the chapter on geostatistics (Chap. 13.2).

Gold is ductile and therefore during milling it is only flattened but not comminuted. The Fire Assay method is the most reliable of the analytical techniques for gold, and for practical reasons a sample of only 50 to 200 g is used, and not 500 g, 1000 g or even 10 kg. This clearly introduces an important problem with respect to sample size.

The problem is double-edged: firstly the problem of obtaining analyses that are as reliable as possible; and secondly that of calculating a meaningful average for high values. The second problem has been addressed in Chapters 8 and 9. The present chapter on sampling describes another method of deriving a reasonable estimate for the average grade. Many gold prospects have a long history of exploration and evaluation, and sometimes minor exploitation might have occurred. Even if only a few tens or hundreds of tonnes were mined, the remains of the tailings are usually still present. The various processes in the treatment plant result in the best homogenisation that can, in practice, be achieved. The grade of the tailings often provides a better indication of the earlier average grade of the ore than the grade derived from resampling the ore.

Example: A gold prospect is submitted for purchase. Ore grading 3 oz/long ton (corresponds to 92 g Au/t, see Economic Evaluations in Exploration, Chap. 1.2.3) is supposed to have been mined on the prospect, and was treated by a stamp mill and then mercury amalgamation. The previous miners are supposed to have worked to a cut-off grade of about 1 oz/long ton (30.6 g Au/t). Since the structure is known to extend along strike and to depth, it is proposed that there must be a considerable potential with a grade of at least 30 g to 50 g Au/t remaining in the ground.

During a visit to the property it was ascertained that most of the tailings from the stamp mill are still present. Ten samples of this material were collected from a regular grid. The grades of these ten samples range from 1.6 g Au/t to 3.4 g Au/t and have a mean value of 2.4 g Au/t, the range of the sample grades is thus relatively small.

If the original information is correct that the ore mined earlier had an average grade of about 90 g Au/t, then the recovery, ε, from the stamp mill was:

$$\varepsilon = \frac{90 - 2.4}{90} = 97.3\%.$$

This is totally unrealistic. A good recovery from that sort of plant would be 85%. Assuming a tailings grade of 2.4 g Au/t and a recovery of 85%, of which the other 15% must be in the tailings, then the grade of the ore was:

$$G_{ore} = \frac{2.4}{0.15} = 16 \text{ g/t}.$$

Even taking the most optimistic case of a 90% recovery, then the average grade was only

$$G_{ore} = \frac{2.4}{0.10} = 24 \text{ g/t}.$$

The old miners were able to mine more selectively than is possible today in countries with high labour costs, and it is clearly impossible that the tenor of the occurrence is greater than 1 oz/t.

The Screen Fire Assay method was developed by the Australian gold mining industry, and at the sampling stage it provides a method of overcoming the problems associated with irregularities in the gold mineralization.

About 1 kg of sample material is ground to a nominal -80 mesh (i.e. -0.77 mm, see Economic Evaluations in Exploration, Appendix Table 4). Nominal means that the sample is ground to this mesh size until there is a residue of about 80 to 150 g of coarser material containing the coarse gold. The coarse residual material in total is then analysed by fire assay. Two samples are taken from the -80 mesh fraction (about 20 - 50 g in each sample), and are also analysed by fire assay. These two samples provide an indication of the variability. Normally, the gold values are very similar to each other, and significant differences occur only for very high gold grades.

The total gold content of the sample is then calculated by using a weighting factor derived from:

a) proportion of the residual coarse fraction and its gold content,

b) proportion of the -80 mesh fraction and the mean value of the two -80-mesh samples.

Example:

Table 35. Assay results for an example of a screen fire assay

Sample no.	+ 80-mesh fraction		- 80-mesh fraction		
	Mass portion (g)	Au grade (g Au/t)	Mass portion (g)	Au grade (g Au/t)	
				Sample 1	Sample 2
LRC 245-1	102.5	99.2	645.6	11.2	11.4

The calculation of the average grade of sample LRC 245-1 is carried out as follows:

a) The mean value of the two -80 mesh samples is 11.3 g Au/t.

b) The total mass of the sample is 102.5 + 645.6 = 784.1 g.

c) The gold values are now weighted according to their relative masses (cf. Chap. 3.1.2):

$$G_\varnothing = \frac{99.2 \cdot 102.5 + 11.3 \cdot 645.6}{748.1} = 23.3 \text{ g Au/t},$$

thus 23.3 g Au/t is the average grade of sample LRC 245-1.

The technique is relatively laborious and expensive because three gold analyses are required for one gold sample, and thus it is only applied in the later stages of exploration.

This technique is also ideally suited for carrying out check analyses (Sect. 11.2). It has been shown that the sampling effect is often extreme for gold and, in order to minimize it as much as possible, only the -80 mesh fraction is dispatched to the check laboratory.

In anticipation of Part B of this textbook (Exploration Statistics), it is mentioned here that the nugget effect is clearly also of importance for the collection of geochemical samples. If the usual 50 - 100 g stream sediment sample is collected during gold exploration, it is found that the anomalous values can be repeated only with difficulty. The presence of a small flake in the sample can yield a highly anomalous value. If this flake is then not in the repeat sample, then the value of the replicate stream sediment sample is obviously low (the 1st example in Sect. 11.1.1 is also valid). The bulk cyanide leach stream sediment method, which is based on the normal CIP (carbon in pulp) method of gold recovery, was developed in an attempt to solve this problem of the nugget effect. Samples of fine sediment are collected from the active zone of the stream or river bed (not from traps!), and are then sieved to -80 mesh (some companies sieve to -40 mesh or -2 mm). A -80 mesh sample should weigh about 2 kg. The sample is then leached in cyanide for 24 h, after which the gold is then absorbed onto charcoal. The detection limit is about 0.05 ppb and, in the first assessment, values greater than 2 ppb are considered to be anomalous.

11.2
Check Analyses

11.2.1
Discussion of the Problem

Errors or differences between analyses are always found if repeat analyses are undertaken at one or more analytical laboratories. Two types of error must be distinguished:

a) random errors and
b) systematic errors.

Random errors occur in every analysis, even with the same analytical procedures. The random error is recognised if, in spite of using the same analytical procedures, the analytical results vary from each other by small, irregular differences. The true content of the sample is usually not known, and lies somewhere within this range of variation. The random error is quantified by calculating the average error (Chap. 7.2).

Systematic errors always affect the measurements in the same way, and thus the true value can also lie outside the range of variation.

Example: The length of an adit is measured ten times with a measuring tape. Each measurement varies by a few cm, which is the random error. The average error is calculated according to the procedures in Chapter 7.2, and the length of the adit is determined to be 28.93 m ±7 cm.

It is subsequently found that the measuring tape has shrunk and is not 2 m long but only 1.90 m. All the measurements therefore have a systematic error of 5%. Regardless of how accurately the measurements are made, each measurement is 5% too long.

There are two other terms that are closely associated with random error and systematic error, and they are:

precision is associated with the random error, and
accuracy with the systematic error.

Reputable analytical laboratories, and only such laboratories should be used, include standards in a series of samples so that they can check themselves for systematic errors. The exploration or mining geologist should not rely on this, however, and every 10th or 20th sample should be sent to another laboratory for a comparative analysis. Alternatively, it is possible to prepare large samples that cover the range of expected grades, and to have them analysed by several laboratories. In this way one can generate one's own individual standard samples. Three or four of these standards are included in every shipment of samples. They are numbered in series with the other samples so that they cannot be identified by the laboratory.

It is important that the samples for check analyses are divided after the fine grinding, i.e. from the pulps of about -200 mesh (-0.074 mm). This procedure minimizes the grain size effect (previous section).

In general, the use of blind samples is of paramount importance and cannot be over-emphasized.

11.2.2
Mathematical Comparison of Two Series of Analyses Using the Student's t-Factors

Random errors cannot be avoided, but it is important to determine if a systematic error is present. The check analyses described above are undertaken so that this error can be identified. The statistical question then raised is: are the differences

Table 36. Check assay results for a lithium deposit

Colum 1	Column 2	Column 3	Column 4	Column 5	Column 6	
Sample no.	Laboratory K % $Li_2O = x_K$	Laboratory L % $Li_2O = x_L$	Laboratory K-laboratory L % $Li_2O = (x_K - x_L)$	$\dfrac{x_K + x_L}{2}$	$\left\| x_K(x_L) - \dfrac{x_K + x_L}{2} \right\| = \left\| \dfrac{x_K - x_L}{2} \right\|$	
22 - 5	2.67	2.85	- 0.18	2.760	0.090	
22 - 23	3.05	3.12	- 0.07	3.085	0.035	
23 - 8	3.18	3.06	+ 0.12	3.120	0.060	
23 - 9	4.82	4.67	+ 0.15	4.745	0.075	
25 - 1	0.58	0.55	+ 0.03	0.565	0.015	
25 - 11	2.08	1.99	+ 0.09	2.035	0.045	
27 - 5	1.43	1.62	- 0.19	1.525	0.095	
27 - 7	2.78	2.53	+ 0.25	2.655	0.125	
27 - 21	1.89	2.02	- 0.13	1.955	0.065	
28 - 9	3.42	3.33	+ 0.09	3.375	0.045	
28 - 13	3.98	3.54	+ 0.44	3.760	0.220	
29 - 5	2.26	2.25	+ 0.01	2.255	0.005	
32 - 11	0.98	1.22	- 0.24	1.100	0.120	
32 - 12	0.77	0.88	- 0.11	0.825	0.055	
33 - 14	2.35	2.16	+ 0.19	2.255	0.095	
34 - 15	0.73	1.89	- 1.16	1.310	0.580	
	$\overline{x}_K = 2.42$	$\overline{x}_L = 2.39$	$\overline{x}_{diff} = 0.03$ $s_{diff} = 0.19$			

between the laboratories significant or not, in other words are the errors only random, or also systematic? This question is tested by use of the Student's t-distribution (e.g. Moroney, 1970).

The procedure is best explained by an example.

Example: During the evaluation of a spodumene deposit, 16 check samples were shipped to a laboratory for comparison (laboratory L). The results from laboratory L are compared to those from laboratory K, which is the laboratory that is usually used, as follows (columns 1 to 3, Table 36):

Question: Are the differences between the two analytical laboratories significant or not?

Solution:

1st Step: The differences between the analyses (laboratory K - laboratory L) are listed in column 4, Table 36, and are critically reviewed. It is noted that the last value is a particularly high difference, and it is suspected of being an outlier (cf. Chap. 8.3.3).

The outlier test is therefore undertaken according to the procedure described in Chapter 8.3.3.1. According to Eq. (1), Chapter 8.3.3.1, the value x_A is considered to be an outlier if the following condition is satisfied:

$$x_A \geq \bar{x} + s \cdot g \, ,$$

where $\bar{x}$ is the mean, s is the standard deviation and g is the outlier threshold, which is determined from the diagram in Fig. 20. Since it is an analytical problem that is under consideration, the level of significance Si = 1% is selected.

If there is no systematic error, then theoretically the sum of the differences should be zero. Therefore it is irrelevant, with respect to the outlier test, if it was laboratory K that analysed 0.73% LiO_2 and laboratory L that analysed 1.89% LiO_2, or if it was the other way round. The absolute value x_A is being tested, not x_A itself.

The mean and the standard deviation are now calculated for the differences listed in column 4, Table 36, by using the standard formulae [Chap. 3.1.1, Eq. (2) and Chap. 3.2.1, Eq. (1)] omitting the value suspected of being an outlier. This gives the following values:

$$\bar{x}_{diff} = 0.03 \, ,$$

$$s_{diff} = 0.19.$$

The value g is selected from the diagram in Fig. 20 for Si=1% and n=15:

$$g = 5.8 \, .$$

If the final value is correctly suspected of being an outlier, then the above equation must be true:

$$|x_A| \geq \bar{\bar{x}} + s \cdot g \; ,$$

$$|-1.16| \geq 0.03 + 0.19 \cdot 5.8 = 1.13 \; .$$

The condition is satisfied, and therefore the last value for sample 34-15 is rejected, and is not taken into any further consideration. Only the remaining 15 values are included in the subsequent steps. (The last sample 34-15 has to be examined again to find out which assay value can be accepted, best by resampling and reassaying.

2nd Step: The test to determine if the difference between the laboratories is significant or not is undertaken with the following formula for calculation of the Student's t-factor (Moroney, 1970):

$$t = \frac{|\bar{x}_K - \bar{x}_L| \cdot \sqrt{n-1}}{s_{diff}} \; . \tag{1}$$

The absolute difference between the mean of the analyses from each laboratory x_K and x_L is in the numerator. Clearly this difference must be the same as the mean of the differences in column 4 of Table 36, which is 0.03. Once more this is an absolute value, and it is thus irrelevant if the difference is derived from laboratory K - laboratory L, or, conversely, from laboratory L - laboratory K.

If the calculated values are substituted in the above equation, then:

$$t = \frac{|\bar{x}_K - \bar{x}_L| \cdot \sqrt{n-1}}{s_{diff}} = \frac{0.03 \cdot \sqrt{14}}{0.19} = 0.59 \; .$$

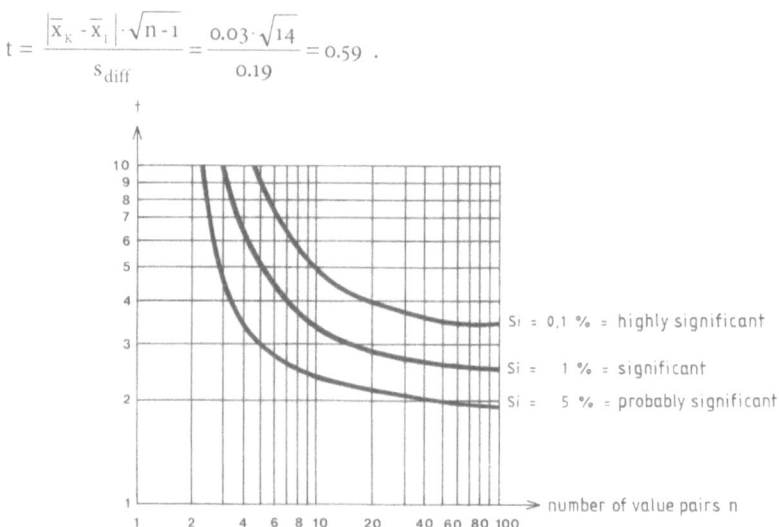

Fig. 31. Student's t-factors for the analyses of significance at the 0.1, 1 and 5% level. (After Moroney 1970)

The value $t = 0.59$ is now plotted on the diagram in Fig. 31, and it is seen that this value lies well below the "probably significant" line. This means that no significant difference can be identified between the laboratories, even if there is one. The analyses from laboratory L, which was used for the check samples, are relatively lower by an average of 1.2% , but there is no necessity to reduce the analyses from laboratory K by, for example, 1.2%. The sample that was rejected as an outlier must be reanalysed.

Finally, a weakness of this test must be mentioned. The standard deviation of the differences, s_{diff}, is in the denominator, which means that the larger the value for s_{diff}, the smaller t becomes, and therefore the chance of detecting a significant difference between the two laboratories is that much less. If the situation arises whereby the means of the sample series from both laboratories x_K and x_L are approximately the same, but there is an anomalously large difference between the individual samples so that s_{diff} is a large value, then it is recommended to engage two new laboratories.

11.2.3
Comparison of Two Series of Analyses by Regression Analysis

If it is found that there is a significant difference between the routine laboratory and the check laboratory, then the values from the routine laboratory must either be repeated or corrected. The check laboratory is known to provide reliable analyses on the basis of previous sample series that were compared with other reputable laboratories. A typical example would be:

A silver deposit is offered for sale. During a visit to the property, 80 splits were collected from the remnants of analysed samples and were submitted for analysis at a check laboratory. This yielded significant differences in the analytical results. Time and other constraints prevent any resampling at the prospect. What is to be done with the analytical results from the routine laboratory?

It is incorrect, and even dangerous, to compare the average values from both analytical laboratories and then to derive just one single correction factor. For example, it is frequently necessary to apply disproportionately larger corrections to the higher values, which are also the values of decisive importance for the economic evaluation.

Therefore the best method is to carry out a regression analysis, and to recalculate each analytical value either by the regression equation, or by graphical means. The procedure of a regression analysis was described in Economic Evaluations in Exploration, Chapter 5.2, and will not be repeated in detail here.

The most important equations for the regression analysis are summarized in Appendix Table 8, as they will be required again later in this book.

The following Table 37 provides the results of the 20 check analyses, and compares them to the results from the routine laboratory.

The differences are tested by the Student's t-factor using the procedures described in Section 11.2.2. This demonstrates that there is a highly significant difference between the analyses. The difference of the mean values between the two analytical laboratories is 31.9 g Ag/t. The standard deviation of the differences is

$$s_{diff} = 24.79.$$

Table 37. Routine and check assay results for an Ag deposit

Routine analysis x_i (g Ag/t)	Check analysis y_i (g Ag/t)	Routine analysis x_i (g Ag/t)	Check analysis y_i (g Ag/t)
15.1	38.2	131.5	103.2
46.3	27.6	137.1	83.2
51.1	54.9	145.0	117.4
67.2	68.8	154.2	106.8
71.1	54.6	164.4	127.1
85.0	50.3	187.3	130.7
93.6	87.2	192.2	144.9
98.3	67.0	205.1	141.0
114.4	90.1	219.3	145.2
127.2	84.3	220.8	165.3

$\bar{x} = 126.3$ g Ag/t $\bar{y} = 94.4$ g Ag/t $\bar{x} - \bar{y} = 31.9$ g Ag/t

Therefore by using Eq. (1) from Sect. 11.2.2:

$$t = \frac{|\bar{x} - \bar{x}| \cdot \sqrt{n-1}}{s_{diff}} = \frac{31.9\sqrt{19}}{24.79} = 5.61.$$

Apply the values t=5.6 and n=20 to the diagram in Fig. 31, then it is seen that this value lies above the "highly significant" line. Thus there is a significant difference between the check analyses and the routine analyses.

A regression analysis is now carried out by using the Eq. (1) to (4), Appendix Table 8, and this yields a regression line defined by the equation

$$y = 14.9 + 0.63 \cdot x \quad \text{[Fußnote 22]}$$

The coefficient of determination r^2 is $r^2=0.92$, which indicates a high correlation that can also be displayed graphically (Fig. 32).

If only the mean values had been used, then

$\bar{x} = 126.3$ (routine laboratory) would be compared with

---

[22] The question of how to identify the analyses of the check and the routine laboratories, with x or y, arises during the regression analysis. There is a difference if a regression is made from y to x or from x to y (see statistical test books, e.g. Kreyszig, 1968 or Wonnacott and Wonnacott, 1985). The following concept must be made clear: there are numerous analyses from the routine laboratory that are to be recalculated by using the regression line derived from the comparison with analyses from the check laboratory. The regression line is defined by the equation $y = a \cdot x + b$; so the routine analyses must be x and the check analyses must be y (see Table 37).

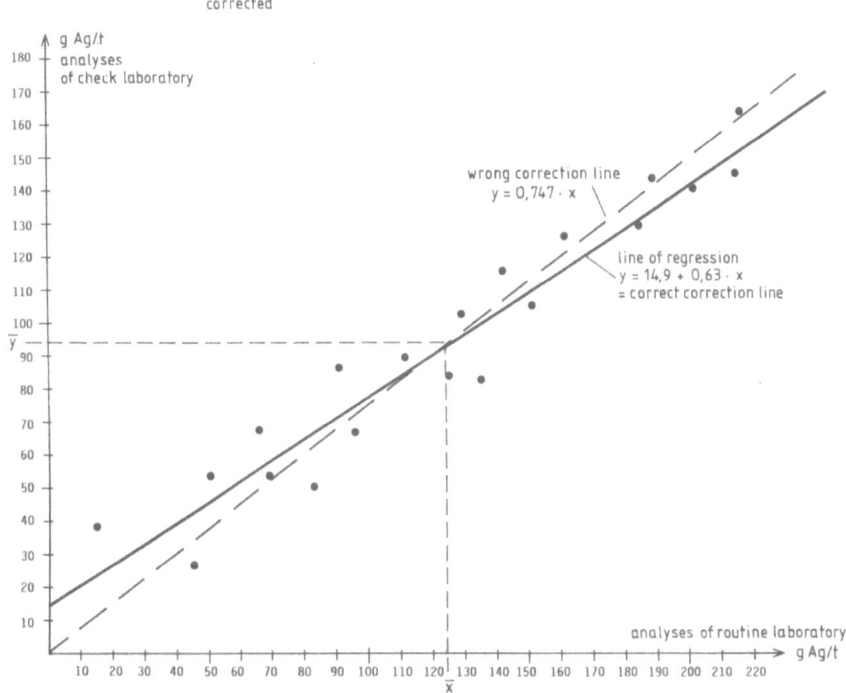

Fig. 32. Graphic comparison of 20 Ag analyses from Table 37

$\bar{y}$ = 94.4 (check laboratory).

The difference is 31.9, which relative to the routine laboratory is:

$$\Delta \, \text{rel} = \frac{31.9}{126.3} = 0.253, \text{ or } 25.3 \, \%.$$

If every value from the routine laboratory is corrected by this factor, so that every value is multiplied by $(1-0.253)=0.747$, then the correction line in Fig. 32 would be:

$$y = 0.747 \cdot x \, .$$

Figure 32 clearly shows the mistakes that might be made:
All values less than the mean $\bar{x}$ would be too strongly corrected, and thus if the cut-off was in this range, then in reality the value selected for the cut-off would be too high and therefore the tonnage would be reduced.

The correction for values above the mean $\bar{x}$ would be too small. A value from the routine laboratory of 220 g Ag/t ought to be corrected to 153.5 g Ag/t, but would be corrected wrongly to 164.3. This means that the grade is overestimated by 7 %, which could be a decisive error for an economic evaluation.

11.2.4
Graphical Comparison of Analytical Laboratories

During longer exploration programmes, check analyses should be dispatched regularly with every shipment of samples, as discussed above in Section 11.2.1. If this is done, then one wants to detect any variations in the differences from shipment to shipment, and wants to recognise trends in the deviations and therefore systematic errors or a reduced repeatability. This can be done with the help of the graphical method proposed by Thompson and Howarth (1973). The mean value for each value pair is plotted on the x-axis of log-log paper, and the analytical deviation from the mean is plotted on the y-axis. The lines of the same

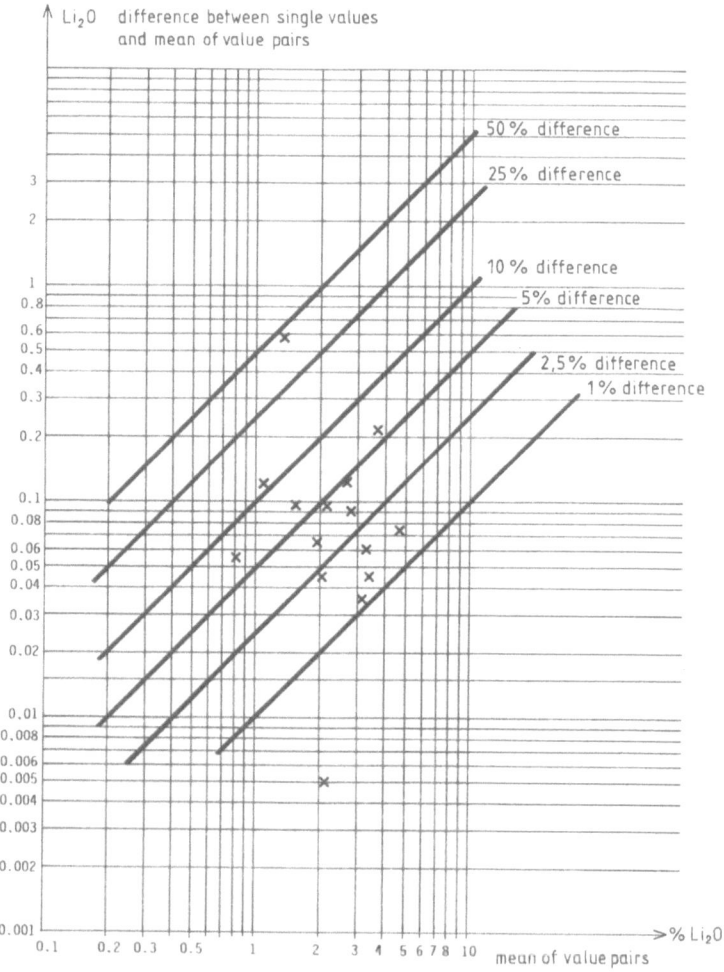

Fig. 33. Diagram for comparison of two laboratories after Thompson and Howarth (1973) with values of Table 36

relative deviation can be easily drawn on the diagram (Fig. 33). The mean value of the value pairs and the difference between the analyses and the mean value are listed, respectively, in columns 5 and 6 in Table 36 for the example described in Section 11.2.2. These values are also plotted on Fig. 33.

11.3
Comparison of Sample Series with Different Support

The problem of different sample support was already introduced in Chapter 1.1 at the beginning of this book. It is easy to imagine that the larger the volume of the sample, the lower the range of the grades or variance should be. The sample-volume-variance relationship, the theory of which was derived by Hazen (1967), suggests that the volume-variance product should be constant. Although this does not strictly occur in nature (cf. Chap. 3.3), there is a definite tendency. The problem of comparing sample series with differing support is commonly encountered during exploration projects. Two examples:

A lateritic nickel deposit has been drilled. In the soft saprolite it is difficult to check accurately the core recovery and therefore, as a further control, pits are dug and sampled.

Most of the mineralization in porphyry copper and molybdenum deposits is generally associated with fractures. During drilling, the core breaks preferentially along the healed fractures. There is therefore a danger that some of the mineralization will be washed out with the sludge, and that the grade of the cores will thus appear to be less than it really is. Therefore, on some porphyry copper deposits, a 10% "upgrading factor" is applied to the drill results as a correction for this effect. Shafts are often sunk and adits driven in order to obtain large bulk samples that can be compared to the drill results.

Before determining whether such an "upgrading factor" is justified or not, it must first be understood how two or more series of samples with different support can be compared, even in the ideal case of each sample type such as drill core or bulk samples being completely free of error.

11.3.1
Theoretical Basis for the Comparison of Sample Series with Different Support.

In his classic study, Krige (1962) discussed this problem in detail by comparing the grades of samples with the grades of mine blocks in the South African mining industry. The problem is illustrated in Fig. 34. The diagram in the centre shows the grades of the samples plotted against the grades of the mining blocks, and they can be seen to have a broad range. Histograms constructed for the rows and columns, as has been done for some cases in Fig. 34, yield approximately normal distributions.

The regression line, which is the best correlation between the two variables, passes through the peaks of each of the normal distributions of the columns (line KL). This correlation line has a gradient that is markedly shallower than the 45° line, which reflects the correlation to be expected if on average the grade of a sample was equivalent to the grade of the enclosing block. The 45° line and the

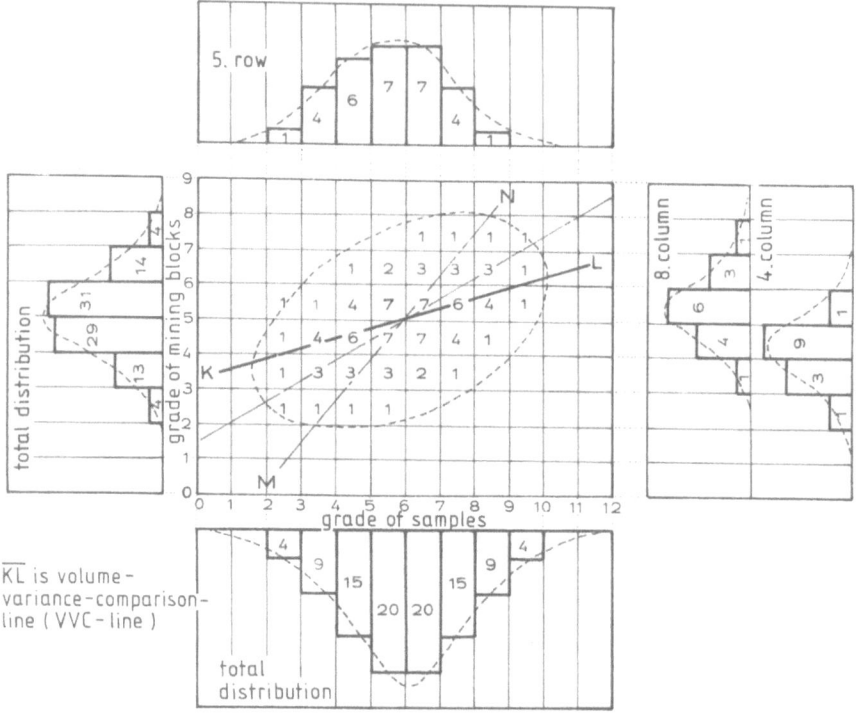

Fig. 34. Diagram with example from South African gold mining, illustrating the correlation between grades of samples with grades of corresponding mining blocks. (Krige 1962 with permission of the author)

optimal correlation line (regression line) intersect at the mean $\bar{x}$ which, assuming that the sampling was truly representative, should theoretically be the same for both the samples and the mining blocks (Fig. 35).

The relatively shallower gradient of the regression line with respect to the 45° line means:

1. Samples, for which the grades are less than the mean $\bar{x}$, are enclosed by blocks that on average have higher grades than the samples.
2. Samples, for which the grades are greater than the mean $\bar{x}$, are surrounded by blocks that on average have lower grades than the samples.

This is a consequence of the regression, and has nothing to do with the quality of the sampling, as can be simply illustrated in Fig. 36. Assume that a deposit has been drilled and that the frequency distribution of the drill results is as shown in the upper part of Fig. 36. The deposit is subsequently mined and the grades of the mining blocks around each of the drill holes were determined. The range, or the variance, of the grades for the mining blocks is lower since the volumes are orders of magnitude greater. The frequency distribution of the grades of the mining blocks

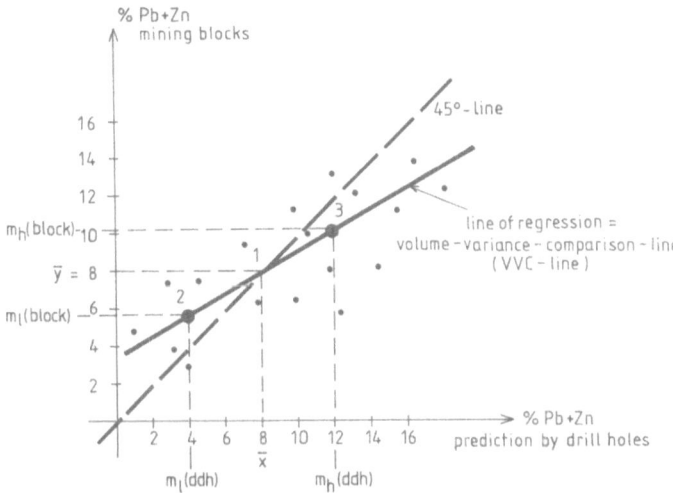

Fig. 35. Comparison between grades of drill hole samples and of corresponding mining blocks

is shown in the lower half of Fig. 36. The means $\bar{x}$ and y are the same for both distributions.

The next step examines the values in both distributions that lie above and below the same means $\bar{x}$ and $\bar{y}$. Each half is treated on its own and, if the mean m_l is derived for that half of the values below the mean $\bar{x} = y$, then it can be seen that the mean of the mining block distribution has shifted to a higher value as compared to that of the drill hole distribution:

$$m_l(ddh) < m_l(\text{mining block}) \cdot$$

The mean value m_h is now determined for that half of the values above the mean value, and it can be seen that the mean for the mining block distribution has shifted to a lower value as compared to that for the drill hole distribution:

$$m_h(ddh) > m_h(\text{mining block}) ,$$

and thus, during the transition from the frequency distribution of the drill holes to that of the mining blocks, the means of both halves shift towards the overall mean value $\bar{x} = \bar{y}$.

A correlation line (regression line) can now be constructed from these three points (Fig. 35):

1. the overall mean $\bar{x}$, at the coordinates $\bar{x} = \bar{y}$;

2. the mean of the lower halves with the coordinates

$$x = m_l(ddh), \quad y = m_l(\text{block});$$

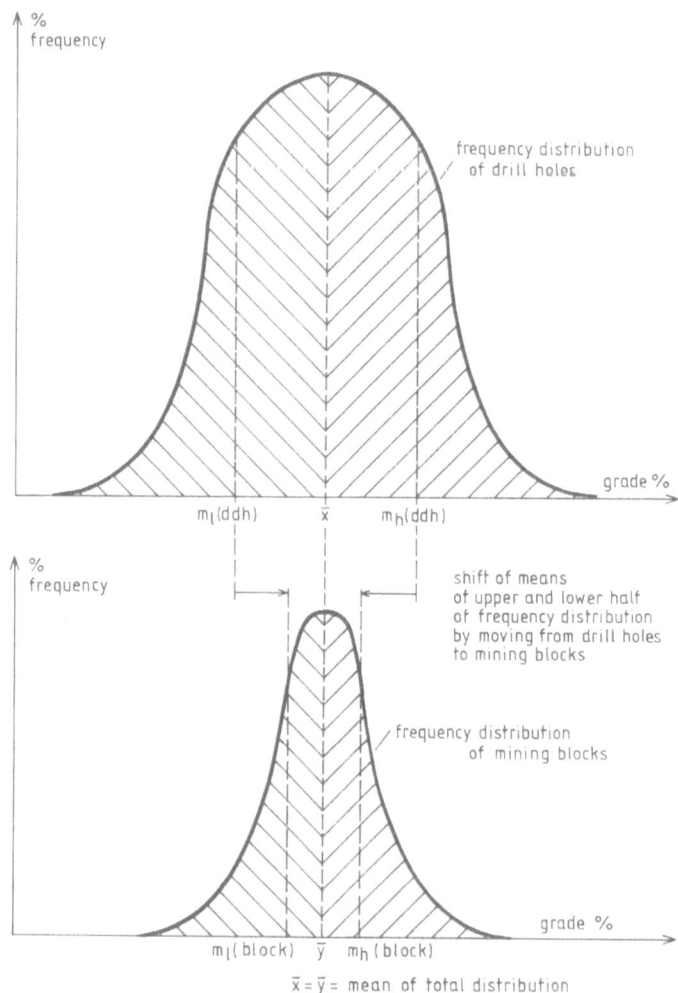

Fig. 36. Hypothetical frequency distributions for drill hole samples and corresponding mining blocks

1. the mean of the upper halves with the coordinates

$$x = m_h(ddh), \quad y = m_h(block).$$

It is now clear that this regression line must have a gradient shallower than the 45°
line.

This principle is valid for each comparison of sample series with different
volumes (i.e. different support) or, in more precise terms, sample series with
different ranges or variances (Fig. 36). Thus this regression line, which is of

fundamental importance, is known as the volume-variance-comparison line, abbreviated to VVC-line. This VVC-line is, of course, directly related to the volume-variance rule that was introduced in Chapter 3.3. The application of this principle is explained by an example:

Example: Assume that drilling results must be checked by pitting, as often happens during the exploration for lateritic nickel deposits. It is important that the channel samples collected from the walls of the pit have the same volume as the drill core samples, and that these samples are collected from one side or wall of the pit. If channel samples are collected from all four walls of the pit, and are then combined into one sample, then this is essentially averaging four values. The quality of this averaged sample, even if it has the same volume as the corresponding drill core samples, is obviously different from the quality of a sample taken from one single site (Chap. 1.1). This is because the variance of a sample that is averaged from four individual samples is lower than that of individual samples (Fig. 37). The improved sample quality that is obtained by combining samples has been cleverly adapted for use in selective gold mining in Australia (cf. Sect. 11.5.2.2).

Fig.37. Examples for the reduction of the variance and the standard deviation. *Top* Comparison of drill core samples with mining blocks. *Centre* Comparison of drill core samples of centre of test pit with four channel samples at wall of test pit. *Bottom* Comparison of drill core samples with mean of random grab samples from ore stockpile

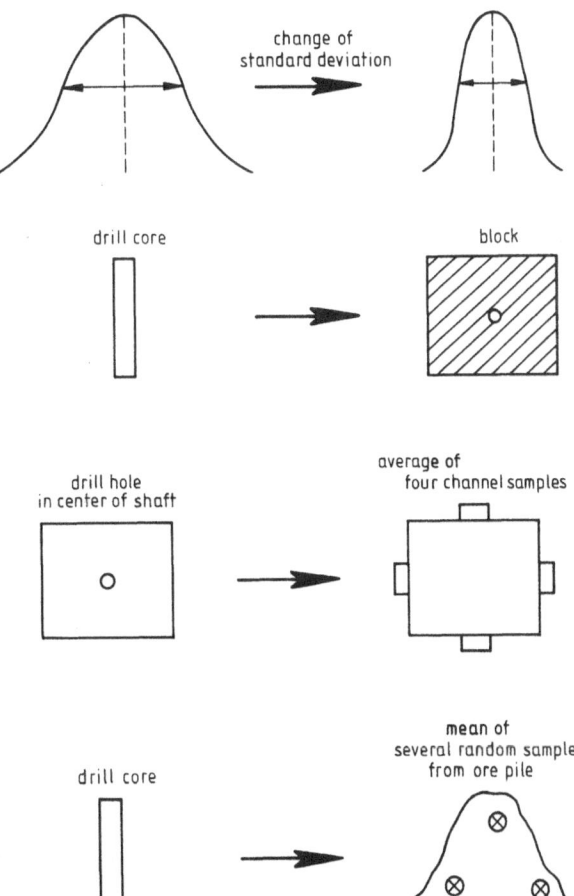

It is surprising how often these relationships are incorrectly understood in bulk sampling programmes and, as a consequence, totally misleading conclusions are made. The following example, which is taken from the prospectus for the sale of a uranium mine, provides a good illustration of this:

The average grade of the uranium deposit was determined by drilling and trenching (channel samples) to be 0.55% U_3O_8. A bulk sample several tonnes in weight was taken from one location, and yielded an average grade of 0.38% U_3O_8. The corresponding drill core grades were 0.29% U_3O_8. It is concluded that the bulk sample indicated that the drilling had not provided reliable results, and had thus undervalued the deposit. If the bulk samples were 25% higher at low grades, how much higher were they at higher grades?

It is quite clear that the author of this prospectus had not realised that, if the bulk samples are less than the mean, the average grades of bulk samples are shifted to higher values and that the converse is true for average grades greater the mean.

Bulk samples collected from those sites where the values are lower than the mean can thus provide a good means of misleading incompetent investors into believing that the deposit is actually of much higher grade, and that an "upgrading factor" could be applied. These theoretical considerations also clearly show that it is meaningless to carry out a bulk sampling programme at only one site. In order to make a serious comparison between drill core results and bulk samples, it is essential to collect the bulk samples from two sites, one of which is expected to yield grades above the mean value and the other below. Ultimately, the straight line, in this case the VVC-line, must be defined by at least two points and there is no rule indicating how shallow the gradient of the VVC-line should be in comparison with the 45° line.

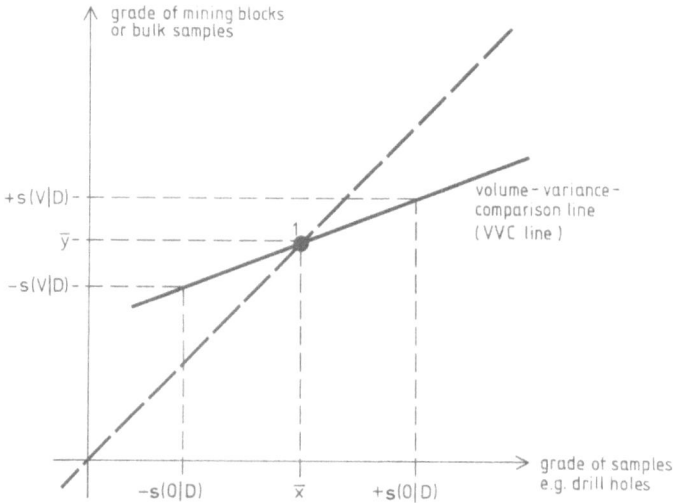

Fig. 38. Schematic illustration for the construction of the volume-variance comparison line (VVC line) out of the variances of the samples in the deposit $\sigma^2(O/D)$ and of the bulk samples or mining blocks $\sigma^2(V/D)$ (σ^2 of population ~ s^2 of sample size)

The problem of the gradient and the location of the VVC-line could theoretically be geostatistically solved, and geostatistical methods will be introduced in Chapter 13.

The variance of the grade of a mining block within the ore deposit, $\sigma^2_{(V/D)}$, can be estimated by calculation of the variogram (Chap. 13.2). The variance of the samples, for example drill holes, in the deposit $\sigma^2_{(O/D)}$ is known anyhow so that the VVC-line can be constructed as shown in Fig. 38. The problems are only:

a) The location of point 1 is not known. The bulk sampling should test that the mean of the samples $\bar{x}$ is truly the same as that of the bulk samples or future mining blocks $\bar{y}$, and

b) if the mean $\bar{x}$ of the samples does not coincide with the mean of the bulk samples or mining blocks $\bar{y}$, then it must be assumed that the drill samples have a systematic error (see introductory remarks to Sect. 11.3), and thus there is a bias. These samples are therefore not a good basis for the calculation of a variogram and the construction of the VVC-line.

11.3.2
Derivation of an Upgrading Factor by Comparing Bulk Samples and Drilling

11.3.2.1
Introduction

Bulk sampling is carried out so that the grade of the mining blocks can be calculated from the drilling results. This assumes that the volumes of the bulk samples can be compared satisfactorily with the blocks of the future mining method. The derivation of the VVC-line is the only purpose of this exercise, and the question of whether the result is a normal VVC-line with a gradient of less than 45° through the mean $\bar{x}$ of the deposit, or if it has a steeper gradient and/or has been parallel shifted, and thus indicating there has been a definite loss of the valuable minerals during the drilling, is only of secondary interest at this stage. This question becomes relevant if these results are applied to other sections of the ore deposit where there might have been either similar or different core recoveries.

If there really has been a systematic loss of the metal values during drilling, then a comparison of the drilling and bulk samples will show that the regression line gradient is generally steeper than 45°. There is no reason to assume that the absolute values of the differences for high-grade value pairs will be any smaller. On the contrary, the higher the proportion of valuable material, then the greater the probability that there has been some loss. The proportion of metal loss might be somewhat smaller, but that ought not to be the case for the absolute amount of the difference.

For a perfect or model data set, the regression line for drill and bulk samples could be compared with the theoretical relationship, or with the VVC-line, for which the gradient is shallower than 45°, as defined in Section 11.3.1. From this, an "upgrading" factor could be derived for the drill data, using the principle illustrated in Fig. 39:

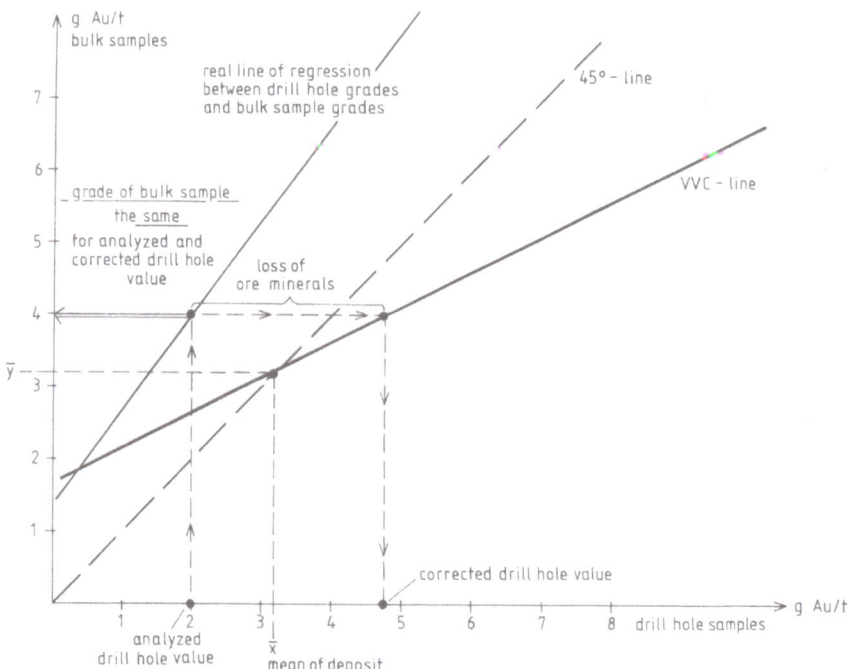

Fig. 39. Theoretical examples for the comparison of the grade of small samples (e.g. drill hole samples) and the corresponding grade of bulk samples and the correction of the grade of small samples with the volume-variance-comparison line (VVC line)

Draw a horizontal line from the intersection of the analysed value of the core section (e.g. 2 g Au/t) with the regression line for drill/bulk samples to the VVC-line. The corrected drill value is then read from this intersection point (4.75 g Au/t in this example). The metal loss was therefore:

$$\frac{4.75 - 2}{4.75} = 0.58, \text{ i.e. } 58\%.$$

Figure 39 demonstrates that this has no effect on the recalculation of the bulk grades, since the bulk grade is 4 g Au/t for both the original analysed value as well as for the corrected value.

(This is, of course, an extreme case that is shown to explain the principle. If a metal loss of over 50% was really encountered, then the drilling was useless and must be repeated!).

In reality, the theoretical VVC-line is not known, and other approximations must be derived. This is now demonstrated by a real example.

11.3.2.2
Standard Derivation of an "Upgrading" Factor

Example: A porphyry copper deposit, with a gold zone within the supergene enrichment zone, has been evaluated by diamond drilling on a regular grid. Most of the gold is associated with the pyrite, which is oxidised to limonite in the zone of gold enrichment. The pyrite, or now the limonite, is located in small veinlets. During the drilling, it was seen that the core preferentially fractured along these limonite veinlets. It is suspected that some of the limonite, together with gold, is washed out with the sludge and that this has resulted in a lower gold grade for the drill core. An adit is driven in order to investigate this possibility. A horizontal hole had been previously drilled along the heading of the adit. The broken ore from the adit is used as a bulk sample.

The horizontal hole is drilled with HQ core (i.e. a diameter of 63.5 mm; cf. Economic Evaluations in Exploration, Appendix Table 2). The adit has a cross-section of 2 x 2.5 m, so there is an area ratio of approximately 1:1600, meaning if the same length of drill core and adit are compared also the volume ratio is 1:1600, 1:1600,

Drill results and bulk sample results are each combined to 10 m sections.

Two homogeneous areas are recognised in the zone of gold mineralisation:

a) high gold, low copper.

b) high gold, high copper.

These homogeneous zones were treated separately, since the drill cores from the high gold, low copper area appear to be more fractured than those from the other homogeneous area (b). The data from area (a) (high gold, low copper) is discussed first, and then the data from area (b) (high gold, high copper).
The results are as follows:

Column 4 in Table 38a shows that the differences between the bulk and the drill core samples are always positive, except for two values. The negative differences for the two exceptions are only very small. It is therefore suspected that the regression line between the bulk samples and drill hole samples is systematically shifted to higher bulk sample values. It has been shown in the previous section that there would otherwise be a more balanced distribution of positive and negative differences.

A comparison of the variances provides an indication that the drill grades must be in error. The plot of the VVC-line shows that the variance of the drill hole results $\sigma^2_{(O/D)}$ must be greater than the variance of the bulk samples or mining blocks $\sigma^2_{(V/D)}$, as in Fig. 38:

$$\sigma^2_{(O/D)} > \sigma^2_{(V/D)}.$$

Table 38a. Drill hole and bulk samples in the homogeneous area a: high Au, low Cu

Section	Drill hole = x (g Au/t)	Bulk sample = y (g Au/t)	Difference Δ bulk-drill hole (g Au/t)
1	0.93	1.63	+ 0.70
2	1.20	1.55	+ 0.35
3	1.68	2.37	+ 0.69
4	7.70	8.59	+ 0.89
5	3.03	3.99	+ 0.96
6	4.05	5.52	+ 1.47
7	3.75	3.58	+ 0.10
8	3.57	4.78	+ 1.21
9	1.77	1.63	- 0.14
10	1.58	2.36	+ 0.78
11	1.30	2.62	+ 1.32
12	3.08	3.02	- 0.06
13	1.03	1.36	+ 0.33
14	0.96	1.06	+ 0.10
15	1.70	1.93	+ 0.23
	$\bar{x} = 2.49$ $s_x^2 = 3.26$	$\bar{y} = 3.08$ $s_y^2 = 4.05$	Sum of differences $\Sigma\Delta = + 8.93$

These conditions are now compared to the results in Table 38a, where it is found that the converse is true:

$$s_x^2 < s_y^2.$$

The mean values of the bulk samples and the drill samples should be approximately the same, assuming that these samples cover zones with grades that are both higher and lower than the mean value. The mean value for the gold zone (a) in the deposit was calculated from the drilling results to be 2.80 g Au/t (uncorrected). The mean value of the drill hole in the adit is 2.49 g Au/t, which is about 10% lower. In practice, this is an acceptable approximation.

The average of the bulk samples is 3.08 g Au/t, which is 24% higher than that of the drill samples. Theoretically the mean value from the drill samples should be the same as the mean value from the bulk samples (Fig. 35), and as a simple procedure an "upgrading" factor could be derived from

$$Fu = \frac{3.08 - 2.49}{2.49} = 0.24, \text{ i.e. } 24\%,$$

and then every drill value is corrected accordingly.

A similar problem was discussed in Section 11.2.3, and the danger was noted of universally applying corrections to grades, without considering the dependence between the correction factor and the grades.

This problem is solved in the following way:

1st Step: The regression line is calculated for the drill and bulk samples. The necessary relationships are provided as Eq. (1) to (4) in Appendix Table 8. This results in a regression line with the equation

$$y = 0.39 + 1.08 \cdot x. \tag{1}$$

As is to be expected for a true increase in metal values, which compensates for the material washed out during the drilling, the gradient of the line is steeper than 45°, and not shallower, as is the case for the VVC-line. A gradient of less than 1 (i.e. a slope angle shallower than 45°) would be expected if there had been no loss of metal values during the drilling (cf. Sect. 11.3.2.1). The individual values and the regression line are plotted on Fig. 40a.

2nd Step: The VVC-line is not known as it was in Fig. 39. What can be done? The gradient of the VVC-line should be shallower than 45° but, as previously explained, there are no rules defining how much shallower. It can only be stated that:

$$\alpha_{VVC} = 45°\text{-}\delta.$$

and

$$d > 0.$$

From this, the limiting value can be derived:

$$\alpha_{VVC} = 45°$$

To be on the safe side, the average value of the homogeneous area (a) (high gold, low copper) was determined from the drill holes to be $\bar{x} = 2.80$ g Au/t. According to the above regression equation [Eq. (1) above in the 1st step] this corresponds to $\bar{y} = 3.4$ g Au/t. Theoretically the VVC-line should intersect the 45° line at this value (point A).

3rd Step: The fields less than A and greater than A are now treated separately.
a) Less than A: This is the field where the values are on average lower than the mean value of the deposit $\bar{x}$. In this field the mining blocks have, on average, higher grades than the corresponding samples (cf. Sect. 11.3.1 and Fig. 35). There are two possible solutions:
    i)   To be completely on the safe side, then the values in this field are not corrected at all.
    ii) If this appears to be too conservative, then point A, through which the VVC-line should pass, is joined with point B, which is the y-value for the regression line where $x = 0$

In the illustrated example, the coordinates for point B are (0/0.39). The line AB is then selected as the VVC-line for this field (Fig. 40a). The horizontal difference between the regression line $y = 0.39 + 1.08 x$ and the line AB is the "upgrading" factor for this field.

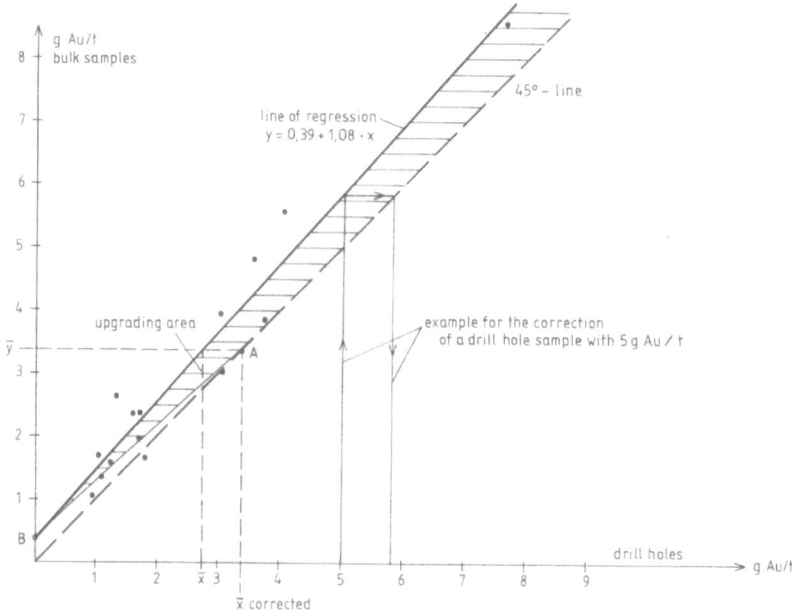

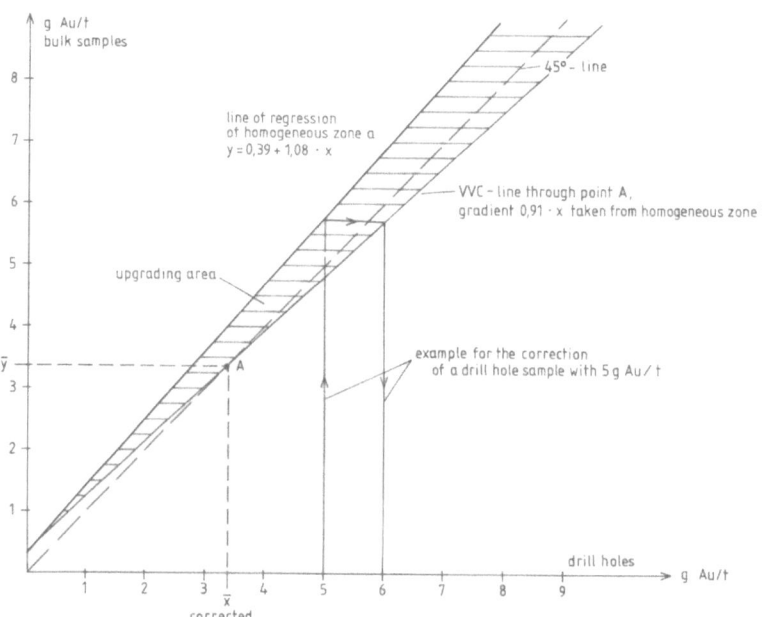

Fig. 40.a (top) Comparison of drill hole and bulk samples of Table 38a and construction of corresponding VVC line. **b** (bottom) The upgrading area (*horizontally hatched*) in the homogeneous zone a = high Au, low Cu

b) Greater than A. This is the field where the values are on average higher than the mean value for the deposit $\bar{x}$. In this field the mining blocks have, on average, lower grades than the corresponding samples (cf. Sect. 11.3.1 and Fig. 35). As stated in the 2nd step, the gradient of the VVC-line is not known.

To be on the safe side, the limiting value of 45° is selected, so that the VVC-line is the same as the 45°-line. The difference between the regression line y=0.39+1.08 x and the 45°-line is then the "upgrading" factor (Fig. 40a).

Figure 40a illustrates an example of how a drilling grade of 5 g Au/t was corrected to 5.8 g Au/t.

11.3.2.3
Derivation of an "Upgrading" Factor by Comparing Zones

The adit in the example outlined in Section 11.3.2.2 was driven further into the 2nd, high gold, high copper homogeneous zone. The results are listed in Table 38b.

An analysis of the data indicates that the mean values for the drill hole data and the bulk samples are relatively close to each other, and the sum of the differences is much less than it is for the high gold, low copper, homogeneous zone as shown in Table 38a. It is therefore assumed that in this zone there has been no metal loss during the drilling.

A comparison of the variances also supports this conclusion. In contrast to the data in Table 38a, the following condition is met (cf. Fig. 38):

$$s_x^2 > s_y^2 .$$

Table 38b. Drill hole and bulk samples in the homogeneous area b: high Au, low Cu

Section	Drill hole = x (g Au/t)	Bulk sample = y (g Au/t)	Difference Δ bulk-drill hole (g Au/t)
17	2.13	2.61	+ 0.48
18	1.90	1.85	- 0.05
19	2.75	2.47	- 0.28
20	2.08	2.44	+ 0.36
21	2.75	3.15	+ 0.40
22	4.74	4.61	- 0.13
23	2.80	2.74	- 0.06
24	5.42	4.19	- 1.23
25	5.55	4.25	- 1.30
26	7.14	8.05	+ 0.91
	$\bar{x} = 3.73$ $s_x^2 = 3.36$	$\bar{y} = 3.64$ $s_y^2 = 3.24$	Sum of differences $\Sigma\Delta = 0.90$

1st Step: The regression line is calculated by the Eqs. (1) to (4), Appendix Table 8. The regression line has the equation:

$$y = 0.26 + 0.91 \cdot x \ .$$

The gradient of the line is shallower than 1 (i.e. the slope angle is less than 45°). This is compatible with the assumption that no metal values have been lost during the drilling, and that there is a true VVC-correlation between the drill samples and the bulk samples.

2nd Step: This VVC-line is now used to determine an "upgrading" factor for the data in Table 38a and Fig. 40a. Since the mean values are different, only the gradient will be used. Thus a line with gradient 0.91 is constructed through the corrected average value of 3.4 g Au/t (Point A in Fig. 40a). This line is now considered to be the VVC-line. This VVC-line and the regression line y = 0.39+1.08 x for the data set in Table 38a are presented again in Fig. 40b. The difference between the regression line y = 0.39 + 1.08 x and the VVC-line is then the respective "upgrading" factor.

Figure 40b illustrates an example of how the drill grade of 5 g Au/t is corrected to 6 g Au/t.

The examples of Chapters 13.2.2.2 and 13.2.2.3 in Tables 38a and b show how important it is to have good estimates of the samples of a deposit with different support. It has been stated already several times how advisable it is to visit similar deposits, possibly in the same area to collect economic and statistical data, which can be used as a-priori information in the statistical and economic evaluation of deposits. This applies also to variances. A-priori information can help to avoid overestimation (and underestimation) of deposits.

Krige et al. (1990) showed with a case history from the Hartebeestfontein mine in the Klerksdorp goldfield, Witwatersrand Basin, South Africa, the advantage to be gained to introduce a-priori information on the variances. In an experiment nine borehole values were drawn 60 times from an 1 x 1 km area in the Hartebeestfontein mine and compared with "true" values based on a large number of data within the relevant 1 x 1 km area. Figure 41 shows the result.

Where abnormally high variability is encountered – usually due to the inclusion of one or more very high values – a positive conditional bias can be expected in the estimate, i.e. a significant overevaluation. The reverse position applies where the variability is abnormally low. According to Krige et al. (1990), this feature can be expected to occur in all grade estimates for ore-bodies with highly variable skewed distributions of grade data where the data are limited.

11.3.2.4
Safety Margin for an "Upgrading" Factor

The application of an "upgrading" factor is always somewhat problematic. Because of the expense, bulk samples can be collected only at one or a few locations on the ore deposit, and are then generalised for the whole of the homogeneous zone. The number of samples is always limited, and the range is relatively wide (Fig. 40a). On the other hand, with few samples one can be suspected of engineering an

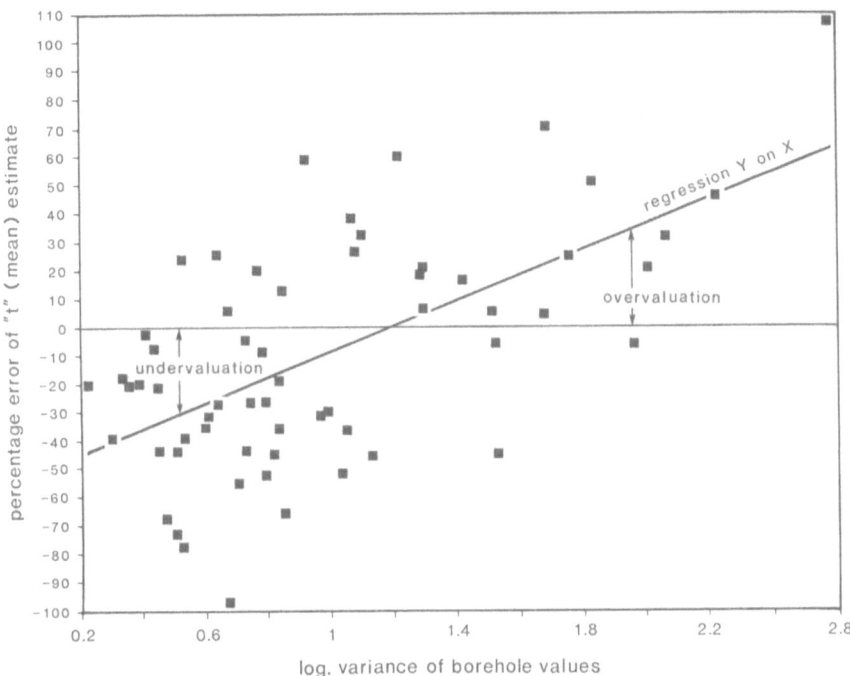

Fig. 41. Correlation of errors of borehole values for Hartebeestfontein mine case history of Krige et al. (1990) (with permission of the authors)

economically favourable calculation. Even if all the indications suggest that metal values have been lost to the drilling sludge, and that the application of an "upgrading" factor is justified, a careful geologist will always build a safety factor into his calculations.

The simplest method is to halve the "upgrading" factors derived in Sections 11.3.2.2 and 11.3.2.3. However, it is better to calculate a confidence limit for the values on the regression line, and then to use a lower confidence limit for the determination of an "upgrading" factor.
This can be illustrated by the example of the previously described gold deposit.
The relationships for the determination of confidence limits are (cf. for example, Kreyszig, 1968):

$$l = \frac{t \cdot h \cdot \sqrt{d}}{\sqrt{n-2}} \, . \tag{1}$$

In this equation, t is the corresponding factor from the Student's t-distribution (cf. Chap. 7.1 and Appendix Table 4) and n is the number of value pairs, which in Table 38a is 15.

The dimensions h and d are defined as

$$h^2 = \frac{1}{n} + \frac{(x-\bar{x})^2}{(n-1)\cdot s_x^2} \, , \tag{2}$$

$$d = (n-1)\cdot(s_y^2 - b^2 \cdot s_x^2) \, . \tag{3}$$

s_x^2 and s_y^2 are the variances of the variables x and y, b is the gradient of the regression line.

A confidence limit is now calculated for the regression line of the data set in Table 38a, which has the equation (cf. Sect. 11.3.2.2):

$$y = 0.39 + 1.08\cdot x \, .$$

1st Step: The factor d is calculated.
The gradient b is b = 1.08.
The variances of the variables x (drill hole samples) and y (bulk samples) are derived from Table 38a as:

$$s_x^2 = 3.26,$$

$$s_y^2 = 4.05 \, .$$

The number of value pairs is n = 15. Therefore d is:

$$d = (n-1)(s_y^2 - b^2 \cdot s_x^2) = 14\cdot(4.05 - 1.08^2 \cdot 3.26) \, ,$$
$$d = 14\cdot(4.05 - 3.80) = 3.47$$

2nd Step: The factor h^2 is calculated.
It can be seen from Eq. (2) above that this expression has a minimum of

$$h^2 = \frac{1}{n} \, ,$$

if $(x-\bar{x})$ is zero (i.e. at the value $\bar{x}$). This illustrates how the confidence limit moves away from the regression line with an increase in the distance of the x-value from $\bar{x}$.

In order to draw the confidence limit, h^2 is calculated for four points:

i) $x_1 = \bar{x} = 2.49$ (Table 38a)

ii) $x_2 = 1.3$

iii) $x_3 = 5.0$

iv) $x_4 = 7.5$.

i) for $x_1 = \bar{x}$: $h^2 = \dfrac{1}{n}$

i.e. $h = \sqrt{\dfrac{1}{15}} = 0.26$;

ii) for $x_2 = 1.3$: $h^2 = \dfrac{1}{15} + \dfrac{(1.3 - 2.49)^2}{14 \cdot 3.26}$

$h^2 = \dfrac{1}{15} + \dfrac{1.42}{45.64} = 0.098$

$h = 0.313$;

iii) for $x_3 = 5.0$: $h^2 = \dfrac{1}{15} + \dfrac{(5.0 - 2.49)^2}{14 \cdot 3.26}$

$h^2 = \dfrac{1}{15} + \dfrac{6.30}{14 \cdot 3.26} = 0.205$

$h = 0.452$;

i) for $x_4 = 7.5$: $h^2 = \dfrac{1}{15} + \dfrac{(7.5 - 2.49)^2}{14 \cdot 3.26}$

$h^2 = \dfrac{1}{15} + \dfrac{25.1}{14 \cdot 3.26} = 0.617$

$h = 0.785$.

3rd Step: The confidence limits l are now calculated from Eq. (1) above.
A level of confidence must initially be selected in order to determine the Student's t-factor. A higher level of confidence, e.g. 95%, is selected for the safety factors. Therefore for 15 value pairs (according to Table 4, Appendix, columns 3 and 4):

$t = 2.16$.

In the equation for the confidence limits [Eq. (1) above]:

$$l = \frac{t \cdot h \cdot \sqrt{d}}{\sqrt{n-2}}$$

only h varies for the variable x, so that the expression

$$\frac{t \cdot \sqrt{d}}{\sqrt{n-2}}$$

needs to be calculated only once:

$$\frac{t \cdot \sqrt{d}}{\sqrt{n-1}} = \frac{2.16 \cdot \sqrt{3.47}}{\sqrt{15-2}} = 1.121 \,.$$

Hence for the following confidence limits:

i)   for $x_1 = \bar{x}$: $l_{\bar{x}} = h_{\bar{x}} \cdot \dfrac{t \cdot \sqrt{d}}{\sqrt{n-2}} = 0.26 \cdot 1.121$

$l_{\bar{x}} = 0.29$;

ii)   for $x_2 = 1.3$: $l_{1.3} = h_{1.3} \cdot \dfrac{t \cdot \sqrt{d}}{\sqrt{n-2}} = 0.313 \cdot 1.121$

$l_{1.3} = 0.35$;

iii) for $x_3 = 5.0$: $l_{5.0} = h_{5.0} \cdot \dfrac{t \cdot \sqrt{d}}{\sqrt{n-2}} = 0.452 \cdot 1.121$

$l_{5.0} = 0.51$;

iv) for $x_4 = 7.5$: $l_{7.5} = h_{7.5} \cdot \dfrac{t \cdot \sqrt{d}}{\sqrt{n-2}} = 0.785 \cdot 1.121$

$l_{7.5} = 0.88$.

This confidence limit means that, for example, with a 95 % level of confidence for the y-value (i.e. the bulk sample value) corresponding to the mean of the drill core samples of $\bar{x} = 2.49$ lies within the range $\bar{y} = 3.08 \pm 0.29$, which is between 2.79 and 3.37.

The upper limit of the confidence interval is clearly of no interest for the safety factor. The curve representing the lower confidence limit is plotted on Fig. 42 together with the regression line for the data in Table 38a and the VVC-line. By taking the safety factor into account, the "upgrading" factor is now the difference between the lower confidence limit and the VVC-line.

Figure 42 illustrates an example of how a drilling grade of 5 g Au/t is corrected to 5.4 g Au/t.

11.4
Comparison of Sample Series with Different Sample Character

The problem of comparing samples with different character occurs if, for example, most of the drilling was carried out by percussion drilling and a few check holes were done by core drilling. Even then, the sample support can be the same or similar.

Usually, a diamond drill hole is not located immediately adjacent to a percussion drill hole, but the cored diamond drill holes are also sited according to the previous grid. It is therefore not possible to make a direct comparison between the diamond

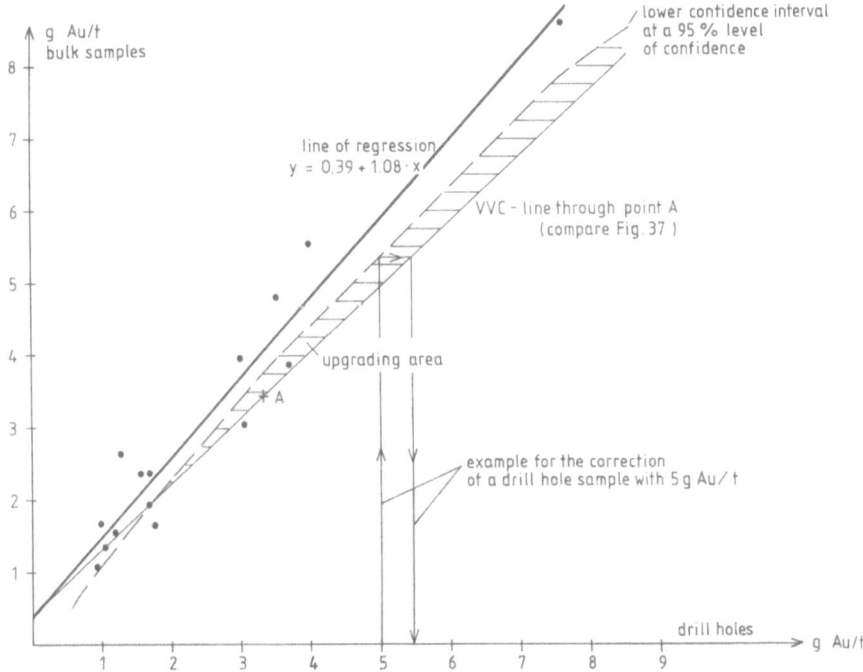

Fig. 42. The upgrading area (*horizontally hatched*) in the homogeneous zone a: high Au, low Cu, taking into account the confidence interval of the line of regression between drill hole samples and bulk samples

drill holes and the percussion drill holes, as was the case for the comparison between drill and bulk sample results described in Section 11.3.2.

The following possibilities for making comparisons can be used:

1. Comparison by means of variograms.
The variogram calculation will be explained in Chapter 13.2. A variogram is calculated with the bulk of the data, such as the percussion drill holes, for example. Then the variogram γ-values are calculated for each of the diamond drill holes. The calculated γ-values should conform with the first variogram. If they are systematically larger or smaller, then there is probably a significant difference.

2. Comparison by point kriging.
Point kriging is a geostatistical method and will be described in Chapter 13.4.5.2. This method uses weighting factors to derive a value from the percussion hole values for the point where the diamond drill hole is located. This calculated value is then compared with that actually obtained in the diamond drill hole.

3. Comparison by Isoline Maps.
An isoline map is drawn for the grade or the grade x thickness product, the

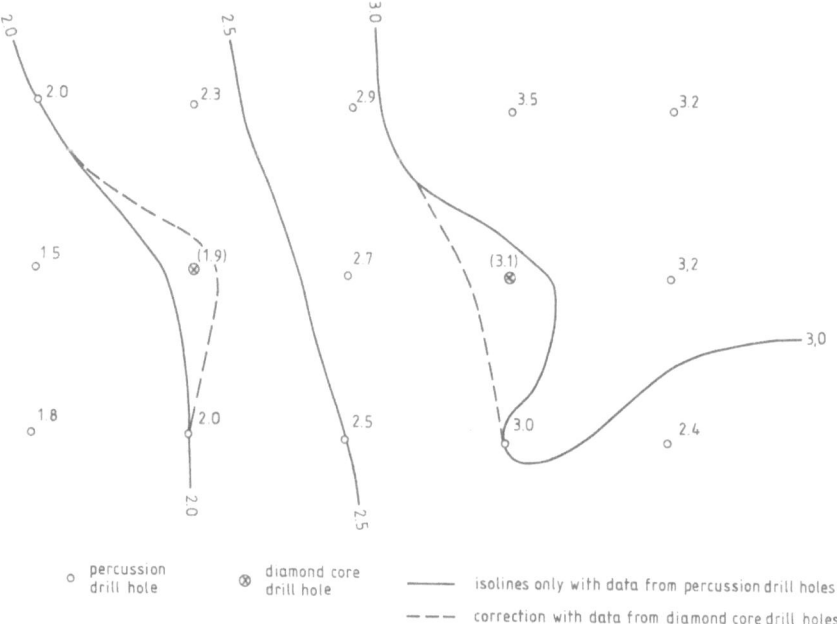

Fig. 43. Accumulation values (grade · thickness) of a uranium deposit (% U_3O_8 m)

accumulation values, from the percussion hole data. The diamond hole data are then matched with this map. The resulting correction of the isolines should result in approximately equal changes towards higher and lower values (see the example in Fig. 43).

If the isolines shift in only one direction, then there is probably a systematic difference between the two types of sample.

4. Comparison of different mining levels.
A comparison of different levels can also be helpful. For example, the ore deposit is divided into 10-m sections and the average values for all diamond drill and percussion hole analyses are calculated separately as composites. This is similar to calculating the grade for open-pit mining levels with a 10-m bench interval. The average values are then plotted on a diagram as a function of the level. The trend of the change in grade with depth should be similar for both types of sample. If this is not the case, then the differences between the two types of sample are probably significant.

Figure 44 illustrates, as an example, this type of test for a porphyry copper deposit. Native copper occurs in this deposit. It can be clearly seen that the diamond drill holes show only a weak trend with depth, while the percussion holes show a well-defined trend. Apparently, the native copper, which is much denser than the silicate matrix, was not flushed out sufficiently well by the percussion drilling and was therefore enriched at the bottom of the drill hole. In purely

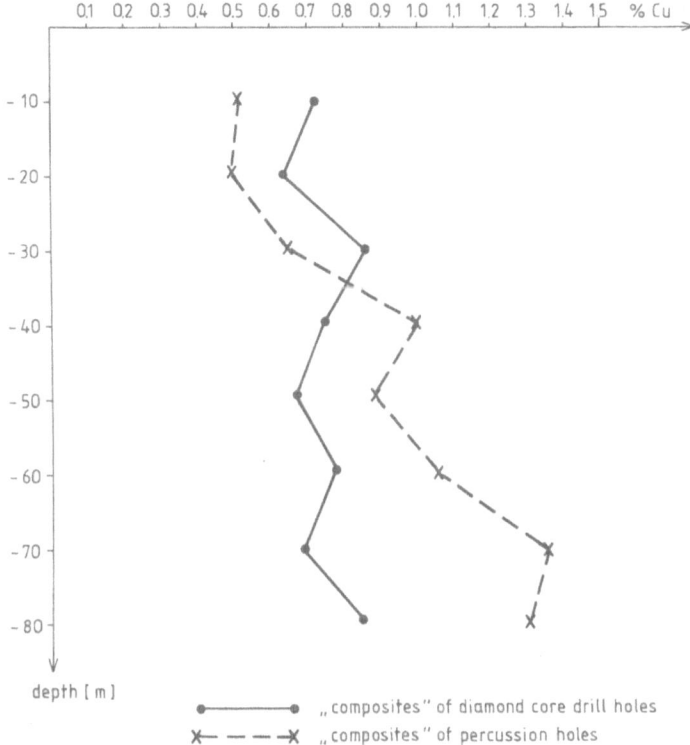

Fig. 44. Comparison of mean grades of 10-m mining levels in a porphyry copper deposit

mathematical terms, the upper levels of the deposit are relatively diminished and the lower levels enriched. The average can remain unchanged. However, the economic feasibility of the project often depends on the highest possible grades being exploited in the early years of mining (cf. Economic Evaluations in Exploration, Chap. 10.2.3.1). Thus, if this trend was not previously identified, this relative enrichment and depletion has a significant effect on assessing the economics of the operation.

11.5
Treatment of Sample Series with Different Sample Qualities

The most common problem of evaluating sample series with different sample qualities (Chap. 1.1) arises if the core recoveries vary during diamond drilling.

11.5.1
Assessment of Core Loss

11.5.1.1
Introduction

Core loss is a relatively common occurrence during diamond drilling. The question is how to assess core losses so that undesirable bias tendencies, to either lower or higher values, are avoided during the final evaluation, including the calculation of grades and tonnage. In the first case, the mineralization would be undervalued and in the second case, it would be overvalued, and of course both effects are unwanted.

Good geological observation and judgement are very important if core loss does occur. Does the mineralization mainly occur on fractures, or is the core recovery lower in particularly strongly fractured or intensely broken zones? Is the mineralization zone softer than the surrounding host rocks? Observations of this type can assist in avoiding bias in the final evaluation.

If the core recovery is mostly in the range between 100 and 85%, then it is generally assumed that the bias is negligible. However, if the core recovery is sometimes less than this, then the possible effect of the core loss on the average grades must be checked.

11.5.1.2
Sampling in the Event of Core Loss

An example was described in Chapter 2.3 of a volcanogenic Cu-Pb-Zn deposit in which the core sections were selected for analysis according to their geology, and were therefore of different lengths. If there is a difference in the core recovery, then the homogeneous zones should be selected primarily according to the same degree of core recovery. These zones can, of course, then be subdivided according to their geology. It is essential to avoid combining a zone of, say, 100% recovery with a zone of 60% recovery into one analytical sample. These zones of different core recovery must be separated.

11.5.1.3
Statistical Treatment of Core Loss

There are two extreme cases for the calculation of reserves and grades from drill core if the recovery is less than 100% :

a) Only the recovered core is taken into account. For example, if the recovery of a 1.5-m section is 60%, then this is calculated to be 1.50 · 0.6 = 0.90 m. It is then assumed that the mineralised zone is 90 cm thick and that the average grade is as analysed. As a consequence, the tonnage is reduced (because of the reduced thickness), but there is no correction for a possible "upgrading" or "downgrading" as a result of the reduced core recovery. For the potential economic feasibility of a project, a change in the grade is much more significant than a change of similar magnitude in the tonnage (see sensibility studies, Economic Evaluations in Exploration, Chap. 10.6). The above mentioned correction has therefore been applied to a second-order dimension that can also be relatively easily estimated from the geology. In reality, this type of correction assumes a 100% core loss from the host rock or within a barren interval, although it is rare for the host rocks to be so soft that absolutely no core is recovered.

b) It is assumed that the core loss is distributed equally throughout the section, and there has been no "upgrading" or "downgrading". This is the normal procedure. If a systematic "downgrading" or "upgrading" is suspected, then the drill hole must either be repeated with a better core recovery by using a larger diameter or better technology (for example, careful drilling with wire-line equipment or triple core barrels), or a different type of sampling must be carried out, such as bulk sampling from pits or adits. The statistical treatment of data obtained in this way was discussed in Section 11.3. A simple regression analysis can be undertaken to test if the suspicion for grade bias is in fact justified. The regression analysis is carried out by using Eqs. (1) to (4) in Appendix Table 8 for the grades on the one hand and the core recovery on the other, and the square of the correlation coefficient is calculated as the coefficient of determination [Eq. (4) in Appendix Table 8]. Ideally the gradient of the regression line should be zero, in which case there is absolutely no correlation (Fig. 45 top). If the gradient is negative, then there is a negative correlation (cf. Economic Evaluations in Exploration, Fig. 11a and b), and the core loss is suspected of causing an artificial "upgrading", in other words the grades of mineralisation appear higher than they really are. If the gradient is positive, then there is a positive correlation, and "downgrading" is suspected (Fig. 45 centre and bottom)

The coefficient of determination r^2 indicates the per cent value of the variation that can be explained by the linear regression (cf. Appendix Table 8). If there is a positive or negative correlation and if r^2 plots in the significant field on the diagram in Appendix Table 8, then the suspicion of bias is justified and additional sampling is recommended. If r^2 does not plot within the significant field, then possible errors are tested by estimating risk from an economic sensibility analysis (cf. Economic Evaluations in Exploration, Chap. 10.6). If necessary the threat of an "upgrading" error can be rectified by applying a safety factor.

It is emphasized again that this is not a conclusive test and thus the term "suspect" is used accordingly. It is quite possible that the grades correlate with the rock mechanical characteristics, and that the softer sections with a poorer core recovery do indeed have higher primary grades (Fig. 45, centre).

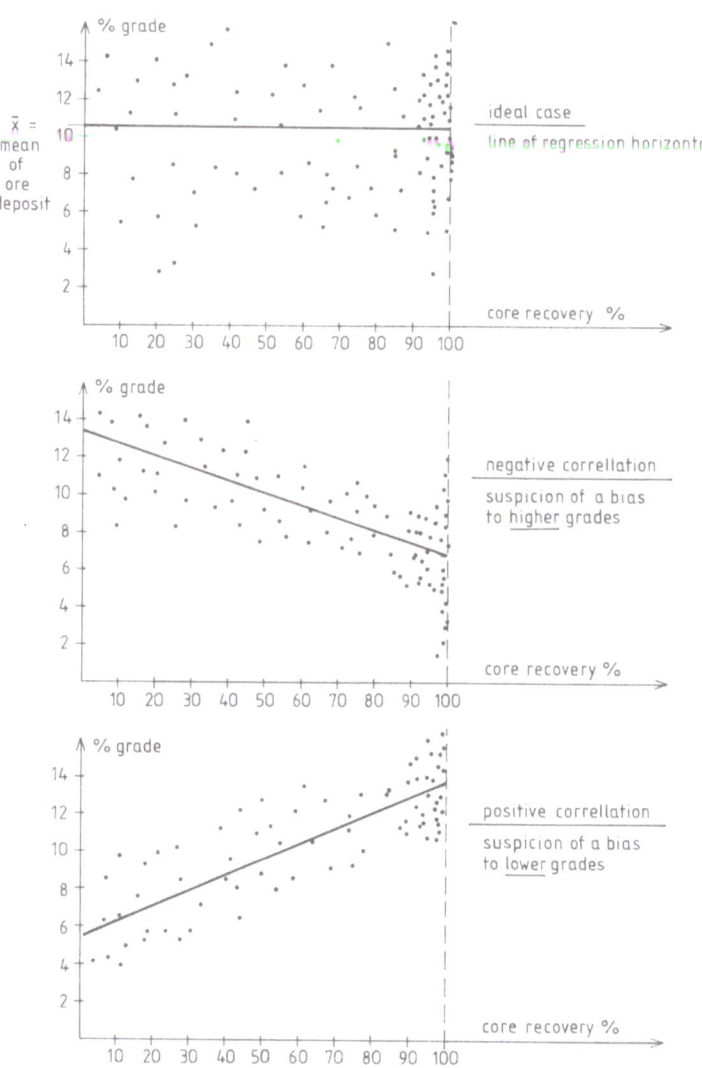

Fig. 45. Comparison of core recovery and grades of drill hole samples

11.5.2
Other Problems with Different Qualities of Sample

11.5.2.1
Channel Sampling

Channel sampling is often associated with a problem of different sample quality, although this may not be realized as such. It is rare that the same volume of sample

Fig.46. Channel sampling of a gold quartz vein in andesitic host rock

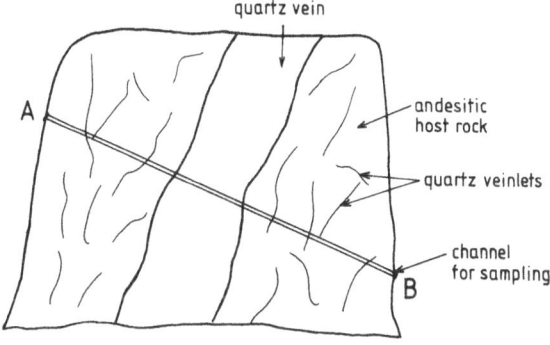

Section through channel $\overline{AB}$

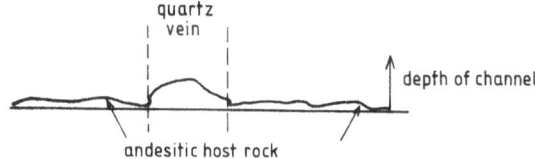

can be hammered from several homogeneous sections, and there is always a tendency favouring those sections from which it is easiest to collect the samples. If these sections are also those that are preferentially mineralized, then the grade will be overvalued. The sampling of a gold-bearing quartz vein in andesitic host rocks provides a good example (Fig. 46). The andesitic host rock contains a stockwork of quartz veinlets (box work), which are also mineralized. The main mineralization occurs in the massive quartz vein. Experience indicates that it is easier to hammer chips from the quartz vein than it is from the andesite. If a profile is drawn along the channel, then it will appear as in the lower part of Fig. 46. This shows that the quartz vein, and therefore also the grades, are clearly overestimated. It is therefore more appropriate to collect two separate samples, each with a different quality, one from the quartz vein and one from the andesitic host rock. The gold grades are then calculated with respect to the thickness by a weighting factor (Chap. 3.1.2), and in some cases are weighted not only by the thickness – but also by the different densities. The same rule as for sampling in cases of different core loss is also valid, namely never combine sections with different core recovery into one sample.

In a mine, where the style of mineralization is known, only the two homogeneous sections, quartz vein and host rock, are separated. If the style of mineralization is not known, then each phase should be sampled separately – in this case the quartz vein, andesitic host rock without the boxwork quartz veining, and the quartz veinlets themselves. The proportion of the quartz veinlets in the host rock can be estimated by making the appropriate measurements along profiles, and the gold grades are recalculated accordingly. If the gold grade is concentrated by the nugget effect into only one small quartz veinlet (Chap. 13.2), it is quite common that the grade is then erroneously extrapolated over a large volume of rock, because the different phases have not been separately sampled. In principle a systematic study of any deposit should be made prior to undertaking the routine sampling so that the sampling itself can be technically and economically optimized.

11.5.2.2
Sampling for Selective Mining

Samples with different qualities are successfully used for grade control in selective gold mining in Australia. The principle is illustrated in Fig. 47. Trenches are dug at 5-m spacing on the open pit mining level, and are sampled over 1-m sections. Three categories are then marked on the mining level that are defined, on the basis of this first sampling, as high-grade ore, low-grade ore and waste.

High- and low-grade ore are extracted selectively by hydraulic shovel excavators, usually to a depth of 1.5 m, and then transported to two separate holding dumps. At

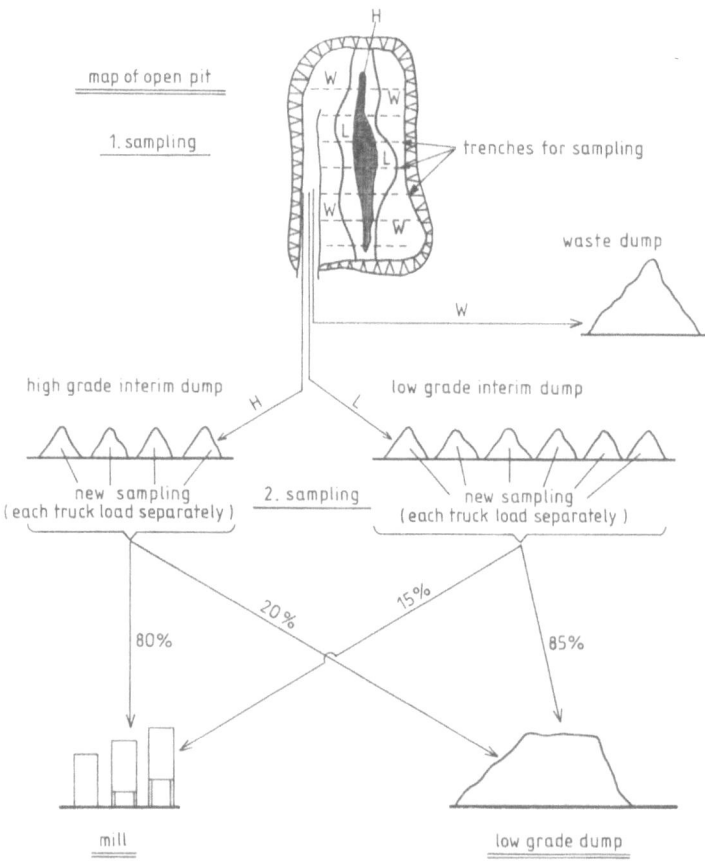

Fig. 47. Principle of the selective gold mining open pits in Western Australia

the holding dump each truck load is sampled by two random grab samples. On the basis of the mean of these two samples the truck load is classified either as ore and taken to the plant, or as low-grade material for the stockpile, which may be treated later in the plant during periods of high gold prices.

In most cases, the grade prediction on the basis of the samples taken at the holding dumps correlates well with the yield from the plant. Experience shows that about 20% of the high-grade truck loads are reclassified at the holding dump as low-grade, and, conversely, about 15% of the truck loads from the low-grade holding dump are reclassified as high-grade and directed to the treatment plant. What happens during the sampling? The quantities for both the 1st and 2nd sampling are approximately the same (2 - 3 kg). Blending occurs as a result of the excavation, loading into the trucks and the dumping onto the holding dumps, and thus the quality of the 2nd sample is quite different. The variance of these samples is then much lower as compared to the 1st set of samples (cf. Fig. 37). The lower the variance, the lower the probability of incorrectly classifying high-grade ore as low-grade, and the converse (cf. Fig. 36).

12
Problems Related to Cut-Off Levels

Statistical questions may arise when cut-off levels are applied to the evaluation or mining of an ore deposit. These questions are closely associated with the problem discussed in Chapter 11.3 Comparison of Samples Series with Different Support, in particular the comparison between the grades of samples and the grades of the enclosing mining blocks (cf. Sect. 11.3.1). The economic implications of cut-off levels were discussed in Economic Evaluations in Exploration, Chapter 10.1, and the statistical implications are the subject of this chapter. Figure 34 again acts as the basis for the discussion. This figure summarizes Krige's (1962) fundamental concepts about the relationship between the grades of samples and those of mining blocks. For each grade interval for the samples there is a population of mining block grades that has an approximately normal distribution (e.g. examples for the 4th and 8th columns at the right side of Fig. 34), whereby the mean of the normal distribution for each interval lies exactly on the regression line between the sample grades and block grades. This line is also the VVC-line (KL). It has been shown in Chapter 11.3.1 that the regression line, or VVC-line, intersects the 45° line at the mean value $\bar{x} = \bar{y}$ for the deposit (Fig. 35), so that the mean of all the sample values is the same as the mean of all the block values. However, this is only true if all the blocks are really mined, without the application of a cut-off grade. If individual blocks are selected for mining and more detailed considerations are taken into account, then the universal concept is no longer valid and the statistical problem is changed. The curve on the diagram that includes all the points, the sample grades plotted against the block grades, approximates an ellipse (Fig. 48, upper). If a cut-off x_C is plotted on the diagram (Fig. 48, upper), then the blocks within the elliptical curve can be divided into four fields:

Field I   : Correctly classified as economic; the blocks are mined as ore.

Field II : Incorrect classification, since the blocks are classified as uneconomic (below the cut-off limit x_C) and are not mined, although their real grades lie above the cut-off grade.

Field III : Correctly classified as uneconomic; the blocks are not mined.

Field IV : Incorrect classification, since the blocks are classified as economic (above the cut-off x_C) and are mined, although they are uneconomic.

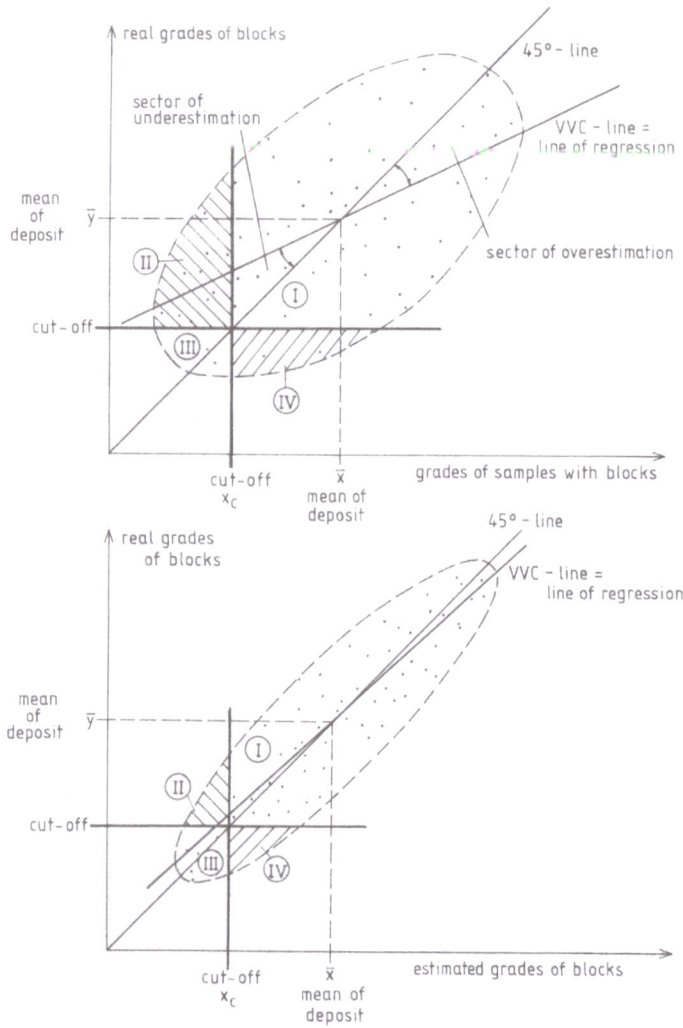

Fig. 48. Correlation between the grades of samples and the corresponding mining blocks and the effect of the application of a cut-off grade

There is, without any doubt, a diminishment of the grade as a result of the incorrect classification in fields II and IV, because a certain number of economic blocks will not be mined and a number of uneconomic blocks will be mined. There are numerous technical factors, such as the quality of the grade control during the mining, as well as the selectivity and flexibility of the mining method, that have to be taken into account if the blocks are to be mined as they have been classified (e.g. Carras 1986). It is very rare that mining plans are made only on the basis of the exploration sampling, which is usually only drilling, and a closer-spaced sampling is always required for a more detailed mining plan.

Examination of the principles illustrated in Fig. 48 shows clearly that the problem of incorrectly classifying the mining blocks becomes even more serious if the cut-off value x_C is shifted towards the "belly" of the enveloping ellipse, towards the mean value $\bar{x}$. By comparing the upper and lower diagrams in Fig. 48, it is also obvious that this problem of incorrect classification becomes that much less as the correlation between the block and sample grades improves or, in other words, as the enveloping ellipse becomes narrower (Fig. 48, lower). Geostatistical techniques are applied to improve the classification, and therefore also the correlation, and these are described in Chapter 13.

There are several practical rules of thumb to estimate how "risky" a cut-off value is (Table 39):

In this context an observation can be helpful, which can be made relative frequently in ore deposits: the difference between the cut-off grade and the mean grade of the deposit above cut-off grade remains more or less constant over a certain grade range. Taylor (1972), for example, has shown this for North American porphyry copper deposits.

Example: The grade of a porphyry Cu deposit is 0.4% Cu at a cut-off of 0.2%, meaning a grade difference of 0.2% Cu. To obtain a grade 0.7% Cu, one would need to apply a cut-off grade of about 0.7-0.2 = 0.5% Cu.

This means, of course, the more the cut-off is pushed up, the relatively smaller the difference between cut-off and grade above cut-off becomes and the "riskier" the application of such a cut-off according to Table 39.

As David (1977) has shown, the effect of a constant difference between cut-off and mean grade is the consequence of a lognormal grade distribution and the validity of Lasky's law. Lasky's law states for the usual range of values of the standard deviation of the grades: when the cut-off is varied, the resulting grade G and tonnage To are in linear dependence.

$$G = K_1 - K_2 \cdot \ln T_0 ,$$

Table 39. Cut-off grades and the degree of danger of overestimating the achievable selectivity

Position of cut-off x_C relative to mean on deposit $\bar{x}_C$ above cut-off	Degree of danger of overestimating achievable selectivity
Cut-off x_C about 50% or less than mean value x_C	General practice in most mining operations
Cut-off x_C >50%, but less than 70% of mean value x_C	Still tolerable, but the more critical, the further away from 50 %
Cut-off x_C >70% of mean value x_C	In most cases the calculated selectivity will not be achieved in practice

Fig.49a-c Three different types
of ore deposit with respect to
the relationship between ore
and waste blocks. a Compact
ore body. b Ore blocks isolated
in waste. c Waste blocks isolated
in ore

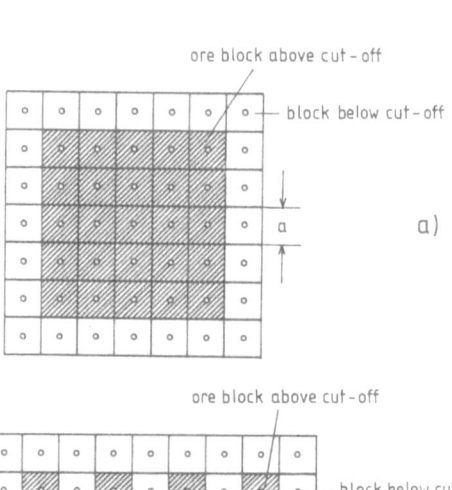

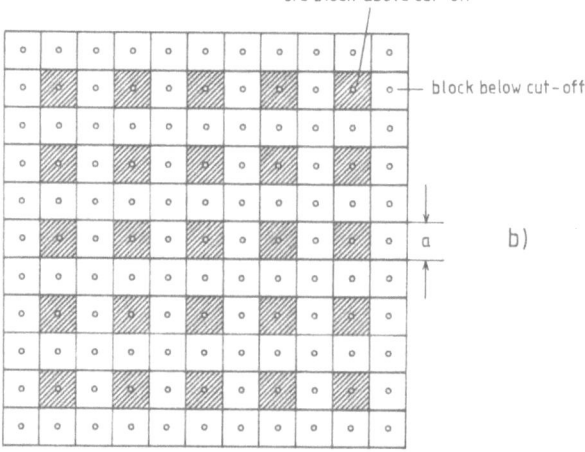

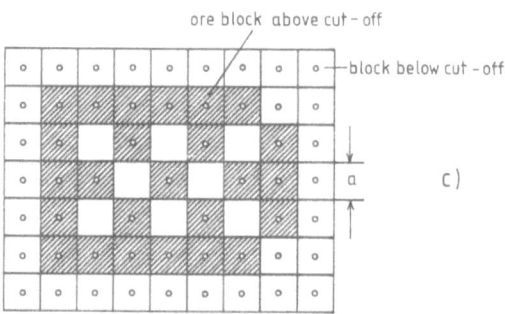

whereby K_1 and K_2 are deposit-specific constants. Lasky's law also means: As grade increases (decreases) linearly, tonnage inversely decreases (increases) exponentially.

The determination of a cut-off level, which corresponds to about 50% of the average ore grade (after application of the cut-off), is usually made by using a cut-off based on the operating costs (see Economic Evaluations in Exploration, Chap. 10.8.2.2). This is because the operating costs are mostly covered by about half of the contained metal value of the deposit (see Economic Evaluations in Exploration, Chap. 9.3.2.1).

If the cut-off x_c approaches the mean of the remaining ore blocks $\bar{x}_c$, then this usually results either in numerous uneconomic blocks (internal waste) between the mining blocks (illustrated schematically in Fig. 49c), or in isolated ore blocks that are surrounded by uneconomic areas (Fig. 49b). The mining method that is used in practice is generally not as selective as that which is required by the theoretical concept. This, together with the incorrect classification, often result in production grades that are considerably lower than the theoretically calculated grades.

In this context, the results of an Australasian Institute of Mining and Metallurgy working group, which studied the judgement of ore reserve calculations, are quoted here (King et al., 1982):
"In Australia, in the last generation, some 50 new mining ventures (coal excluded) reached the production stage. Of these, 15 were based on large, good grade deposits, relatively easily assessed. Of the remainder, 10 suffered ore reserve disappointments more or less serious and some mortal."

Simple and good ore deposits are only offered for sale for huge premiums. As a consequence, the problem of internal waste, or the isolated ore blocks, is often encountered during the evaluation of marginal or difficult deposits.

Only scarce attention is paid in the literature on geostatistics to the problem of internal waste. In practice, however, there is a major difference if the application of a cut-off grade results in a continuous zone of ore (Figs. 49a, 50) or numerous individual ore blocks surrounded by waste blocks (Fig. 49b and c). Therefore the simple diagrams (e.g. the well-known example of Formery, see David 1977), from which the potential amount of metal output can be determined in relation to the cut-off grade and the variances of the blocks, are not discussed in this textbook. In practice, a block-by-block calculation proves to be better[23].

If one or two different cut-off grades are applied, and the economic zone of the ore deposit remains consolidated as one compact area (Figs. 49a and 50), then there is normally no serious mining problem in using a normal grade control procedure. The average grades in the mining area generally correlate, within the limit of error, to the calculated grades. Grade variations are more important than tonnage variations in determining the economic feasibility of a mining operation (e.g. Wellmer 1986). Overestimates and underestimates that are made in delineating the cut-off boundaries in the mine generally cancel each other out (Fig. 50, enlargement). Accurate definition of the cut-off grade boundaries is made much simpler if the cut-off grade follows a natural geological contact, such as was the case for the Nanisivik Zn-Pb deposit in Canada (Fig. 51; Wellmer 1976).

---

[23] This viewpoint is true for blocks that have approximately the same dimensions as the drilling grid. If the mining blocks are much smaller than the size of the drilling grid, as, for example, the blocks for a detailed mine plan, then the problem of internal waste can only be treated probabilistically. This will not be discussed further.

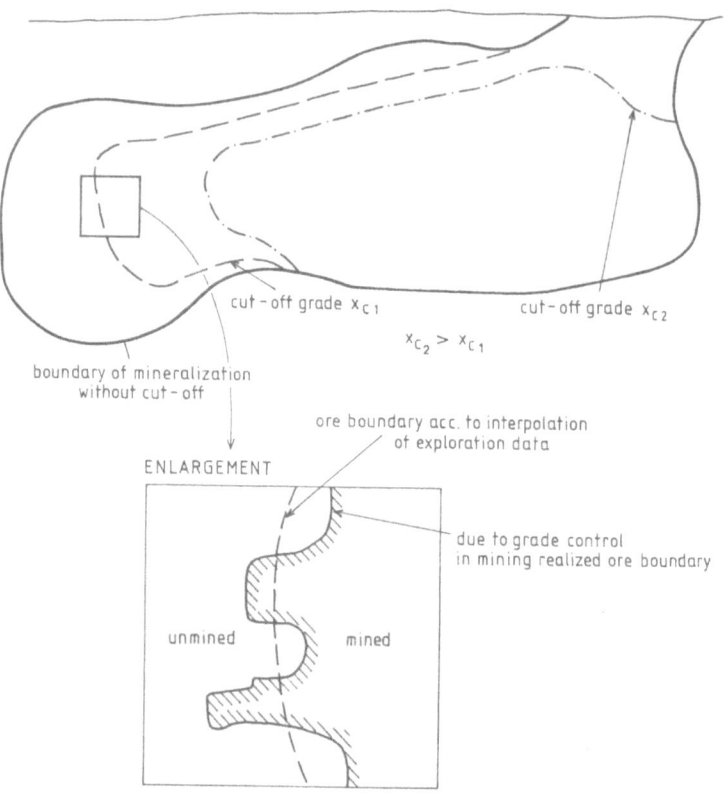

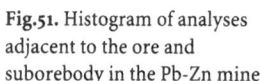

Fig. 50. Example of a compact orebody

Fig.51. Histogram of analyses adjacent to the ore and suborebody in the Pb-Zn mine

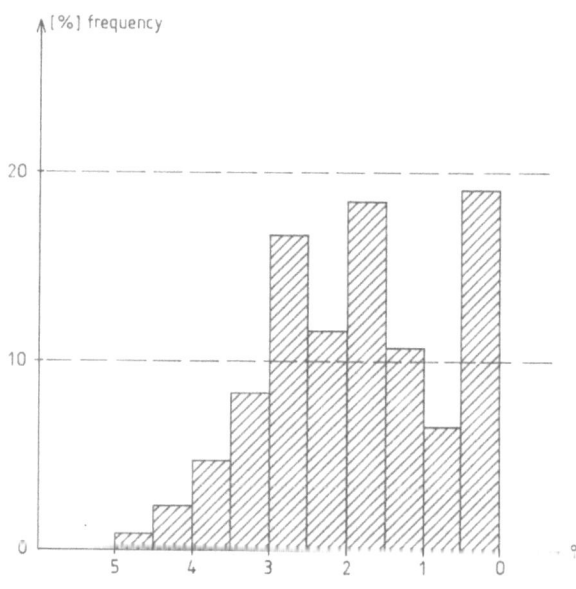

However, the effect of the deposit being broken up by the application of a cut-off grade is schematically illustrated in Fig. 52. The x-axis represents the main drive through the deposit, which leads to the ore blocks numbered from 1 to 13. The grades of the mining blocks are plotted on the y-axis. Line a marks the drilling grades, without being corrected geostatistically or converted by the VVC-line (Chap. 11.3.1), into the grades of the mining blocks, as has been shown in Chapter 11.3.1:

- below the mean grade of the deposit $\bar{x}$, the blocks generally have higher grades as compared to the samples from those blocks, and
- above the mean grade of the deposit $\bar{x}$, the blocks generally have lower grades as compared to the samples from those blocks.

Therefore the true grades of the mining blocks are not those shown by line a in Fig. 52, but are those shown by line b. Evidently the cut-off x_{C_1} does not affect anything, since the average grade of all the blocks is greater than x_{C_1}. However, if the average grade is to be increased to x_c, then the cut-off grade must be increased quite considerably to x_{C_2}. A test, which indicates whether the ratio between the cut-off grade x_{C_2} and the increase mean value x_c lies within the "high risk zones" shown in Table 39, should now be applied. This is, of course, a theoretical concept. In practice the blocks, which have been selected using the x_{C_1} cut-off, will have a higher average cut-off that might be approximately the same as x_{C_2}, and therefore under certain circumstances it could lie within the "high risk zone" referred to in Table 39.

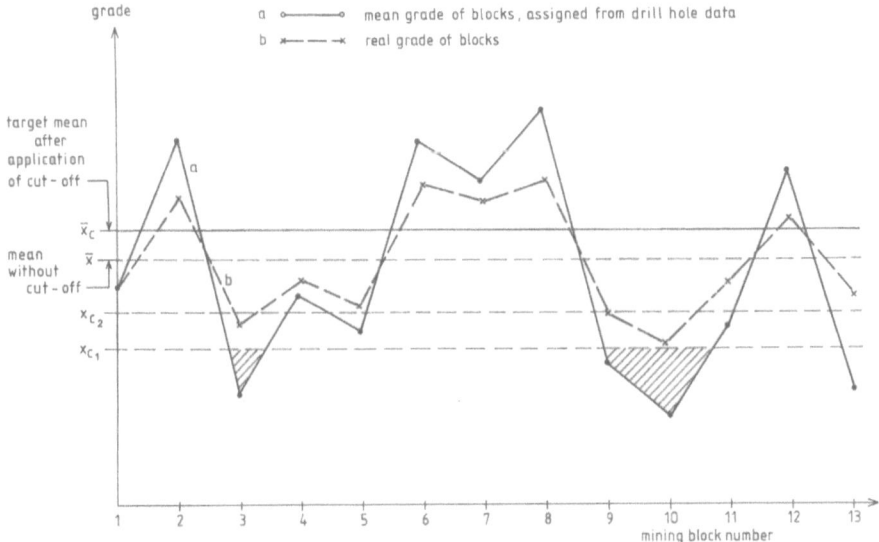

Fig. 52. Diagram for the comparison of the grades of mining blocks and of corresponding drill hole data and the effect of the application of the cut-off grades x_{c1} and x_{c2}

Fig.53.a-b Creation of a skewed distribution by application of a cut-off grade

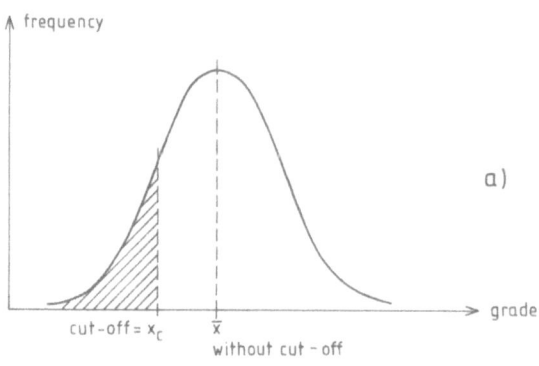

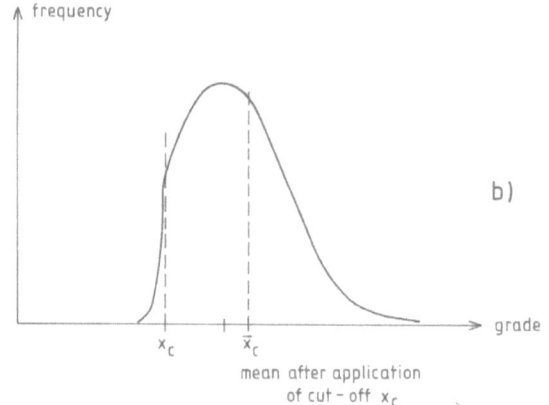

This example demonstrates how important it is, when quoting a cut-off grade, to also mention the sample support (drilling, mining blocks, etc.) on which that grade is based.

The statistical problems encountered by the application of a cut-off grade can also be considered from another point of view. Assume there is an ideal orebody, and the grades of its mining blocks have a normal distribution (Fig. 53a). If a cut-off x_C is now applied, then the lower tail, which are the values less than x_C, will be truncated. The remaining values will now have a skewed distribution, and it will be more difficult to estimate a reliable mean value, as was fully described in the Chapters 8 and 9. Thus some "lognormal" distributions of analytical values or other ore deposit data are primarily caused by the truncation of the lower values (e.g. Attanasi and Drew 1985). In addition to the classification problems mentioned above, in practice it will also prove impossible to strictly apply the cut-off during mine planning. For engineering reasons, some small areas below the cut-off grade must always be included in the mine plan, and this results in the skewed distribution shown in Fig. 53b.

So far the use of the cut-off has been considered only with respect to the statistics of samples with different support. In practice, however, this is closely associated with the question of dilution by adjacent blocks of uneconomic material (see Economic Evaluations in Exploration, Chap. 6.1)[25] , and it is often difficult to separate the two problems. Is the grade of a mining block lower than previously calculated because of the statistical variations, or because there is too much dilution? The dilution is not only very dependent on the selectivity of the mining method (cf. Economic Evaluations in Exploration, Chap. 6.1), but also on the geometry of the ore deposit. The risk of dilution is lower for a compact and consolidated orebody than it is for isolated blocks of ore. The greater the length of the contact with the uneconomic host rocks, the greater the risk of dilution. The principle is illustrated in Fig. 49; Figure 49a shows 25 ore blocks, each with an edge-length a, of a compact orebody. The boundary between the individual blocks should be marked as a contact. If for the sake of simplicity the third dimension is ignored, then the length of the contact with the host rocks is 20 a. Figure 49b shows the same number of ore blocks, all of which are isolated in the host rock. The length of the contact with the host rocks is now 100 a, or five times that of Fig. 49a. Figure 49c shows isolated blocks, which have grades less than the cut-off limit, within the orebody and thus it illustrates an ideal case of the internal waste problem. The length of the contact between the ore blocks and the uneconomic material now amounts to 56 a, which is nearly a threefold increase compared to the original case in Fig. 49a.

These comparisons clearly show that, for a compact orebody such as Fig. 49a and 50, it is more important to correctly estimate the total area of the mineralisation rather than the individual blocks, in other words to make a "global estimate" such as will be described in Chapter 13.3. In the case of isolated ore blocks (Fig. 49b) or isolated uneconomic blocks (Fig. 49c), it is important to evaluate each of the blocks separately in order to derive a reasonable estimation of the average grade and the tonnage. The global estimation can be made by a simple averaging of all the relevant blocks, but the individual calculation of the blocks requires the application of geostatistical methods that are described in Chapter 13.

In the assessment of ore deposits a yes/no decision is often required in the early phases (e.g. should one "farm-in" into the project or not?), rather than a detailed economic evaluation. However, in the event of the deposit containing numerous blocks of internal waste or isolated ore blocks, then seriously erroneous decisions can be made if the individual blocks are appraised by a global estimate rather than by geostatistical methods.

Finally, grade tonnage curves will be mentioned briefly. These are curves showing the grade of a deposit as a function of the applied cut-off grades or the remaining tonnage as a function of the applied cut-off grades respectively. In practice, especially in the exploration stage, one operates just with one or two cut-off grades based on operating costs (see Economic Evaluations in Exploration, Chap. 10.1). Therefore, grade tonnage curves are not dealt with in this book in detail.

---

[25] David (1988) treated this geostatistically.

13
Geostatistical Calculations

13.1
Introduction

The classical statistical methods have been described so far, and they are based on the assumption that the individual samples, such as sample values from drill holes or channel sample values, are statistically independent of each other. This condition is satisfied by, for example, the throwing of a die; if a six is thrown, then the chance of the next throw being another six is exactly the same as if a one, or any other number between a one and a five, had been thrown – actually 1/6. ("chance has no memory.") This type of independence is rarely found in mineral deposit data. Every geologist knows that the chance of drilling another high-grade hole is greater near a previous high-grade intersection than it is near a low-grade intersection. There is, therefore, a certain spatial interdependence. Geostatistics is statistics by which this spatial association is taken into consideration, and where the variables are known as regionalised variables. Spatial dependence has already been mentioned a couple of times in previous chapters, for example Section 8.3.3.2.

D.J. Krige, from South Africa, and the school led by Matheron at Fontainebleau in France were pioneers in the discipline of geostatistics. In the meantime, several centres for geostatistical studies have become established in, among other countries, Canada, the USA, the United Kingdom and the Federal Republic of Germany. There are now numerous publications on geostatistics, particularly in French and English (e.g. Clark 1979; Cressie 1991, David 1977 and 1988, Dimitrakopoulos 1994, Isaaks and Srivastava 1989, Journel and Huijbregts 1978, Krige 1978, Matheron 1971, Rendu 1978). German publications include the book by Dutter (1985) and Akin and Siemes (1988), as well as the GDMB publication *Classification of Mineral Deposit Reserves by Geostatistical Methods* (Series 1, Vol. 39). The regular APCOM conferences (Application of Computers in the Mineral Industry) are the most important symposia for geostatisticians, and the conference publications are an excellent source of case histories. The books by Matheron and Armstrong (1987), Armstrong (1989), Annels (1992) and Whateley and Harvey (1994) also provide a good selection of case histories. Mathematical Geology is the main journal for geostatisticians. The articles, however, stress the theoretical aspects and there is a relative lack of case history material. The Teacher's Aid articles on the principles of geostatistics (e.g. Barnes 1991) are also helpful. A geostatistical multilingual dictionary and glossary has been published by Olea (1991).

Geostatistical calculations can be very complex, and many must be carried out on a computer. This textbook is designed to be a vade mecum for quick reference, and thus it only describes calculations that can be carried out by hand or with diagrams. More complicated computations are provided in the book by David (1977) and Journel and Huijbregts (1978).

There are two areas where geostatistical calculations can be important, even in the early phases of evaluating a mineral deposit:

1. the calculation of errors or uncertainties in reserve estimates, and therefore the possibility of classifying resources and reserves (Sect. 13.3);
2. the determination of grades for single mining blocks, particularly if the application of a cut-off grade (Chap. 12) results in individual blocks of ore or waste occurring within an ore deposit (Sect. 13.4).

Other geostatistical aspects become important at a later, slightly more advanced, phase of mineral deposit evaluation. In Economic Evaluations in Exploration, Chapter 11, the calculation of dynamic profitability is discussed and it is demonstrated that the profitability of a mineral deposit can be enhanced by extracting those blocks with the highest possible grades during the early years of a mining operation. An average grade for each year is then required for the calculation of the profitability, as well as for mine planning. The grades must

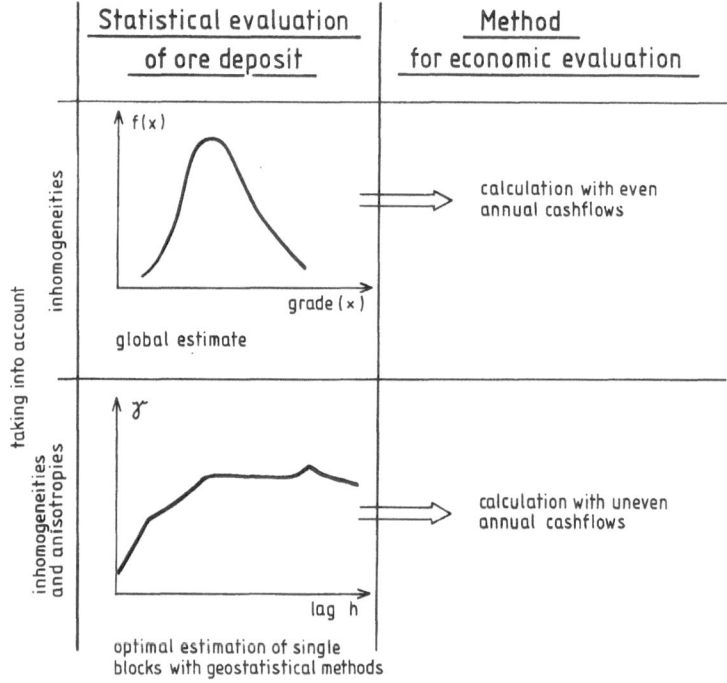

Fig. 54. Relationship between methods for the statistical evaluation of an ore deposit and the methods for economic evaluations

therefore be separately calculated for each and every annual production area, perhaps even for subareas. Overestimates and underestimates of the individual ore blocks should balance each other out for a global estimate of the whole of the deposit (see Chap. 11.3.1, Fig. 34). However, this is definitely not the case for all those blocks that comprise the mining reserves for a specific year, since they will be the blocks with the best possible grades. The dynamic profitability calculations use compound interest formulae (cf. Economic Evaluations in Exploration, Chap. 11.2.3.1), and therefore the information on how much cash flow will be generated in any particular year (and that implies the grades that can be mined in each year) is a critical factor in the calculation. If, for example, the grade for the first mining year is overestimated and this will be numerically balanced by an underestimation of the average grade for the ninth year, then this is clearly quite different from a balance of the profitability. The present value of the cash flow in the first year is quite different from that in the ninth year.

During the advanced prefeasibility and feasibility studies, it is therefore necessary to estimate the grades of each block as accurately as possible. The so-called kriging method is the best technique to make these estimations, and it will be discussed in Section 13.4.5. There is thus a relationship between the two dynamic methods for calculating profitability, specifically the calculations with equal and variable annual cash flows (Economic Evaluations in Exploration, Chap. 11) on the one hand and, respectively, the statistical and geostatistical methods of calculating mineral deposit data on the other. This relationship is schematically illustrated in Fig. 54.

13.2
The Variogram

13.2.1
Fundamental Principles for Calculating the Variogram

The previous section demonstrated that geostatistics considers the spatial dependence of the sample or analytical values between each other. The variogram is the basic means of quantifying this spatial dependence. In practice, it describes the average of the squared differences of two values that are calculated as a function of the distance between them (Fig. 55). The variogram is the basis of all geostatistical calculations. The following section should explain just what a variogram is, as well as its applications.

Consider the example in Fig. 55. In the first step the differences are calculated for the pairs spaced 50 m from each other. In the second step the differences are calculated for pairs spaced 100 m from each other, and so on. The values for the variogram, or the semi variogram function[25], are calculated experimentally from the following formula:

---

[25] Originally, the variogram was defined by the values 2γ and the semivariogram by the value γ. However, since the value γ is used for the calculations, the name variogram is now used for γ; and the name semivariogram has become outdated (e.g. David 1977, p. 94).

Fig.55. Principle of the calculation of an experimental variogram

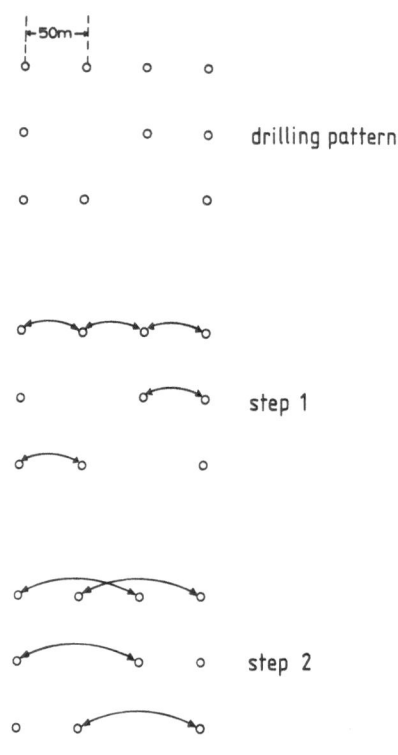

$$\gamma_{(h)} = \frac{1}{2n_{(h)}} \cdot \sum_{i=1}^{n} (x_i - x_{i+h})^2 \, , \tag{1}$$

Where x_i are the data values (e.g. ore grades), x_{i+h} is the data value at a distance h from x_i, and $n(h)$ is the total number of value pairs that are included in the comparison. On the variogram γ is plotted as a function of the spacing or lag h[26].

Variograms are ideal for identifying anisotropic features in a mineral deposit, such as structures, as well as any spatial interdependence (see below). In the first stage the γ-values are only calculated for one (e.g. horizontal) direction (see Fig. 55), and in the second stage the γ-values are calculated for the perpendicular direction. However, this assumes that the distribution of the data is relatively dense, which is usually not the case in the early stages of an evaluation. One generally requires more than 30 drill holes in order to calculate meaningful variograms in several orientations. If the data distribution is not sufficiently dense, as is quite typical in the early stages of exploration, then an average or overall variogram is used that is independent of orientation. An average variogram is the basis for all the following calculations.

---

[26] Variograms can be used not only for ore deposits modeling but for any spatially arranged data, be it geologic, topographic, meteorological, etc. data. A good example for other applications would be seafloor classification by directional variograms (Herzfeld 1993).

The variograms for stratabound or tabular ore deposits are usually not calculated from the values, x_i, of the grade, but from the grade-thickness product (GT). For example, if there are two drill intersections with the same average grade of, say 0.8% MoS_2, then the difference of the grades would be zero, even if the intersected thicknesses were only 5 m in one drill hole and 25 m in the other. Using the grade-thickness product, or GT, values the differences would now be 0.8 x 25 - 0.8 x 5 = 16 (% m). Where the thicknesses remain approximately constant, such as in the benches of an open pit mine, then the variogram can be calculated only from the grades, and this is known as a graded variogram.

Example: Figure 56 shows six drill holes and their GT values from a potash deposit. The unit for the GT values is % $K_2O \cdot$ m thickness. An average variogram will be calculated.

1st Step: All possible differences are derived for the smallest interval, which is 200 m. There are five such differences (Fig. 56, upper part). The y-value is derived from Eq. (1) above:

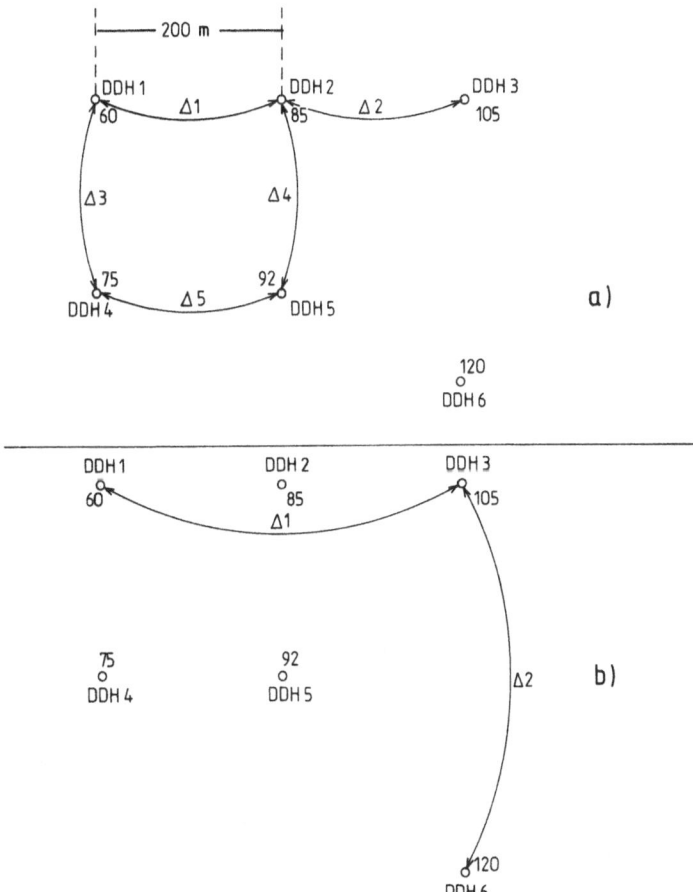

Fig. 56. Example for the calculation of γ-values from six drill holes of a potash deposit

Table 40. Example for the calculation steps for a variogram

Difference no. Δ	$x_i - x_{i+h}$	$(x_i - x_{i+h})^2$
$\Delta 1$	60 - 85 = -25	625
$\Delta 2$	85 - 105 = -20	400
$\Delta 3$	60 - 75 = -15	225
$\Delta 4$	85 - 92 = -7	49
$\Delta 5$	75 - 92 = -17	289
		Sum = 1588

$$Y_{(200)} = \frac{1}{2 \cdot 5} \cdot \sum_{i=1}^{n} (x_i - x_{i+h})^2 \ .$$

According to Table 40, the sum of the squares of the differences is 1588.

Table 40

Therefore:

$$Y_{(200)} = \frac{1}{2 \cdot 5} \cdot 1588 = 158.8 \ .$$

2nd Step: The diagonal spacing of h = 283 m, could be calculated in this step. However, for the sake of the example, the calculation for a spacing h = 400 m is carried out. There are only two differences.
The value is therefore:

$$Y_{(400)} = \frac{1}{2 \cdot 2} \left[(60 - 105)^2 + (105 - 120)^2 \right] \ ,$$

$$Y_{(400)} = \frac{1}{4} (2025 + 225) \ ,$$

$$Y_{(400)} = \frac{2250}{4} = 562.5.$$

However, in practice, only two values for the differences are not definitive, and this was calculated only as an example. There should be at least five difference values in each step for the construction of a meaningful variogram.

The γ-values are then plotted as a function of the spacing or lag h, and the result is an experimental variogram.

The manual calculation of the variogram becomes very time consuming for large data sets. Appendix Table 9 provides a simple BASIC programme for the calculation of the variogram on a microcomputer.

There are several types of variogram that can be distinguished (e.g. David 1977). Many deposits yield a so-called transitive type of variogram (Fig. 57). If the function $\gamma(h)$ increases (within the range, a), then there is a spatial interdependence of the grades and the so-called covariance (Fig. 57) is not equal to zero. In the horizontal orientation, beyond the range a, the sample values are statistically independent, the covariance is zero and the γ-value is the same as the variance.

A variogram is therefore ideally suited for clarifying the problem of whether the sample values are statistically independent of each other, or if they are spatially interdependent. This procedure has already been applied in Chapter 10.

The theoretical relationships between the γ-value, the so-called covariance and the variance σ^2 should be briefly outlined. At the same time, the factor 2 in the denominator of Eq. (1) above, and therefore the outdated term semivariogram, can also be explained.

The variance of a sum or a difference of variables can be generated in the same way as a quadratic equation (e.g. David 1977). For the variance of the difference between x_i and x_i+h, $\sigma^2_{x_i-(x_i+h)}$, is derived from:

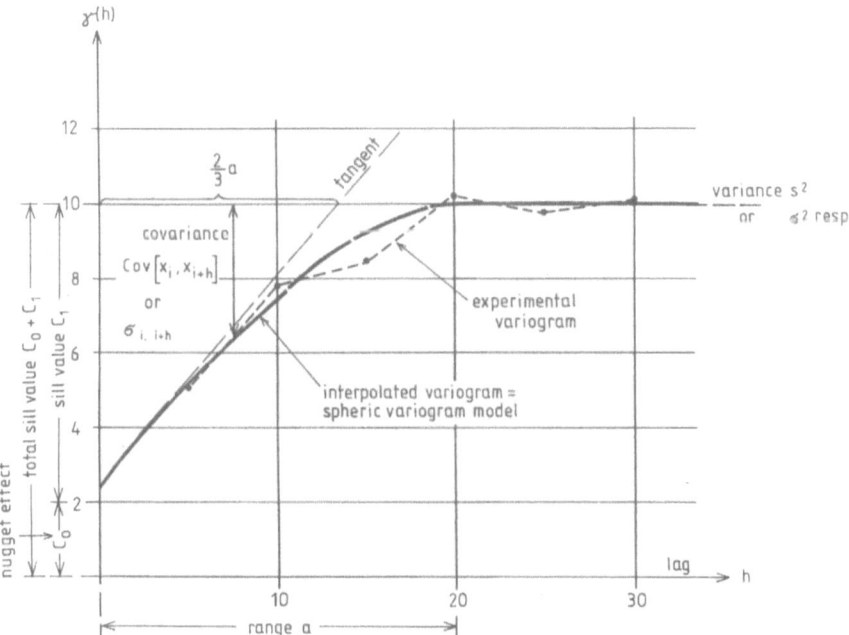

Fig. 57. Example of a variogram of the transitive type with a spherical model fitted to it

$$\sigma^2_{x_i-(x_{i+h})} = \sigma^2_{x_i} + \sigma^2_{x_{i+h}} - 2\sigma_{x_i} \cdot \sigma_{x_{i+h}} \tag{2}$$

$\sigma_{x_i} \cdot \sigma_{x_{i+h}}$ is now denoted the covariance between x_i and x_{i+h}:

$$C_{OV}\left[x_i, x_{i+h}\right] (\text{or } \sigma_{i, i+h}).$$

Since x_i and x_{i+h} are derived from the same population, they must have the same variance, or:

$$\sigma^2_{x_i-(x_{i+h})} = 2\sigma^2_{x_i} - 2Cov\left[x_i, x_{i+h}\right] \tag{3}$$

   or

$$\frac{1}{2}\sigma^2_{x_i-(x_{i+h})} = \sigma^2_{x_i} - Cov\left[x_i, x_{i+h}\right].$$

This is the same as $\gamma_{(h)}$, therefore:

$$\frac{1}{2}\sigma^2_{x_i-(x_{i+h})} = \gamma_{(h)} = \sigma^2_{x_i} - Cov\left[x_i, x_{i+h}\right]. \tag{4}$$

The covariance is zero if x_i and $x(i+h)$ are independent variables, or if there is no spatial interdependence. This results from the addition rule for variances of independent random variables (e.g. Kreyszig 1968). If the distance h is greater than the range a, then the values are statistically independent. Therefore for $h \geq a$:

$$\gamma_{(h)} = \sigma^2_{x_i} \quad [27], \tag{5}$$

so that the horizontal value, towards which the normal, transitive type of variogram approaches, has the same magnitude as the variance. By drawing a horizontal line at the interval s^2 for the γ-value on the variogram, this approximation can be used for the interpolation of the horizontal component of a model variogram from an experimental variogram (Fig. 57). As explained above, however, the condition has to be satisfied that most of the samples for calculating the variance lie outside the range a. According to Barnes (1991), a rule of thumb for such an approximation is that the data must be somewhat evenly distributed over an area with dimensions greater than three times the range (a) of the variogram.

If the variogram is generated by using the variance, then the covariance can be derived from the variogram by applying Eq. (4) above. This will be referred to again for the kriging calculations (Chap. 13.4.5).

---

[27] From the equations in the footnote 20 (Section 11.1.2.3.2) it can be derived that the γ-value divided by 2 is the same value as the variance outside of the range of spatial interdependence.

A mathematical model, a so-called spherical variogram model, can be interpolated from an experimental, transitive-type of variogram (Fig. 57). It has the equation:

$$\gamma_{(h)} = C_1 \left(\frac{3}{2} \frac{h}{a} - \frac{1}{2} \frac{h^3}{a^3} \right) + C_0 \qquad h \leq a \qquad (6)$$

$$\gamma_{(h)} = C_1 + C_0 \qquad h > a .$$

The geostatistical textbooks provide tables for this function (6). For example, in Akin and Siemes (1988) the values for $C_0=0$ and $C_1=1$ are listed for $h/a=0$ to $h/a=1$.

These tables are not used in this book. Instead, spherical variograms are interpolated by eye. Variograms are a robust statistical technique, and small deviations do not significantly affect the final results (e.g. Krige 1978). An "eyeball" interpolation is good enough for the first-order estimates that are dealt with in this book. The geostatistical diagrams and tables in the following chapters are all based on the spherical model.

In addition to one characteristic of the transitive variogram that has already been mentioned, specifically that its horizontal component has the same value as the variance of the samples s^2, another characteristic of the function (6) above can be used for manual interpolations – the tangent of the curved component of the variogram intersects the extension of the horizontal component at $2/3$ a (a = the range; Fig. 57).

The transitive type of variogram, or the spherical model variogram, can be described quantitatively by three parameters (Fig. 57):

- range[28] a,
- nugget effect C_0, and
- sill C_1. C_1 is also known in the geostatistical literature as the pitch. The notation C is sometimes used instead of C_1.

The sum of the nugget effect plus the threshold $C_1 + C_0$ is known as the total sill value.
The total sill value $C_1 + C_0$ is the same as the variance σ^2 [Eq. (5) above, and Fig. 57].
Every universe, and therefore every mineral deposit (Chap. 1.1), will have a typical variogram. The respective γ-functions are not known but, as in the previous example, they must be estimated or approximated by applying an experimental variogram to the available data. This approximation is comparable to estimating the distribution. The derivation of a, C_0 and C_1 is therefore comparable to estimating the mean μ and the variance σ^2 of the population from the mean $\bar{x}$ and the variance s^2 of the sample size, as was discussed at the end of Chapter 3.2.1.

---

[28] This range a obviously has no connection with the other range, R (Chap. 3.1.3).

The range a has already been mentioned. Values for the range are provided in the Appendix (Table 10) for several types of mineral deposits. The nugget effect represents a variability at zero, or practically zero spacing. For example, one sample is collected, and then others at a separation of $\Delta x \approx 0$ or at a spacing of nearly zero. (If it were exactly zero, the difference would have to be zero, of course.) C_0 is the average of the squared differences. These differences are extreme if nuggets occur in the mineralization, and hence the name for C_0. The best possible estimate for the nugget effect C_0 is of the greatest importance when establishing the variogram parameters. Great care is required, and the geological information must be taken into consideration.

The three parameters – range a, nugget effect C_0, and sill C_1 – characterise each type of mineral deposit. A very irregular deposit, such as a gold or pegmatite deposit, will have a high nugget effect and/or a small range a; relatively uniform deposits such as stratiform, sedimentary Pb-Zn occurrences have a low, or even zero, nugget effect and a large range a (Appendix Table 10). In Fig. 58 for example, Mississippi Valley-type, sedex-type and vein-type Pb-Zn deposits are classified in a diagram with the range a as the x-axis and the ratio C_0/C_1 as the y-axis. Information from other deposits of the same type preferably neighbouring deposits or deposits in the same geological region can help as a-priori information, for example for the estimation of the range a or the relative nugget effect Co/C, if in the early stage of the exploration only limited data are available to calculate a variogram (see, e.g. Krige et al. 1990).

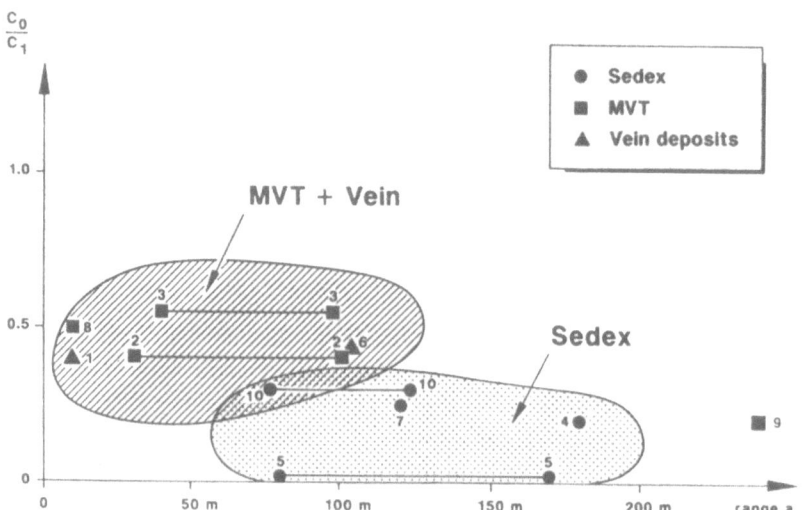

Fig. 58. Plot of the ratio nugget effect C_0/sill value C_1 against range a of variograms of Zn-Pb deposits for the product of grade x thickness. 1 Ramsbeck, Germany; 2 Nanisivik, Canada; 3 Pine Point, Canada; 4 Rampura-Agucha, India; 5 Mt. Isa, Australia; 6 Bad Grund, Germany; 7 Meggen, Germany; 8 Bleiberg, Austria; 9 Song Toh, Thailand; 10 Tynagh, Ireland. (Wellmer et al. 1994)

"Absolute" γ-values are calculated by Eq. (1) above, for the computation of γ-values for the variogram. Relative γ-values can also be calculated by dividing every γ-value by the square of the mean $\bar{x}$:

$$\gamma_{re} = \frac{\gamma_{abs(h)}}{\bar{x}^2} \, .$$
(7)

The following calculations for estimating the error are always based on the relative variogram according to Eq. (7) above, so that the error is quoted relative to the mean value.

Another formula for the relative variogram, the so-called pairwise relative variogram (Isaaks and Srivastava 1989) shall be mentioned:

$$\gamma_\pi = \frac{1}{2_{n(h)}} \cdot \sum_{i=1}^{n} \frac{\left(x_i - x_{i+h}\right)^2}{\left(\dfrac{x_i + x_{i+h}}{2}\right)^2} \, .$$
(8)

It often helps to produce clearer display of spatial continuity and is more outlier-resistant.Section 13.2.3 also deals with the problem of outliers and variograms.

As already mentioned above, variograms in different orientations can also identify the presence of anisotropic features in mineral deposits, and in some circumstances this can be allowed for in the calculations. Anisotropic features are reflected by the range a and sill C_1, which are dependent on the orientation. The nugget effect is generally an isotropic quantity (see above).

In addition to the spherical model of the transitive type, there are also other models of variogram (cf. for example, David 1977; Isaaks and Srivastava 1989) which are briefly discussed in Section 13.2.2. The spherical model is most frequently fitted to experimental variograms, and the following discussion will be almost entirely restricted to this type. Another type of variogram (the de Wijs variogram) will be briefly introduced in Section 13.4.7.

Finally, the stationarity condition will be mentioned, since this is one of the conditions that must be satisfied for the generation of the variogram. The difference between x_i and x_{i+h} (or, in the case of anisotropies, the vector between the two points) must be independent of the location. Akin and Siemes (1988) discuss stationarity in more detail. In practice, it implies that geologically inhomogeneous zones, such as richly and poorly mineralised zones, should not be treated together. Instead, the mineral deposit should be divided into geologically homogeneous zones, which are then treated as separate geostatistical units with different variograms.

The geology must always be taken into consideration in the generation of experimental variograms. h, for example, is the distance between two points at the time of mineralization. The subsequent tectonic deformation, such as folding and faulting, can change this distance. It is therefore incorrect to identify the intersections with x, y and z coordinates in a computer programme, and then blindly to calculate the distance between the intersections for the distance h. A

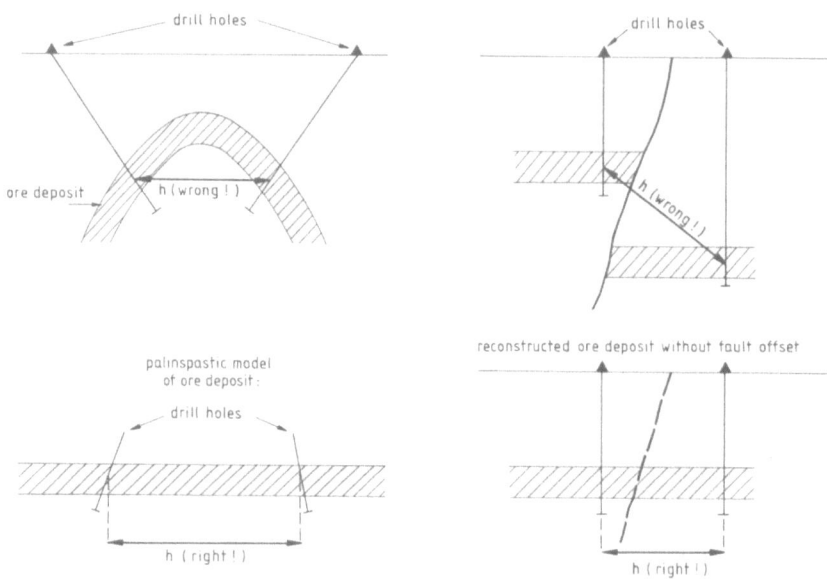

Fig. 59. Examples for the correct and incorrect determination of lag h for calculating a variogram

palinspastic model must be constructed from the folded deposit for the correct determination of the distance h, and faulted deposits must be corrected by the relevant displacement (Fig. 59).

13.2.2
Variogram Models

Various models of variograms are described in the literature (see, e.g. David 1977; Isaaks and Srivastava 1989). Some shall briefly be mentioned. For a transitive type of variogram, e.g. a variogram with a finite sill (C_o+C_1; see Fig. 57) which is reached at a finite distance, the range a, or a sill which is reached aymptotically, in which case the practical range a is given by the distance at which the variogram model is equal to 95% of the value of the sill (Olea 1991), the spherical model has already been described in Section 13.2.1 (Fig. 60A,a).

The exponential model has the formula

$$\gamma(h) = C_1 \left(1 - e^{\frac{-3h}{a}} \right).$$

Whereas in the case of the spherical model the tangent at the beginning of the variogram intersects the sill at 2/3 of the sill (see Chap. 13.2.1), in the exponential model the tangent intersects the sill at about 1/5 of the range a (Fig. 60A,b).

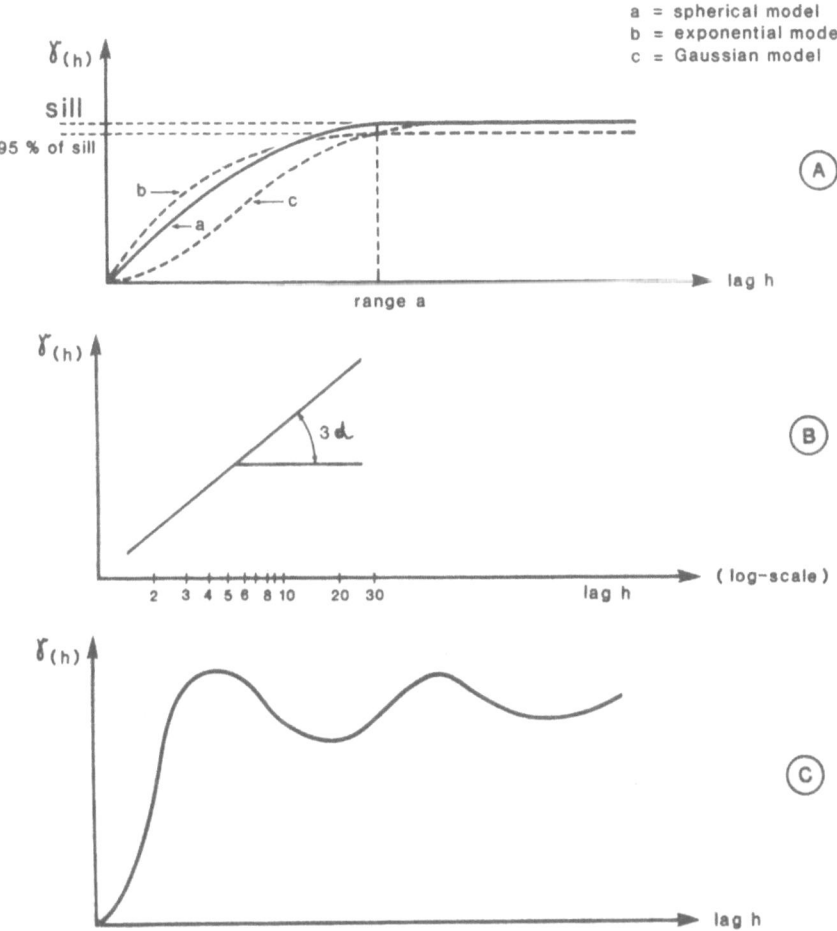

Fig. 60A-C. Types of variogram models

The Gaussian model has the formula

$$\gamma(h) = C_1 \left(1 - e^{\frac{-3h^2}{a^2}} \right).$$

It has a parabolic behaviour at the beginning. It is the only model shown here whose shape has a point of inflexion (Fig. 60A,c).
The de Wijs model is a special case of the logarithmic variogram model. It is a nontransitive model, i.e. the model has no sill. The de Wijs model has the formula

$$\gamma(h) = 3\alpha \cdot \ln h + b.$$

As shown in Fig. 60B 3α is the gradient of the variogram. This variogram will be applied in Section 13.4.7.

Finally, the hole effect model shall be mentioned (Fig. 60c). If an experimental variogram reaches a maximum and then with increasing lag h the γ-values decrease and then increase again, then a hole effect is present. This periodic effect, which has to be distinguished from normal fluctuations of an experimental variogram, is caused by a succession of rich and poor zones. It can be modeled with sin- and cos-functions (see Journel and Huijbregts 1978). This model can help to better understand the structure of an orebody.

13.2.3
Allowing for Outliers in the Calculation of Variograms

Since the square of the differences is included in the calculation of the experimental variogram [Eq. (1), Sect. 13.2.1], isolated high values can seriously distort the shape of the variogram[29]. For example, consider the values of the potash deposit in Fig. 56 and Table 40. If the value in DDH 3 was not 105, but 205, then the difference $\Delta 2$ would be 120 and the square $(\Delta 2)^2 = 14\,400$. The sum of all the differences in Table 40, 3rd column, would then be 15 588, and thus 92% of the γ-value would be caused by only this one difference from DDH 3.

Various methods of dealing with this problem are described in the literature, and several of them, besides the pairwise relative variogram dealt with in Section 13.2.1 [Eq. (8)], are briefly mentioned here:

1. Outlier test
Outlier tests were described in Chapter 8.3.3, and they can be applied to this problem. Values suspected for being outliers are not taken into consideration.

2. Lognormal variograms
Logarithms of the values can be used as the variables, and thus also as the basis for the calculation of a variogram (Krige 1978; Krige and Magri 1982a). Outlier tests can also be carried out on the logarithmic values as described in Chapter 8.3.3.

It must be emphasised that ratios of the original non-logarithmic data, rather than the differences, are now used for the logarithmic variograms, as was already discussed for the variance of the logarithmic distribution (Chap. 9.2.3.1).

3. Transformation
Cressie and Hawkins (1980), as well as Krige and Magri (1982b), concluded that the transformation to

$$\sqrt[4]{(x_i - x_{i+h})^2}$$

yielded a good estimation of the variogram.

---

[29] Champigny and Armstrong (1989) use the variation coefficient, C>1.5, as a criterion (cf. Chap. 3.3).

Fig.61A-C. Calculation of an indicator variogram. (After Akin und Siemes 1988)

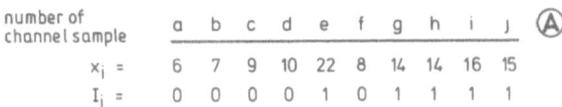

number of channel sample	a	b	c	d	e	f	g	h	i	j
$x_i =$	6	7	9	10	22	8	14	14	16	15
$I_i =$	0	0	0	0	1	0	1	1	1	1

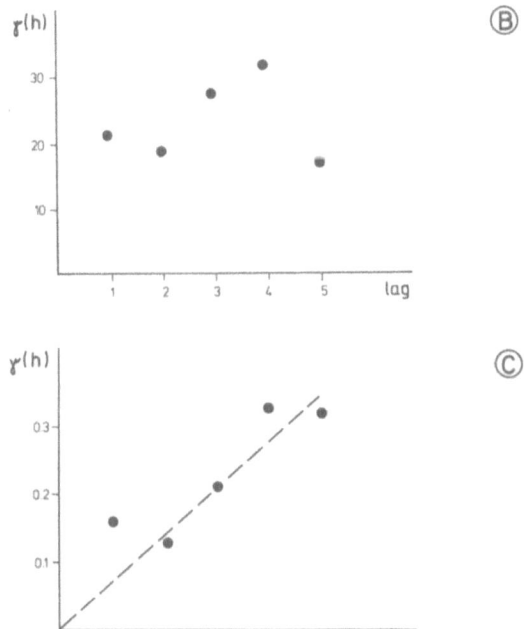

4. Indicator variograms

Journel (1983) introduced the method of using an indicator variogram for very variable sample values. Akin and Siemes (1988) provide an example to illustrate this method, and it is repeated here.

A cut-off xg is selected, for example the median value (Chap. 3.4), and the dependent variables are transformed as follows into a new variable I:

if $x_i \geq x_g$, then $I_i = 1$,

if $x_i < xg$, then $I_i = 0$.

The example provided by Akin and Siemes (1988) is reproduced in Fig. 61

Figure 61a shows values derived from sampling along an axis, for example regularly spaced samples from a vein along the main heading on the vein. Although there are too few values in this example, it does clearly illustrate the effect of an indicator variogram. The normal variogram with individual values is illustrated in Fig. 61b. The points are so broadly dispersed that it is impossible to make an interpretation or interpolation.

The limiting value $x_g = 12$ is now selected, and the data are transformed accordingly (2nd row in Fig. 61a). The calculation is now exactly the same as described in Section 13.2.1. The result is shown in Fig. 61c and, in contrast to Fig. 61b, a distinct trend can be identified.

13.3
Reserve Classification by Geostatistical Calculations

A detailed classification of reserves with geostatistical methods is nowadays standard procedure for feasibility studies and subsequent detailed economic evaluations including sensitivity studies based on errors of grades – in particular the classification of proven and probable reserves as distinct from possible reserves, and the determination of errors of grades. Just as the book *Economic Evaluations in Exploration* was not intended for feasibility stage evalutions, this present book concentrates on calculations for early exploration/prefeasibility studies only for possible and inferred reserves or resources. However, even in the early stages of exploration, such as the design of a drilling grid, it is important to understand the procedures for reserve classification because this grid may be used subsequently to define proven and probable reserves.

13.3.1
Introduction

As already pointed out in Chapter 7.1, various attempts in several countries have been made to define the level of confidence and confidence intervals for the different reserve categories. The results of these attempts are summarized in the Appendix Table 3. Confidence intervals or relative errors of reserves are needed as input in economic sensitivity studies (see Economic Evaluations in Exploration, Chap. 11.6.). Reserve classification, however, has also other purposes, for example informing the shareholders of a mining company as accurately as possible about the reserve assets of that mining company, meaning that the applied reserve classification system has to be understood also by an educated layman. This is one of the reasons why in most countries verbal descriptions and definitions for the different reserve categories are preferred in the official reserve classification systems. At present, two major initiatives exist to standardize the national reserve classification systems for the preparations of an internationally accepted one. The results of this effort in the English speaking mining regions and countries Australasia, the United Kingdom, the USA, Canada and South Africa are summarized by Miskelly (1994). The Energy Council of the UN-suborganization Economic Council for Europe (ECE) sponsors a United Nations International Framework for Reserve/Resource Classification that is based on a three-dimensional classification system, developed in Germany, and takes into account not only the geological knowledge but also the intensity of economic evaluations (Kelter and Wellmer 1993).

For detailed economic studies, however, as pointed out above, errors or intervals of confidence respectively, for the reserve data, have to be taken into account. The following examples, therefore, use the German recommendations for reserve classification.

As already mentioned in Chapter 7.1, in 1983 a GDMB (German Association for Metallurgists and Mining Engineers – the German equivalent to the British Institution of Mining and Metallurgy) working group published recommendations for the geostatistical definition of reserve classifications by applying extension and dispersion variances (Wellmer, 1983a,b). The GDMB recommendations are reproduced in Appendix Table 3.

The standard deviation can be derived from the geostatistical extension and dispersion variances, and then the confidence intervals or error limits for a 90% level degree of confidence can be calculated. These error limits are the basis for the classification of reserves applied to the parameter which in economic evaluations is the most sensitive one. In most cases for a metallic orebody this is the metal content.

The magnitude of the confidence intervals, or the relative error, is the only factor that affects the classification of reserves, and not the average grade. In the following discussion it is assumed that, within the area under consideration, the blocks have the same size. The global estimate of the average grade is then made by simple arithmetic averages weighted by the thickness of the individual blocks, i.e. by means of the grade-thickness (GT) product (Sect. 13.2).

13.3.2
Size of the Blocks

A previously mentioned relationship stated: the larger the volume or support of the sample, the lower the variance of these sample values (cf. Chaps. 3.3 and 11.3). The selection of the size of the blocks, or the zones of the deposit used as a basis for defining the reserves, is therefore critical in calculating the geostatistical variances and standard deviations. The estimate for the whole mineral deposit will have a lower variance (estimation error) as compared to the estimate for only a part of the mineral deposit.

The GDMB working group agreed on the following solution (Wellmer 1983a,b; Fig. 62):

a) The size of the blocks is defined by the sampling grid (e.g. drill holes). These are the "original or basic blocks", and are only an intermediate stage in the calculation of the variances.

b) The original blocks are summed into one block, which corresponds to the reserves for about 3 to 4 years of mining: this is the "definition block".

13.3.3
Drill Grid

Three types of drill grid can be distinguished (Fig. 63):

a) the regular grid, a quasi-regular grid, in which one drill hole is sited precisely in the centre of each block;

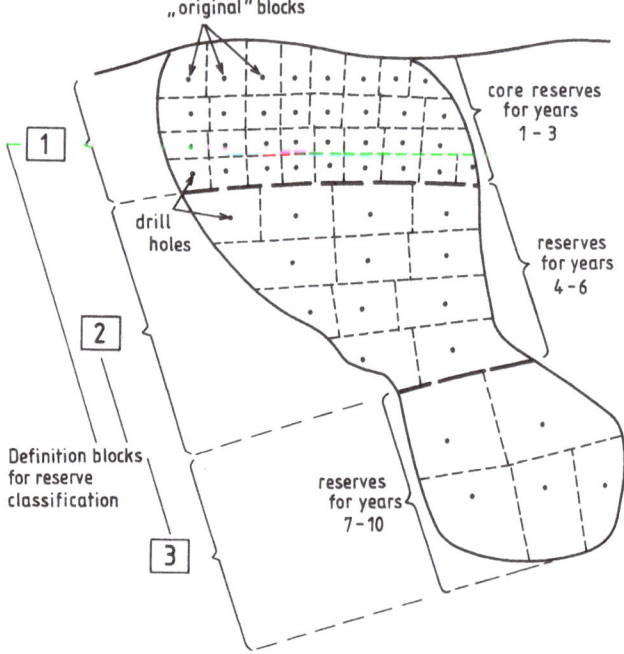

Fig. 62. System of definition blocks according to the reserve classification system of the GDMB. (Wellmer 1983a)

b) the random stratified grid, in which each block contains one drill hole although the drill holes can be sited anywhere within the blocks;

c) the irregular grid, in which both the site as well as the number of drill holes within the blocks can vary. The number of blocks should approximately correspond to the number of drill holes.

These three grids can be rectangular, and do not have to be square.

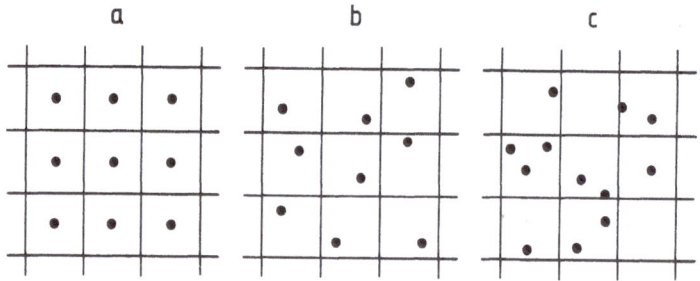

Fig. 63a-c. Classification of drill grids. a Regular; b random stratified; c irregular

According to this definition of the drill grid, and therefore of the "basic blocks" (Sect. 13.3.2), the location of the intersections in the orebody are critical, and not the location of the drill site. Flat-lying orebodies, such as potash seams or the carbonate-hosted Mississippi-Valley type of Pb-Zn deposits, which are drilled on a regular grid, usually have "basic blocks" with drill sites in their centres. In moderately to steeply dipping orebodies, on the other hand, the grids are usually only random stratified because deviation of the drill holes as well as spatial restrictions on the siting of the drill holes rarely make it practically possible for a grid to have regular intersection points.

13.3.4
Calculation of the Geostatistical Estimation Variance

13.3.4.1
Reference Datum

The GDMB working group recommended that the parameter most sensitive in economic evulations should be selected to be the reference datum for the classification of reserves (Appendix Table 3). The metal content of the definition block is the most sensitive parameter for metal deposits. (Sect. 13.3.2 and Fig. 62), and it is derived from the product of the volumes or tonnages and the grades. For practical reasons, however, it is common only to select the grade (see below, Sect. 13.3.4.3),

The metal content of a mineral deposit is Q:

$$Q = G \times V \times \varrho ,$$

where G is the grade, V is the volume of the mineral deposit and ϱ is the density. It has already been mentioned that the accumulation GT value is used in preference to the grade. The metal content Q is therefore calculated by:

$$Q = (\overline{G \cdot T}) \cdot S \cdot \varrho , \tag{1}$$

where $\overline{GT}$ is the mean value for all GT values in the zone of reserve definition, and S is the area of the mineral deposit within the zone of reserve definition. If the estimation variance is calculated for a figure that is derived from several variables (all of which are independent of each other, such as the GT values and the area S), then the variances are summed. The relative variance for the metal content Q is therefore:

$$\frac{\sigma_Q^2}{Q^2} = \frac{\sigma_S^2}{S^2} + \frac{\sigma_{GT}^2}{(\overline{GT})^2} . \tag{2}$$

In this case, ϱ is considered to be a constant. The uncertainty related to the density is usually very small in comparison with that of the GT values and the surface area S.

The relative variances for the surface area S and the GT values must be calculated separately and then added together. The calculation procedure is demonstrated in the following chapters.

The relative standard deviation can then be derived from the square root of the relative variance of the metal content, $\dfrac{\sigma_Q^2}{Q^2}$, or as shown below (Sect. 13.3.4.3) of the grade, $\dfrac{\sigma_G^2}{\overline{G}^2}$. The relative standard deviation must then be multiplied by the Student's t-factor in order to derive the relative error estimate $\dfrac{ki}{\overline{x}}$ for a 90% degree of certainty.

The Student's t-factor can be found in column 1 (number of drill holes) and column 2 of Appendix Table 4. t = 1.83 for ten drill holes, or ten "basic blocks" in the definition block (compare the calculation procedure for the confidence level in Chap. 7.1):

$$\frac{ki}{\overline{x}} = \frac{\sigma_Q}{Q} \cdot t \quad \text{or} \quad \frac{ki}{\overline{x}} = \frac{\sigma_G}{G} \cdot t \; . \tag{3}$$

13.3.4.2
The Relative Estimation Variance for the Area S of the Mineral Deposit

The relative estimation variance for a mineral deposit with area S can be calculated according to a formula proposed by Matheron (1971):

$$\frac{\sigma_S^2}{S^2} = \frac{1}{n^2}(\frac{1}{6}N_2 + 0.061\frac{N_1^2}{N_2}) \text{ for } N_2 \le N_1 \; , \tag{4}$$

where n is the number of ore blocks, $2N_1$ is the number of edges of the ore blocks that are parallel to the direction of the drill grid, and $2N_2$ is the number of edges of the ore blocks that are perpendicular to those used to find N_1 .

Figure 64 provides an example and shows how the edges of the ore blocks are counted in a N-S direction.

It is obvious from formula (4) above, that the relative estimation variance for the mineral deposit area S will be that much lower if the number of edges N is fewer. In other words, the relative estimation variance will be lower either if the orebody is compact and there is a relatively short length of contact with the host rocks, or with blocks containing grades below the cut-off. Similarly, the relative estimation variance will decrease if the total number of blocks n is larger.

This calculation can now be related to the earlier discussion of dilution in Chapter 12 and Fig. 49. In Fig. 49 there are three cases, each of which has n = 25 ore blocks:

Fig.64. Determination of number of edges of ore blocks N_1 and N_2 for the calculation of the estimation variance of the ore deposit area S. (After Didyk and Tulcanaza 1970)

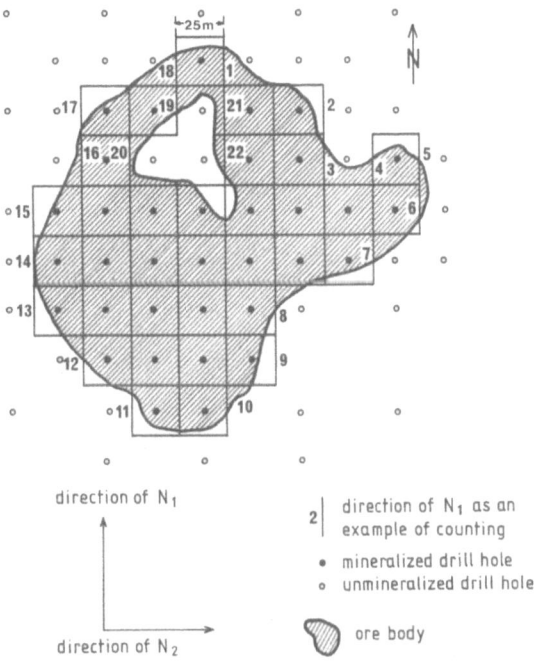

direction of N_1

direction of N_2

$2 |$ direction of N_1 as an example of counting

• mineralized drill hole
○ unmineralized drill hole

ore body

1. Fig. 49a: compact orebody, contact to uneconomic zones is 20 a;

2. Fig. 49b: isolated ore blocks, contact to uneconomic areas is 100 a;

3. Fig. 49c: isolated blocks of internal waste in the ore, contact between the ore blocks to uneconomic blocks is 56 a.

The orebodies are isometric in cases 1 and 2, and thus $N_1=N_2$:

in case 1: $N_1 = N_2 = 10$, and

in case 2: $N_1 = N_2 = 50$.

The orebody in case 3 is longer in the east-west orientation as compared to the north-south orientation. Since N_1 must be greater than N_2 in Eq. (4) above, then for case 3 in Fig. 49c:

$N_1 = 30$, and

$N_2 = 26$.

The relative estimation variance for the mineral deposit area S is derived from these values:

1. Fig. 49a: compact orebody

$$\frac{\sigma_S^2}{S^2} = \frac{1}{n^2}(\frac{1}{6}N_2 + 0.061\frac{N_1^2}{N_2})$$

$$\frac{\sigma_S^2}{S^2} = \frac{1}{25^2}(\frac{1}{6}\cdot 10 + 0.061\frac{100}{10}) = \frac{1}{625}(1.67 + 0.61) = 0.0036 ,$$

and thus 0.36 %;

2. Fig. 49b: isolated orebody

$$\frac{\sigma_S^2}{S^2} = \frac{1}{25^2}(\frac{1}{6}\cdot 50 + 0.061\frac{50^2}{50}) = \frac{1}{625}(8.33 + 3.05) = 0.0182 ,$$

and thus 1.82 %;

3. Fig. 49c: isolated blocks of internal waste:

$$\frac{\sigma_S^2}{S^2} = \frac{1}{25^2}(\frac{1}{6}\cdot 26 + 0.061\frac{30^2}{26}) = \frac{1}{625}(4.33 + 2.11) = 0.0103 ,$$

and thus 1.03 %.

13.3.4.3
The Relative Estimation Variance of the Accumulation Value GT

Three cases can be distinguished according to the type of drill grid, and therefore the location of the drill sites within the "basic blocks" (cf. Sects. 13.3.2 and 13.3.3). A relative variogram for the accumulation value GT is essential for the calculating the relative estimation variance. For the purposes of the calculation it is assumed that this variogram is of the transitive type (cf. Sect. 13.2; see also Akin 1983).

13.3.4.3.1
The Relative Estimation Variance for the Regular Grid

The relative estimation variance for the GT accumulation value can be determined from the extension variance, σ_e^2. The extension variance for the GT value from a particular drill hole can therefore be extrapolated to cover the whole of the "basic block", or extrapolated and then recalculated for the whole of the definition area. In order to do this the extension variance σ_e^2 must be divided by the number of blocks n in the definition area:

Fig.65. Diagram for the determination of the extension variance for a regular drill grid. (Royle 1977 with permission of the author and the Institute of Mining and Metallurgy)

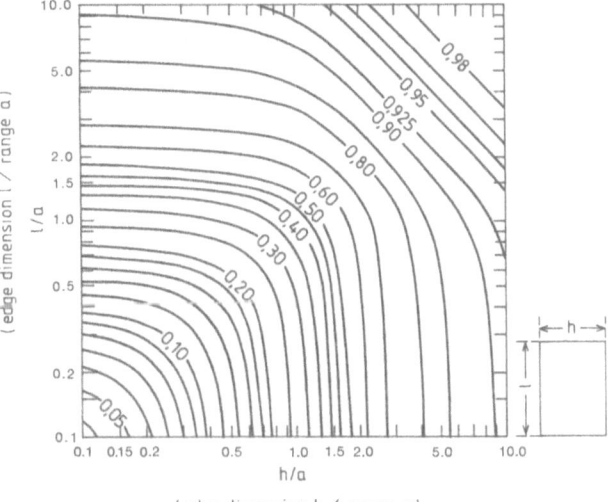

(edge dimension h / range a)

$$\frac{\sigma^2_{GT}}{(GT)^2} = \frac{\sigma^2_e}{n} = \frac{C_0 + C_1 \left\{\sigma^2(h/a, l/a)\right\}}{n} \quad {}^{30} \tag{5}$$

C_0 and C_1 are the nugget effect and sill value for the transitive variogram (Fig. 57), the value σ^2 (h/a, l/a) can be found in a diagram provided by Royle (1977). This diagram is reproduced as Fig. 65.

In the diagram, h and l are the edges of the ore block, so that for square blocks such as in Fig. 63 these values h and l would be the same, a is the range of the transitive variogram (cf. Fig. 57). Royle's diagram (Fig. 65) shows the extension variances for a value $C_1 = 1$ and $C_0 = 0$. The values derived from the diagram must therefore be multiplied by the value for C_1, and then the nugget effect C_0 must be added.

13.3.4.3.2
The Relative Estimation Variance for the Random Stratified Grid

Since, for this type of grid, a drill hole can have any site within the basic block, the relative estimation variance is determined by the dispersion variance of a point (O) within a block (v) or, in other words, by $\sigma^2_{(O,v)}$:

---

[30] In order to emphasize that the $\sigma^2(h/a, l/a)$ value in the brackets is related to a normalized value, the function is often written as $C_1\left\{\sigma^2(h/a, l/a)/C_1\right\}$. The same is true for formula (6) in Section 13.3.4.3.2.

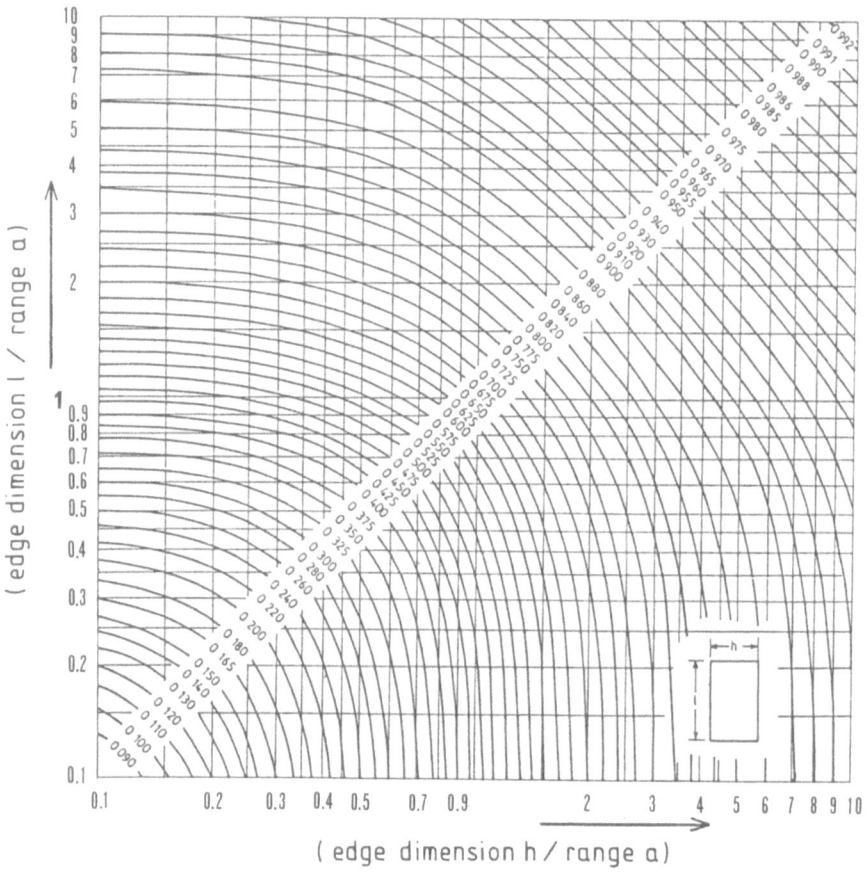

(edge dimension h / range a)

Fig. 66. Diagram for the determination of the dispersion variance of a point within a block (F-chart). (Reprinted from M. David, Geostatistical Ore Reserve Estimation 1977, p. 178, with kind permission from Elsevier Science-NL, Sara Burgerhartstraat 25, 1055 KV Amsterdam, The Netherlands)

$$\frac{\sigma_{GT}^2}{(\overline{GT})^2} = \frac{C_0 + \sigma_{(0,v)}^2}{n} = \frac{C_0 + C_1 \cdot F(h/a, l/a)}{n} \; , \tag{6}$$

where n is again the number of basic blocks in the definition block, and the F-value can be determined from a diagram (Fig. 66). In Fig. 66, h and l are again the lengths of the edges of the basic blocks, and a is the range in the transitive-type relative variogram for the GT values. The so-called F-value must be multiplied by the C_1 value as derived from the variogram and added to the nugget effect C_0 .

13.3.4.3.3
The Relative Estimation Variance for the Irregular Grid

The relative estimation variance for the irregular grid is calculated from the normal relative statistical variance, i.e. from the total sill value $C_0 + C_1$ (cf. Sect. 13.2).

$$\sigma^2 = C_0 + C_1 \, , \tag{7}$$

$$\frac{\sigma^2_{GT}}{(GT)^2} = \frac{\sigma^2}{n} = \frac{C_0 + C_1}{n} \, ,$$

where n is the number of bore holes in the definition zone.

13.3.4.3.4
The Edge Effect

The unknown geometry of the mineral deposit can result in an additional error in the calculation of the estimation variance. This is known as the contour effect or edge effect. Matheron (1971) has shown how a correction factor $\sigma^2 \text{sup}$ can be calculated. If the grid is regular or random stratified, then his factor must be added to the relative variances:

$$\sigma^2_{sup} = \frac{\sigma^2_s}{S^2} \cdot \sigma^2_{D(O/D)} \, . \tag{8}$$

The function $\dfrac{\sigma^2_s}{S^2}$ has already been calculated by Eq. (3) in Section 13.3.4.1. $\sigma^2_{D(O/D)}$ is the relative dispersion variance of the samples (O) within the whole of the mineral deposit (D), and it is derived from the normal relative variogram:

$$\sigma^2_{D(O/D)} = C_0 + C_1 \, . \tag{9}$$

However, in practice, this factor is usually not taken into consideration.

13.3.4.4
The Relative Estimation Variance for Grades

The error is often calculated for the grade, rather than the metal content, as this is the most sensitive of the values. A change in grade has the greatest effect on the economic parameters (cf. Sect. 13.3.4.1 and Appendix Table 3, as well as Economic Evaluations in Exploration, Chap. 10.6).

The estimation variances have so far been calculated for the accumulation GT values, which is the grade x thickness product. The average grade is derived by dividing the mean of the GT values by the mean of the thickness T:

$$\overline{G} = \frac{\overline{GT}}{\overline{T}}.$$ (10)

Again, as in Section 13.3.4.1, the relative variances must be added. In contrast to the area S and the GT values in Section 13.3.4.1, the GT and T values are not independent, but are correlated with each other. This correlation is introduced as a factor containing the correlation coefficient r (i.e. for the correlation between GT and T), see the Appendix Table 8, Eq. (4). The relative estimation variance for the grade is therefore (approximately):

$$\frac{\sigma_G^2}{(\overline{G})^2} = \frac{\sigma_{GT}^2}{(\overline{GT})^2} + \frac{\sigma_T^2}{(\overline{T})^2} - 2r\frac{\sigma_{GT}}{\overline{GT}}\cdot\frac{\sigma_T}{\overline{T}}.$$ (11)

The relative estimation variance for the thickness $\dfrac{\sigma_T^2}{(\overline{T})^2}$ is calculated in the same way as that for the GT value by using Eqs. (5), (6) and (7) in Sections 13.3.4.3.1 to 13.3.4.3.3, depending on the type of drill grid. It is also possible to determine the estimation variance for the tonnage of ore P = S x $\overline{T}$ x ϱ, in the same way as the grade is derived from

$$Q = (\overline{G\cdot T})\cdot S\cdot\varrho \quad [\text{Sect. 13.3.4.1, Eq. (1)}].$$

13.3.5
Example of Using Geostatistical Calculations for Classifying Reserves

The geostatistical formulae mentioned in Section 13.3.4 will be demonstrated by an example.

A zoned pegmatite, in which a large spodumene ore deposit has been discovered, is drilled on a 50 x 60 m drill grid. The average thickness is $\overline{T}$ = 28 m, the average grade is 3.04% Li_2O = G, and the average accumulation value $\overline{GT}$ is therefore 85.16 $Li_2O\cdot m$.

Six blocks are chosen as the core reserves for the first 3 to 4 years of mining. The "basic block" is therefore defined by a drill grid of 50 x 60 m, and the definition block consists of six "basic blocks".

A strict 50 x 60 m grid could not be maintained because of the open pit, and the spacing between drill holes had to be somewhat variable. Each block contains one drill hole so that the grid was of the random stratified type.

Transitive-type variograms were derived for the GT accumulation value and the thickness T (Fig. 67a,b). The correlation coefficient r between GT and T is 0.69 [Eq. (4), Appendix Table 8]. The classification of the core reserves is calculated by applying the error for the grades.

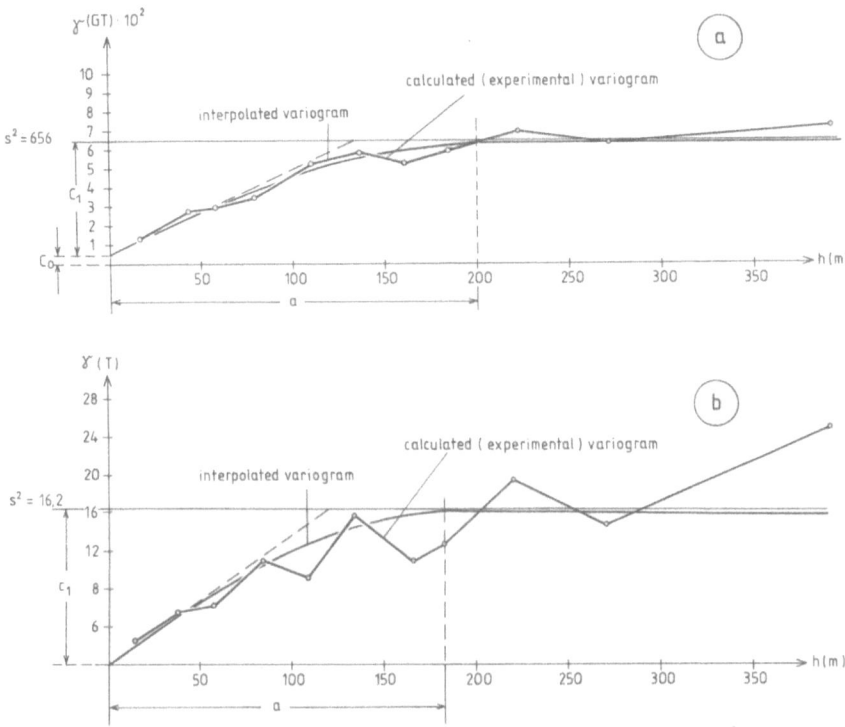

Fig. 67.a Variogram for the accumulation values G·T of a spodumene deposit (grade·thickness in LiO_2·m). b Variogram for thickness of the spodumen deposit of a

1st Step: An idealized variogram is interpolated from the calculated, or experimental, variogram. Since the sum of the nugget effect C_0 and the sill C_1, or the total sill value, is equal to the variance, the known variance values for GT, $s^2 = 656$, and for the thickness T, $s^2 = 16.2$, are plotted on the variograms. The sloping section of the variogram curve is then interpolated using the rule that the tangent at the beginning of the variogram intersects the extension of the horizontal component at 2/3 a (see Sect. 13.2.1 and Fig. 57).

2nd Step: The corresponding values for the range a, the nugget effect C_0 and the sill C_1 are read off Fig. 67a. These values are:

a) for the accumulation value GT:

    range   a = 200 m

    nugget effect C_0 = 60, and

    sill C_1 = 596.

C_O and C_1 are absolute values. The variograms in Fig. 67a and 67b are not relative variograms, and these values must be divided by the square of the mean GT value, which is $(\overline{GT})^2 = (85.1)^2 = 7242$.

This results in:

relative nugget effect $\quad C_O = \dfrac{60}{7242} = 0.0083$, and

relative sill $\qquad\qquad C_1 = \dfrac{586}{7242} = 0.0823$.

b) for the thickness T (Fig. 67b):

range $\qquad\qquad$ a = 185 m

nugget effect $\qquad C_O = 0$, and

sill $\qquad\qquad\qquad C_1 = 16.2$.

Again, this is an absolute value, and it must be divided by the square of the mean value for the thickness ($\overline{T}$ = 28 m), so that the relative sill value is:

relative sill value $C_1 = \dfrac{16,2}{(28)^2} = 0.0207$.

3rd Step: The relative estimation variance is now calculated for the accumulation GT and the thickness T. The appropriate equation is Eq. (6) in Section 13.3.4.3.2, as follows:

$$\frac{\sigma_{GT}^2}{(GT)^2} = \frac{C_O + \sigma_{(o/v)}^2}{n} = \frac{C_O + C_1 \cdot F(h/a, l/a)}{n} .$$

The F-value is obtained from the diagram in Fig. 66.

a) Accumulation value GT:

The quotients h/a and l/a are required for the diagram. h and l are the edges of the "basic blocks" that are derived from the drill grid, in this case 50 and 60 m. a is the range in the variogram, which is 200 m (see 2nd step above). The quotients are therefore:

h/a = 50/200 = 0.25 , and

l/a = 60/200 = 0.30 .

The F-value is derived from the diagram in Fig. 66 for these values, and is found to be between 0.20 and 0.22, although nearer 0.22. The value derived by interpolation is 0.215.

This F-value is based on a C_1-value of 1, so that it must be multiplied by the C_1-value obtained in this example. The number of "basic blocks" within the definition block is n = 6, so that the Eq. (6), Section 3.3, Chapter 4.3.2, is now as follows:

$$\frac{\sigma_{GT}^2}{(GT)^2} = \frac{0.0083 + 0.0823 \cdot 0.251}{6} = \frac{0.0083 + 0.0178}{6} = 0.0043 \, .$$

The relative standard deviation is then:

$$\frac{\sigma_{GT}}{\overline{GT}} = \pm 0.0658, \text{ or } \pm 6.58 \, \% .$$

a) Thickness T:

The same calculation procedure is now used for the thickness T. Since the range a is only 185 m,

$$h/a = 50/185 = 0.27 \, ,$$

$$l/a = 60/185 = 0.32 \, .$$

An F-value of 0.22 is derived from the diagram in Fig. 66 for these values. Since the nugget effect $C_O = 0$, Eq. (6), Section 13.3.4.3.2, is now:

$$\frac{\sigma_T^2}{\overline{T}^2} = \frac{0.0207 \cdot 0.22}{6} = 0.0008 \, .$$

The relative standard deviation is then:

$$\frac{\sigma_T}{\overline{T}} = \pm 0.0275, \text{ or } \pm 2.75 \, \% .$$

4th Step: No correction is made for the edge effect (Sect. 13.3.4.3.4), since the definition block is in contact with ore, and is open both along strike as well as to depth. The relative variance of the grade can now be calculated with Eq. (11) in Section 13.3.4.4, as follows:

$$\frac{\sigma_G^2}{\overline{G}^2} = \frac{\sigma_{GT}^2}{(GT)^2} + \frac{\sigma_T^2}{\overline{T}^2} - 2r \frac{\sigma_{GT}}{\overline{GT}} \cdot \frac{\sigma_T}{\overline{T}} \, .$$

The correlation coefficient between GT and T was 0.69. This figure is then substituted in the equation, which gives:

$$\frac{\sigma_G^2}{G^2} = 0.0043 + 0.0008 - 2 \cdot 0.69 \cdot 0.0659 \cdot 0.0275$$

$$\frac{\sigma_G^2}{G^2} = 0.0043 + 0.00080 - 0.0025 = 0.0026.$$

Therefore the relative standard deviation is now:

$$\frac{\sigma_G}{G} = 0.0510, \text{ or } \pm 5.1\ \%.$$

5th Step: The relative standard deviation must now be multiplied by the corresponding Student's t-factor in order to derive the estimation error at the 90% level of confidence [Eq. (3), Sect. 13.3.4.1]. The Student's t-factor can be obtained from Appendix Table 4. Columns 1 and 2 yield t = 2.02 for n = 6 "basic blocks". Therefore the relative estimation error is $ki/\bar{x}$:

$$\frac{ki}{\bar{x}} = \frac{\sigma_G}{G} \cdot t = 0.0501 \times 2.02 = 0.103, \text{ which is } 10.3\%.$$

A comparison with the requirements for the classification of reserves as recommended by the GDMB (see Appendix Table 3) shows that an error of about 10% lies just within the "proven" category. However, since other possible sources of error have been ignored (e.g. density, analytical precision), it is recommended that the reserves are classified as "probable".

6th Step: The possibility of a regular grid is briefly considered. The diagram in Fig. 65 is used. The quotients h/a and l/a remain unchanged. An extension variance of

$$\sigma_e^2 \quad = 0.0095$$

is then derived for the GT accumulation values from

$$h/a \quad = 0.25\ ,$$

$$l/a \quad = 0.3.$$

For the thickness T, the quotient h/a = 0.27 and l/a = 0.32, so that

$$\sigma_e^2 \quad = 0.10.$$

The further calculation, multiplication with the sill value C_1 and the addition of the nugget effect C_0 continues as described in the 3rd step.

If this calculation is then carried out, then an estimation error of 8.3% is derived. This is obviously an improvement as compared to the result from the random stratified grid. The exploration is clearly most productive if a drill hole is located in the centre of a block.

13.4
Estimating the Grades of Individual Bblocks

13.4.1
Introduction

The regression effect on samples with different support was mentioned in Chapter 11.3.1. This showed how overestimates and underestimates can be caused by the uncorrected transfer of sample grades to block grades (see also Fig. 35). The problems that occur by the application of a cut-off grade, and the consequent occurrence of isolated ore blocks or internal waste, were discussed in Chapter 12. The evaluation of the individual blocks is not so important for compact orebodies, as long as the economic feasibility calculation does not require different annual grades, and therefore annual cash flows (Economic Evaluations in Exploration, Chap. 10.2.4.2). Under these circumstances, it is better to use a *global* estimate of the whole of the mineral deposit. However, if there are isolated blocks of ore and/or waste, then the best possible estimate must be made for the individual blocks, otherwise the application of a cut-off can cause serious errors in the calculated grades and tonnages for the orebody. It is therefore essential that the estimates are as good as possible for *local* blocks.

The problem is schematically illustrated in Fig. 68. It is assumed there are four mining blocks in an orebody, and that very close-spaced samples were collected along a drive, so that it is possible to draw a more or less continuous profile of the

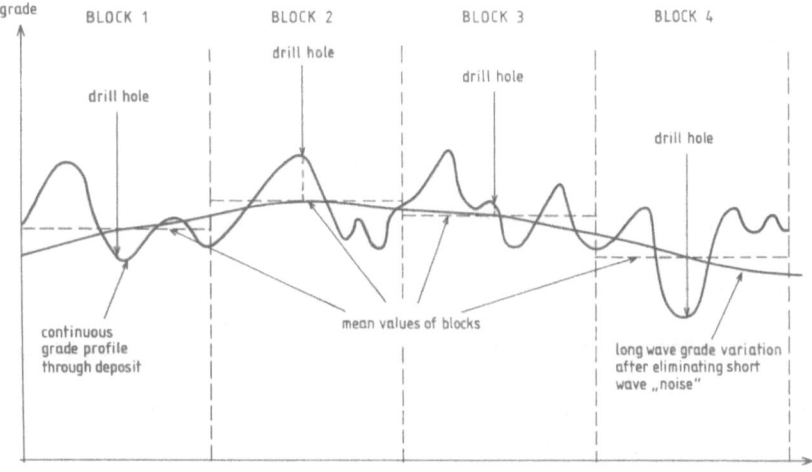

Fig. 68. Grade profile through an ore deposit

grade. Numerous local variations in the grade along this profile result in a high "noise" level. If each block is tested by one drill hole located at its centre, then Fig. 68 clearly illustrates how the results from these holes are affected by the local "noise", and how far the drill hole values can deviate from the average value for the block. The local variations in grade should therefore be eliminated so that the more significant trends in grade, and therefore the correct average grade for the block, can be identified.

The following chapter discusses the methods of estimating the grade of the individual blocks as realistically as possible. Considering that, for mineral deposits with a transitive-type of variogram, there is a spatial interdependence within the range a, it would appear sensible to include the drill holes close to the block together with those within the block when evaluating its grade. In other words, weighting factors should be determined for the drill holes or hole *within* the block as well as for the drill holes *outside* the block.

The symmetry of the block and the distribution of the drill holes is a great help in the calculation of the weighting factors. Section 13.2 demonstrated that *average* variograms are usually used in the early stages of evaluating a mineral deposit. In the early exploration stages it is rarely possible to model variograms in several orientations with sufficient accuracy such that the anisotropy of a mineral deposit, which is present to some degree in all mineral deposits, can be satisfactorily and precisely identified. When using an average variogram, therefore, every point with the same position relative to a reference point or distance from a reference plane must have the same weighting factor.

Weighting factors are always dependent on the distance from the reference point and, because of the role of distance in the geostatistical methods, on the γ-values or covariances that can be derived from the variograms. The weighting factors are independent of the grades and, as a result, they need to be calculated only once for each mineral deposit and each geometric situation. Once calculated they can be used repeatedly for the same problem.

If the estimates for individual blocks are applied to a compact area within a mineral deposit, then the sum of all the blocks should correspond to the global estimate. Indeed, within the limits of error, this is in practice the case. This is easiest to explain with a conceptual model: assume that a mineral deposit is extended to infinity and the drilling grid is regular. The same weighting procedure is then carried out for each of the equal-sized blocks. Whatever the type of weighting factors, the same factors must be applied to each and every drill hole. If all the blocks are added, then all the weighting factors can be excluded from the main addition, and the final total must correspond to the global estimate, which is derived by a simple averaging of all the drill holes.

As a consequence, differences between the global estimate and the total derived from the individual blocks, which were estimated by weighting factors, are caused only by the marginal blocks. However, as mentioned above, in practice these differences for compact ore deposits are generally minor.

Fig.69. Mining blocks as a consequence of the drill pattern

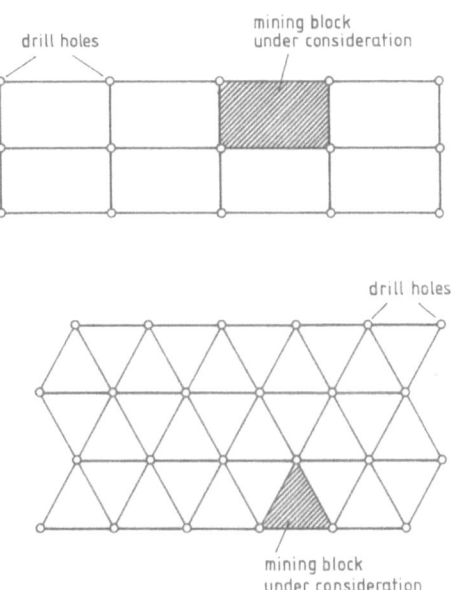

13.4.2
Simple Weighting with the Corner Points of a Block

The simplest and quickest methods for determining the grade of a block by weighting numerous drill holes can be carried out on regular rectangular or triangular drill grids (Chap. 20.1), in which the blocks are arranged so that the drill sites are always located at the corner points (Fig. 69). Since an average variogram is always used, and thus anisotropic features are not taken into account, each drill site has the same weighting and the grade of the block is therefore the average of the drill holes at its corner points.

In practice, this method is improving the sample quality. A similar situation was described in Chapter 11.3.1 for the sampling of a lateritic nickel deposit by pits, and then averaging the channel samples from all four walls of the pits.

This technique is also used in practice, for example in some open pit bauxite mines in Western Australia, where it is referred to as ultra-simplified kriging (see Sect. 13.4.5 for kriging).

13.4.3
The Inverse Squared Distance (ISD) Weighting Method

In many mining camps it is common practice to weight drill holes within a search radius by the inverse of the square of their distance from a reference point. This method is not based on a variogram, and therefore can not be considered to be a geostatistical method but, as in kriging (Sect. 13.4.5), it exploits the relationship between decreasing influence with increasing distance.

In the early stages of evaluating a mineral deposit – for example if a deposit is being offered for sale – as many mineral deposits of similar type in the district should be visited in order to ascertain which mathematical techniques are being applied. This important exercise has already been emphasised in Economic Evaluations in Exploration as well as in this book. Since they allow for the local geological conditions, mining methods, etc., the calculation methods and techniques actually used in practice are always preferable to those that are theoretically derived. This is the reason for introducing the inverse square distance weighting method at this juncture.

Weighting has already been discussed in Economic Evaluations in Exploration, Chapter 4, as well as in this volume, for example Chapter 3.1.2. The grades are multiplied by a weighting factor, the products are then added together, and this figure is then divided by the sum of the weighting factors. Since the grades G_i are weighted by the inverse value of the square of the distance d_i, the average grade G is then:

$$\overline{G} = \frac{\displaystyle\sum_{i=1}^{n} G_i \cdot \frac{1}{d_i^2}}{\displaystyle\sum_{i=1}^{n} \frac{1}{d_i^2}} \qquad \text{[31]} \tag{1}$$

An example, which is illustrated in Fig. 70, is used to explain the method.

Example: A Mississippi Valley-type, carbonate-hosted, zinc-(lead) deposit is submitted for purchase. It has been drilled on a 30-m grid (modified from an earlier 100-ft grid). Acess to several drill sites was prevented by the landowners, and thus the drill grid could not be strictly maintained throughout the program. It is known that the inverse square distance method is applied at another, nearby, mine in the same district, where a search radius of 50 m (with respect to the centre of a block) is used.

The average grade G will be calculated for the block marked on Fig. 70.

1st Step: The search radius is drawn from the centre of the block. There are eight drill holes within the search circle, and they are located at either 21.2 m or 47.2 m from the centre of the block.

2nd Step: The above Eq. (1) is now applied. Since three and five drill holes are located, respectively, at the same distance from the centre of the block, these drill holes can be combined:

---

[31] This is actually a special case. The usual method is with factors $\left(\dfrac{1}{d_i}\right)^n$, by which various values can be obtained that are often based on practical experience from a mining camp.

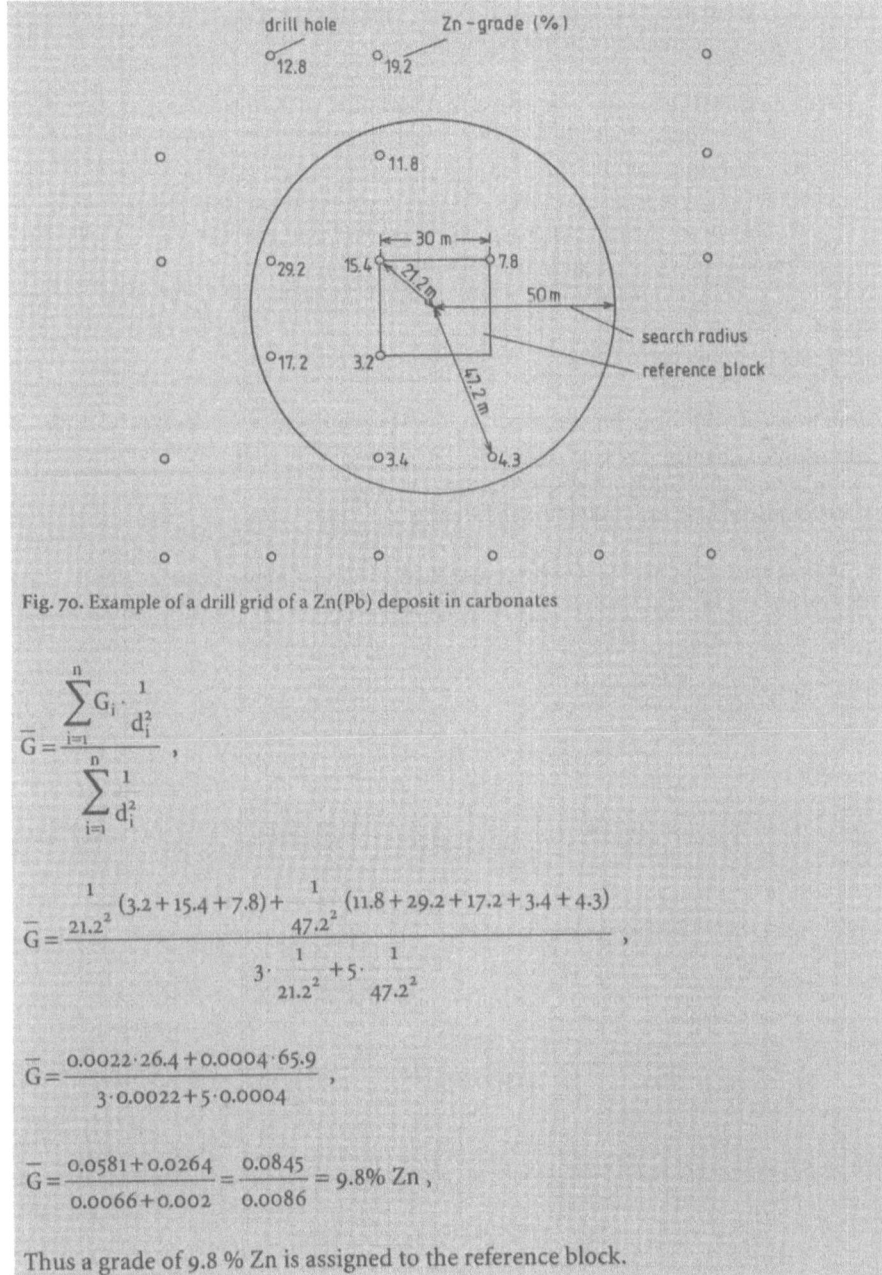

Fig. 70. Example of a drill grid of a Zn(Pb) deposit in carbonates

$$\overline{G} = \frac{\displaystyle\sum_{i=1}^{n} G_i \cdot \frac{1}{d_i^2}}{\displaystyle\sum_{i=1}^{n} \frac{1}{d_i^2}},$$

$$\overline{G} = \frac{\dfrac{1}{21.2^2}(3.2 + 15.4 + 7.8) + \dfrac{1}{47.2^2}(11.8 + 29.2 + 17.2 + 3.4 + 4.3)}{3 \cdot \dfrac{1}{21.2^2} + 5 \cdot \dfrac{1}{47.2^2}},$$

$$\overline{G} = \frac{0.0022 \cdot 26.4 + 0.0004 \cdot 65.9}{3 \cdot 0.0022 + 5 \cdot 0.0004},$$

$$\overline{G} = \frac{0.0581 + 0.0264}{0.0066 + 0.002} = \frac{0.0845}{0.0086} = 9.8\% \ Zn,$$

Thus a grade of 9.8 % Zn is assigned to the reference block.

There are three weaknesses in the inverse squared distance method:
1. There is no relationship between the search radius and the range a of a variogram. An irregular deposit with a small range a would be treated in exactly the same way as a regular deposit with large a.

2. If the reference point is a drill hole, then the weighting factor is infinity. Therefore the method can not be used.

3. The method is basically a point estimate (similar to that which will be described for point kriging in Sect. 13.4.5) and is completely independent of the size of the blocks. Various comparisons between the inverse square distance method and the geostatistical kriging method (Sect. 13.4.5) have been published (e.g. David 1977; Raymond 1979), and they all show that the inverse squared distance method is inferior to kriging to be discussed below.

13.4.4
Weighting with Factors Derived Directly from the Variogram

One way of allowing for the spatial interdependence that is expressed by a variogram is to use the covariances as weighting factors. The covariances can be derived from the variogram (Sect. 13.2 and Fig. 57).
An example will again explain the method:

Example: A gold deposit has been closely drilled on a 10 m grid (Fig. 71a).

Fig.71a-b. Drill grid for a gold deposit

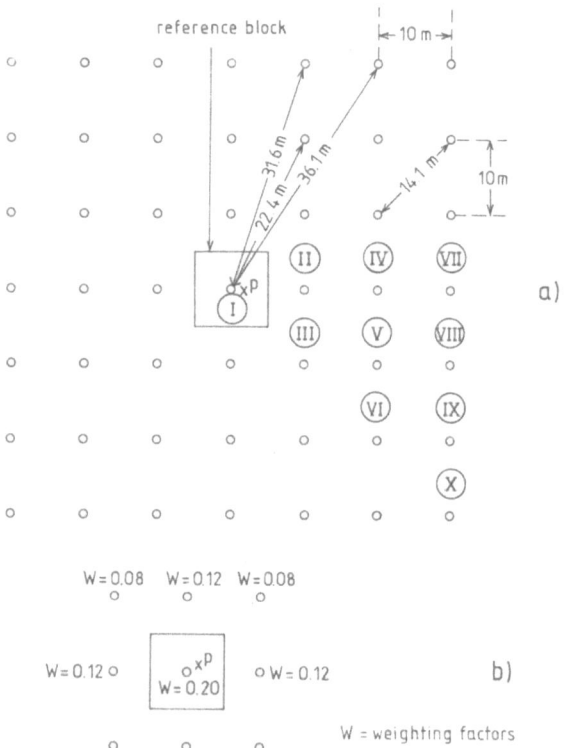

W = weighting factors

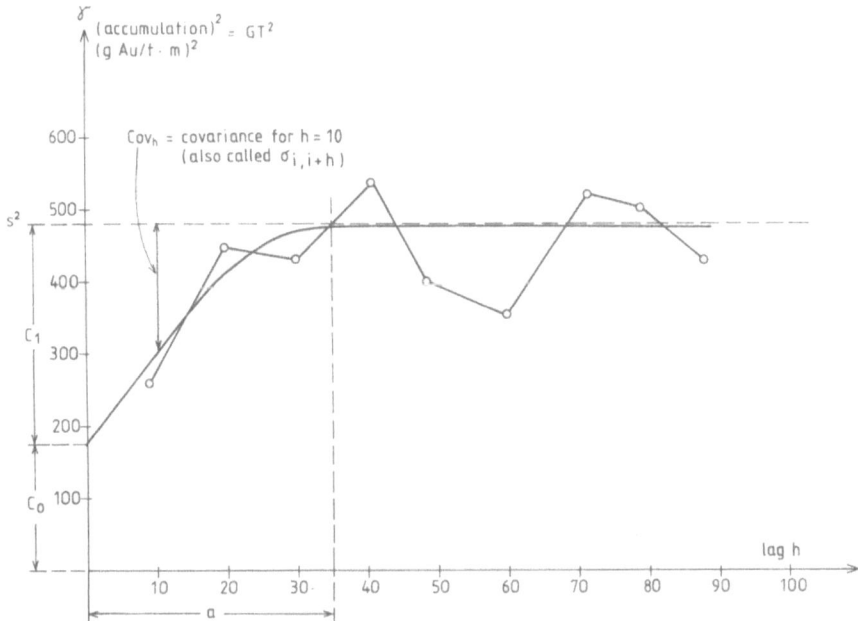

Fig. 72. Variogram for the gold deposit in Fig. 71

The grades of the blocks must be calculated as follows: each drill hole is located in the centre of a block. The reference point P is located next to the central drill hole, from which it is separated by the distance Δx. This difference is so small that, in practical terms, the symmetry with respect to the other drill holes remains the same. The grade calculated by weighting with reference to point P is then assigned to the whole block.

An experimental variogram has been prepared for the GT accumulation value (g Au/t x m), and has been interpolated as shown in Fig. 72.

1st Step: Analysis of the variogram in Fig. 72 indicates that the range a is located at about 36 m. For the sake of the example, however, all the points will be taken into consideration.

2nd Step: The symmetry of the drill grid around the reference point P is analysed. Ten classes of drill sites, denoted with I to X on Fig. 71a, can be distinguished on the basis of their distance from point P. The appropriate distances are also shown on Fig. 71a.

3rd Step: The number of points in each class is now determined, and this figure together with the distance (spacing) of the respective points from the reference point P is listed in Table 41. The covariance for each class is then derived from the variogram in Fig. 72. The value for point I, in the centre of the reference block, is the sill value C_1.

The covariances are zero for the points in classes IX and X, which are located beyond the range a.

4th Step: The covariances in each class are multiplied by the respective number of points in the class, and the results are listed in the 5th column of Table 41, and then summed together. The normalized weighting factors are then calculated in the 6th column by dividing the conversion factors in column 5 by the sum of these factors (2126) and the numbers of points in each class. An analysis of these normalised weighting factors shows that the factors for the points in the 2nd and 3rd rows relative to P, which are the categories IV to VIII, have only a low weighting value; 73% of the total weighting factor is derived from the 9 points in classes I, II and III, and the remaining 27% from the 20 points in classes IV to VIII.

The closer rows of drill sites therefore appear to shield the rows of sites that are further away. This is known as the screen effect, and will be referred to again in the discussion on kriging factors (Sect. 13.4.6).

5th Step: As more points are included in the weighting, the manual computation of the calculation obviously becomes more complex.

In view of the screen effect as well as practical considerations, only the points from classes I, II and III will be considered. These classes include only those drill holes that are in immediate proximity to the reference point P. The computation procedure outlined in the 4th step for the normalisation of the weighting factors must therefore be repeated once again, and the results are listed in columns 7 and 8 of Table 41. The final results are then the weighting factors in column 8, and they are plotted on Fig. 71b.

6th Step: The accumulation values in g Au/t x m are shown on Fig. 73. The value is now calculated for the reference point P using the weighting factors that were derived in Step 5.

The usual weighting equation is [see Eq. (2) in Sect. 3.1.2]:

$$\bar{x} = \frac{\sum\limits_{i=1}^{n} x_i a_i}{\sum\limits_{i=1}^{n} a_i},$$

where a_i are the weighting factors. After taking the symmetry into account (i.e. all points of one class are included in one bracketed function), this yields:

$$\bar{x}_p = \frac{(classI) \cdot 0.2 + (\sum classII) \cdot 0.12 + (\sum classIII) \cdot 0.08}{1}.$$

Fig. 73. Accumulation values for a drill hole within a block and neighbouring drill holes outside the block for the gold deposit in Fig. 71

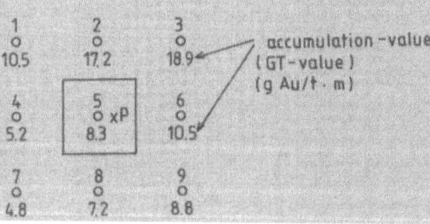

Table 41. Calculation of weighting factors directly from the variogram

Column 1	Column 2	Column 3	Column 4	Column 5	Column 6	Column 7	Column 8
Point category	No. of points n	Distance from P h	Covariance Cov (corresponds to σ_i)	$Cov_x \cdot n$	Normalized weighting factors W_i	$Cov_i \cdot n$	Normalized weighting factors W_i
I	1	x = 0	$C_i = 310$	310	0.15	310	0.20
II	4	10.0 m	185	740	0.09	740	0.12
III	4	14.1 m	125	500	0.06	500	0.08
IV	4	20.0 m	65	260	0.03	Σ 1550	
V	4	22.4 m	45	180	0.02		
VI	4	28.1 m	17	68	0.01		
VII	4	30.0 m	10	40	0.00		
VIII	4	31.6 m	7	28	0.00		
IX	4	36.1 m	0	0	0		
X	4	42.3 m	0	0	0		
				Σ 2126			

Since column 8 of Table 41 has been normalized, the sum of all the weighting factors and therefore the denominator is 1.

$$\bar{x}\,p = 8.3 \cdot 0.2 + (17.2 + 5.2 + 10.5 + 7.2) \cdot 0.12 + (10.5 + 18.9 + 4.8 + 8.8) \cdot 0.08$$

$$\bar{x}\,p = 1.66 + 4.81 + 3.44 = 9.9 \text{ g Au/t x m}.$$

This is the weighted grade for the reference point P, which is now assigned to the whole of the block. If this value of 9.9 g Au/t is compared to the grade of 8.3 g Au/t for drill hole I, then it can be seen that the influence of the drill holes with higher grades (drill holes 1,2,3,6,9) has enhanced the overall value for the block.

The weakness of this method is obviously the same as that for the inverse squared distance weighting method (Sect. 13.4.3). Although it is a method of point estimation, the result is transferred to a block and is completely independent of the size of the block.

13.4.5
Kriging

13.4.5.1
Introduction

The kriging method is named after the South African geostatistician D.J. Krige. It is a technique for determining the best linear unbiased estimator with minimal estimation variance. It can be used on a point as well as on a block. A block is simulated by numerous points that are then integrated.

The kriging technique is usually quite laborious and a computer is required. Kriging is preferably carried out during the later phases of an evaluation, in particular for the computation of individual blocks as required by the detailed mine planning and cash flow analyses. However, a reliable identification of the ore blocks with respect to the cut-off grade is extremely important during the early stages of evaluation of those deposits with isolated single ore and/or waste blocks. Simple kriging techniques are therefore introduced here, since kriging estimates do provide the best weighting factors. The application of the kriging method will be described in two stages.

Stage 1: Point kriging

In the point kriging method, as already explained in Sections 13.4.3 and 13.4.4, a reference point P is selected in a block, and its grade is estimated by the kriging technique. If a value is generalized for the whole block, then this results in the same weakness mentioned in Sections 13.4.3. and 13.4.4, namely assigning a point grade to the whole block regardless of the size of the block.

Stage 2: Block kriging

This technique takes the size of the block into consideration. The theory behind the kriging technique will not be explained here, and the interested reader is referred to David (1977). Instead, the calculation of the weighting factors, known as the kriging factors and denoted by λ_i, will be explained by examples. There are two types of kriging:

Kriging without the mean $\bar{x}$ (kriging with the unknown mean).

a) Kriging using the mean $\bar{x}$ of the deposit or zone within the deposit (kriging with the known mean);

For the calculation of case (a) it is assumed that the sum of the kriging factors is one, or:

$$\sum_{i=1}^{n} \lambda_i = 1 \qquad (1)$$

This assumption is of course not made for case b) (kriging using the mean). The difference between the sum of λ_i and 1 is denoted as the weighting factor of the mean $\bar{x}$ (Sect. 13.4.5.2.3).

In general, if the grade of the deposit exhibits no obvious trends, then kriging with the mean is the preferred method. This method is also easier to do by hand, since one less unknown occurs (the Lagrange multiplier, μ, see below). However, if grade trends are identified or there are clearly definable low- and high-grade zones, then it is less suitable to work with the mean.

13.4.5.2
Point Kkriging

13.4.5.2.1
Equations for the Kriging System Without and With a Known Mean

Kriging uses a linear equation system that can be written using a matrix notation:

1. The kriging matrix K contains the variances σ_{ii} (e.g. σ_{11}) and covariances σ_{ij} (e.g. σ_{12}) of all the points x_i (i = 1, 2 ... n) around the reference point P that are included in the weighting. σ_{ij} is the covariance between the values x_i and x_j, or $\sigma(x_i, x_j)$ or $Cov(x_i, x_j)$ (Sect. 13.2.1 and Fig. 57), σ_{ii} are the covariances of a point with respect to itself (where h = 0), and thus are identical with the variances.

$$K = \begin{bmatrix} \sigma_{11}\sigma_{12} \cdots\cdots\cdots\cdots\cdots\cdots \sigma_{1n} & 1 \\ \sigma_{21}\sigma_{22} \cdots\cdots\cdots\cdots\cdots\cdots \sigma_{2n} & 1 \\ \cdot & \\ \cdot & \\ \cdot & \\ \cdot & \\ \sigma_{n1}\sigma_{n2} \cdots\cdots\cdots\cdots\cdots \sigma_{nn} & 1 \\ 1 \quad\quad 1 \quad\quad\quad\quad\quad 1 & 0 \end{bmatrix} \tag{2}$$

This is the kriging matrix for case (a) (kriging without a mean). The final row and column are missing from the matrix for case (b) (kriging with known mean).

Since $\sigma_{ik} = \sigma_{ki}$, the matrix is symmetrical and, for example, $\sigma_{12} = \sigma_{21}$. Regardless of whether point 1 is viewed from point 2 or the other way round, the covariance between points 1 and 2 must be the same. The spacing between the two points is the same, and it is this dimension that determines the covariance, which can be derived from the variogram (Sect. 13.2.1 and Fig. 57).

2. The weighting factors λ_i describe a vector , λ. In addition, the Lagrange multiplier μ is introduced for the case (a) of kriging without a mean. This parameter does not occur in case (b) (kriging with a known mean).

$$\vec{\lambda} = \begin{bmatrix} \lambda_1 \\ \lambda_2 \\ \vdots \\ \lambda_n \\ \mu \end{bmatrix} \tag{3}$$

3. The third dimension in Eq. (5) see below is the vector $\vec{D}$. This vector contains the covariances of the reference point P to all the other points that are being taken into consideration:

$$\vec{D} = \begin{bmatrix} \sigma_{P,1} \\ \sigma_{P,2} \\ \vdots \\ \sigma_{P,n} \\ 1 \end{bmatrix}$$ (4)

This is also the vector for the case (a) (kriging with an unknown mean). The final value 1 does not occur in case (b) (kriging with a known mean).

The following relationship between the kriging matrix and the two vectors is then true:

$$[K] \cdot \vec{\lambda} = \vec{D} \ .$$ (5)

Consider, for example, the reference point P and two additional points 1 and 2, then the system (5) above for case (a) (kriging without a mean) is:

$$\begin{bmatrix} \sigma_{11} & \sigma_{12} & 1 \\ \sigma_{21} & \sigma_{22} & 1 \\ 1 & 1 & 0 \end{bmatrix} \cdot \begin{bmatrix} \lambda_1 \\ \lambda_2 \\ \mu \end{bmatrix} = \begin{bmatrix} \sigma_{P,1} \\ \sigma_{P,2} \\ 1 \end{bmatrix} \ .$$ (6)

For case (b) (kriging with a known mean), the system (5) above, for the reference point P and two additional points 1 and 2 is:

$$\begin{bmatrix} \sigma_{11} & \sigma_{12} \\ \sigma_{21} & \sigma_{22} \end{bmatrix} \cdot \begin{bmatrix} \lambda_1 \\ \lambda_2 \end{bmatrix} = \begin{bmatrix} \sigma_{P,1} \\ \sigma_{P,2} \end{bmatrix} \ .$$ (7)

The computation of this matrix system is as follows: each value in a row of the matrix is multiplied by the corresponding value for the vector, and then they (row x column) are summed together (Fig. 74).

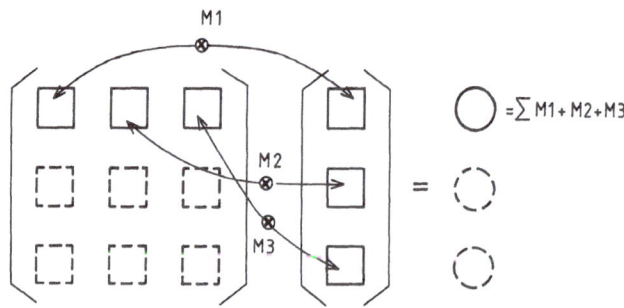

Fig. 74. System of calculation for the multiplication of a matrix with a vector

In case (a) (kriging without a mean) this yields the following system of equations:

$$\sigma_{11} \cdot \lambda_1 + \sigma_{12} \cdot \lambda_2 + \mu = \sigma_{P,1}$$

$$\sigma_{21} \cdot \lambda_1 + \sigma_{22} \cdot \lambda_2 + \mu = \sigma_{P,2} \tag{8}$$

$$1 \cdot \lambda_1 + 1 \cdot \lambda_2 + 0 \cdot \mu = 1.$$

This is a system of three equations with the three unknowns λ_1, λ_2 and μ. In case (b) (kriging with a known mean), the system is as follows:

$$\sigma_{11} \cdot \lambda_1 + \sigma_{12} \cdot \lambda_2 = \sigma_{P,1}$$
$$\sigma_{21} \cdot \lambda_1 + \sigma_{22} \cdot \lambda_2 = \sigma_{P,2} \tag{9}$$

The unknown μ is missing from this system, and therefore it is a simpler system with only two equations with two unknowns.

13.4.5.2.2
Example of Kriging Without a Mean

The system of matrices and vectors that has just been outlined in Section 13.4.5.2.1 will now be applied to the case of the gold deposit that was introduced in Section 13.4.4. In order to emphasize the principle, only those points that are closest to the reference point P will be taken into consideration. These are the point I, with a spacing of Δx, and the four points in class II, which have a spacing of 10 m. These points are shown in Fig. 75, and they are numbered from 1 to 5 (Fig. 75). Needless to say, the variogram shown on Fig. 72 is still valid.

Fig.75a-b. The drill holes taken into account for the point kriging example for the gold deposit of Fig. 71

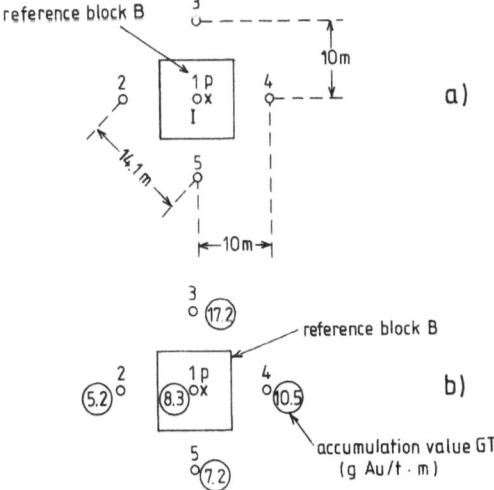

1st Step: Similar to the Eq. (5) and (6) in Section 13.4.5.2, the following system of equations is derived for the reference point P and the five drill sites, 1 to 5:

$$
\begin{bmatrix}
\sigma_{11} & \sigma_{12} & \sigma_{13} & \sigma_{14} & \sigma_{15} & 1 \\
\sigma_{21} & \sigma_{22} & \sigma_{23} & \sigma_{24} & \sigma_{25} & 1 \\
\sigma_{31} & \sigma_{32} & \sigma_{33} & \sigma_{34} & \sigma_{35} & 1 \\
\sigma_{41} & \sigma_{42} & \sigma_{43} & \sigma_{44} & \sigma_{45} & 1 \\
\sigma_{51} & \sigma_{52} & \sigma_{53} & \sigma_{54} & \sigma_{55} & 1 \\
1 & 1 & 1 & 1 & 1 & 0
\end{bmatrix}
\begin{bmatrix}
\lambda_1 \\ \lambda_2 \\ \lambda_3 \\ \lambda_4 \\ \lambda_5 \\ \mu
\end{bmatrix}
=
\begin{bmatrix}
\sigma_{P,1} \\ \sigma_{P,2} \\ \sigma_{P,3} \\ \sigma_{P,4} \\ \sigma_{P,5} \\ 1
\end{bmatrix}
\tag{10}
$$

2nd Step: By using the symmetry of the matrix, this system can be computed as follows:

1. Since the points 2, 3, 4 and 5 are practically all the same distance from the reference point P (the displacement by Δx from point 1 is negligible), then λ_2, λ_3, λ_4 and λ_5 must also be the same. λ_2 is substituted for them in the λ vector.

2. σ_{11}, σ_{22}, σ_{33}, σ_{44} and σ_{55} on the main diagonal of the kriging matrix are all the same. They are the variance of one point within the deposit, and thus the same as the total sill value $C_0 + C_1$ (cf. Sect. 13.2). The variogram in Fig. 72 shows that:

    $$\sigma_{ii} = C_0 + C_1 = 480.$$

3. All distances from point 1 to the other points 2, 3, 4 and 5 are the same, specifically 10 m. Assuming that the difference Δx between the reference point P and the point 1 is negligible, then this is also the case for the reference point P. Therefore the covariances are:

    $$\sigma_{12} = \sigma_{21} = \sigma_{13} = \sigma_{31} = \sigma_{14} = \sigma_{41} = \sigma_{15} = \sigma_{51} = \sigma_{P,2} = \sigma_{P,3} = \sigma_{P,4} = \sigma_{P,5}.$$

    The distance between these points is 10 m, and the corresponding value for the covariances of 185 is derived from the variogram in Fig. 72 (cf. the columns 3 and 4 in Table 41).

4. The spacings between the points 2 and 3, 3 and 4, 4 and 5, and 5 and 2 are the same, specifically 14.1 m. Therefore

    $$\sigma_{23} = \sigma_{32} = \sigma_{34} = \sigma_{43} = \sigma_{45} = \sigma_{54} = \sigma_{52} = \sigma_{25}.$$

    For a spacing of 14.1 m, the corresponding value for the covariance of 125 is derived from the variogram in Fig. 72 (compare the columns 3 and 4 in Table 41).

5. The spacing between the points 2 and 4 and 3 and 5 is the same, specifically 20. The corresponding value for the covariance of 65 is derived from the variogram in Fig. 72.

$$\sigma_{24} = \sigma_{42} = \sigma_{35} = \sigma_{53} \, .$$

6. The only value that has not been determined is the covariance $\sigma_{p,1}$ between the reference point P and point 1, between which the spacing is Δx. This is the sill value $C_1 = 310$ (cf. Sect. 13.4.4).

The matrix Eq. (10) above can now be written as follows:

$$\begin{bmatrix} 480 & 185 & 185 & 185 & 185 & 1 \\ 185 & 480 & 125 & 65 & 125 & 1 \\ 185 & 125 & 480 & 125 & 65 & 1 \\ 185 & 65 & 125 & 480 & 125 & 1 \\ 185 & 125 & 65 & 125 & 480 & 1 \\ 1 & 1 & 1 & 1 & 1 & 0 \end{bmatrix} \cdot \begin{bmatrix} \lambda_1 \\ \lambda_2 \\ \lambda_2 \\ \lambda_2 \\ \lambda_2 \\ \mu \end{bmatrix} = \begin{bmatrix} 310 \\ 185 \\ 185 \\ 185 \\ 185 \\ 1 \end{bmatrix} \qquad (11)$$

3rd Step: The multiplication procedure is illustrated by Fig. 74, and since the matrix K has six rows and six columns, then six linear equations are generated. However, rows 2 to 5 are the same[32].

a) The same values always occur in matrix K, but they are permutated (sequence is changed). Since the same factor always occurs in the vector λ, these permutations are insignificant.

b) λ_2 always occurs in the vector $\bar{\lambda}$.

c) the same value, specifically 185, always occurs in the vector $\bar{D}$.

As a result, the system is reduced to three equations, specifically rows 1, 2 and 6 in the matrix. By multiplication, the following is obtained:

row 1:

$$\lambda_1 \cdot 480 + \lambda_2 \cdot 185 + \lambda_2 \cdot 185 + \lambda_2 \cdot 185 + \lambda_2 \cdot 185 + \mu = 310 \, ,$$

---

[32] The rows 2 to 5 are therefore linearly dependent, i.e. the so-called determinant of the system in Eqs. (2) to (5) above equals zero (refer to mathematics textbooks, e.g. Bronstein and Semendjajew 1965).

row 2:   (12)

$\lambda_1 \cdot 185 + \lambda_2 \cdot 480 + \lambda_2 \cdot 125 + \lambda_2 \cdot 65 + \lambda_2 \cdot 125 + \mu = 185$, and

row 6:

$\lambda_1 + \lambda_2 + \lambda_2 + \lambda_2 + \lambda_2 = 1$.

This can be summarized as:

$480 \cdot \lambda_1 + 740 \cdot \lambda_2 + \mu = 310$,

$185 \cdot \lambda_1 + 795 \cdot \lambda_2 + \mu = 185$,   (13)

$\lambda_1 + 4 \cdot \lambda_2 = 1$.

4th Step: There are three equations with the three unknowns λ_1, λ_2, and μ. μ occurs only in the first two equations, and it can easily be eliminated by subtracting the second equation from the first. By doing this, the following is obtained:

$480 \cdot \lambda_1 + 740 \cdot \lambda_2 + \mu = 310$

$\dfrac{-\left(185 \cdot \lambda_1 + 795 \cdot \lambda_2 + \mu = 185\right)}{= 295 \cdot \lambda_1 - 55 \cdot \lambda_2 = 125}$

5th Step: There are now only two equations with two unknowns:

$295 \cdot \lambda_1 - 55 \cdot \lambda_2 = 125$, and

$\lambda_1 + 4 \cdot \lambda_2 = 1$.

The second row can be rewritten as:

$\lambda_1 = 1 - 4 \cdot \lambda_2$.

Substituting for λ_1 in the first row gives:

$295 (1 - 4 \cdot \lambda_2) - 55 \cdot \lambda_2 = 125$,

$295 - 1180 \cdot \lambda_2 - 55 \cdot \lambda_2 = 125$,

$-1235 \cdot \lambda_2 = -170$,

$\lambda_2 = 0.138$.

6th Step: Assuming that the sum of the weighting factors must equal 1, then from the above:

$$\lambda_1 + 4 \cdot \lambda_2 = 1,$$

therefore for λ_1:

$$\lambda_1 = 0.448 .$$

Therefore, in order to determine the grade or accumulation value for the reference point P, point 1 is weighted by 0.448, and the points 2, 3, 4 and 5 are each weighted by 0.138.

7th Step: The accumulation value for the reference point P is now calculated by:

$$\bar{x}_P = 0.448 \cdot 8.3 + 0.138(5.2 + 17.2 + 10.5 + 7.2) ,$$

$$\bar{x}_P = 3.72 + 5.53 = 9.25 \text{ g Au/t}$$

It must be noted that, because only the five points from classes I and II have been taken into consideration for this example, this value is not directly comparable to that derived in Section 13.4.4. In the earlier example in Section 13.4.4, nine points were used from classes I, II and III.

In order to obtain comparable values the normalized weighting factors for classes I and II must be derived from columns 7 and 8 in Table 41. These are:

for class I : 0.30, and

for class II : 0.18 .

These values can now be directly compared to $\lambda_1 = 0.448$ and $\lambda_2 = 0.138$. This demonstrates that the kriging factors, which are the best linear estimators without a bias, assign a significantly higher value to point 1 in the centre of the block.

The solution demonstrated in Steps 4, 5 and 6 for a linear system of equations with three unknowns can be schematically achieved with the methods of calculating a unit matrix for eigenvalues. The matrix is inverted so that all values, excluding the diagonal values that are 1, are converted to zero. This method can also be applied to larger systems, and therefore more points. This computational procedure is demonstrated in Appendix Table 11 with the above example.

Under normal circumstances linear systems with several unkowns are solved with a programmed calculator or a computer.

13.4.5.2.3
Example of Kriging With a Known Mean

Kriging with a known mean will be demonstrated with the example of the gold deposit in Sections 13.4.4 and 13.4.5.2.2. In Sections 13.4.5.1 and 13.4.5.2.1 it was shown that the only difference between the equation system for kriging with a known mean and that for kriging without a mean is the lack of the condition

$$\sum \lambda_i = 1.$$

Thus the sum of the weighting factors need not be 1. The difference between this sum and 1 is thus assigned the weighting factor of the mean $\bar{x}$. The matrix Eqs. (10) and (11) from Section 13.4.5.2.2 can be applied, and the 6th row and 6th column are then omitted from the kriging matrix, as well as the Lagrange multiplier μ from vector λ and the final value 1 from vector $\bar{D}$. As a result, the equation systems (12) and (13) from Section 13.4.5.2.2 are reduced from three equations with three unknowns to two equations with the two unknowns λ_1 and λ_2:

$$480 \cdot \lambda_1 + 740 \cdot \lambda_2 = 310, \text{ and} \tag{14}$$

$$185 \cdot \lambda_1 + 795 \cdot \lambda_2 = 185 .$$

To solve these equations, the first equation is multiplied by a factor so that it now also begins with $185 \cdot \lambda_1$. This factor is therefore $185 / 480 = 0.39$. This yields the following:

$$(480 \cdot \lambda_1 + 740 \cdot \lambda_2) \cdot 0.39 = 310 \cdot 0.39 \tag{15}$$

$$185 \cdot \lambda_1 + 795 \cdot \lambda_2 \qquad = 185 ,$$

therefore:

$$185 \cdot \lambda_1 + 288.6 \cdot \lambda_2 = 120.9 ,$$

$$185 \cdot \lambda_1 + 785 \quad \cdot \lambda_2 = 185 .$$

The second equation can now be subtracted from the first equation, and one obtains:

$$-506.40 \cdot \lambda_2 = -64.1$$

or $\lambda_2 = 0.127$.

Substituting this value in one of the two equations of the system (14) above gives:

$$480 \cdot \lambda_1 + 740 \cdot 0.127 = 310$$

$$480 \cdot \lambda_1 = 216.0 ,$$

$$\lambda_1 = 0.450 .$$

As mentioned above, the difference between the sum of all the weighting factors and 1 is regarded as the weighting factor for the mean of the deposit $\overline{x}$.
The sum of the weighting factors is:

$$\lambda_1 + 4 \cdot \lambda_2 = 0.450 + 0.508 = 0.958 \text{ , therefore}$$

$$1 - \lambda_1 - 4 \cdot \lambda_2 = 1 - 0.958 = 0.042.$$

The weighting factor of the mean is therefore 0.042.

Assume that the mean of the accumulation values is 12.5 g Au/t $\cdot$ m, then the example in Fig. 75b gives the following accumulation value relative to the reference point:

$$\overline{x}_p = 0.45 \cdot 8.3 + 0.127(5.2 + 17.2 + 10.5 + 7.2) + 0.042 \cdot 12.5 \text{ ,}$$

$$\overline{x}_p = 3.74 + 5.09 + 0.53 = 9.26 = m_K$$

This value m_K is known as the kriging estimator.

13.4.5.3
Block Kriging

Point kriging (Sect. 13.4.5.2) takes into account only the relationships between the individual sample points, which were drill hole sites in the previous example, but does not take the size of the blocks into consideration. Block kriging uses the covariances of the sample points in relation to the reference block.

Block kriging can also be carried out without a mean and with a known mean. Equation (5) from Section 13.4.5.2.1 is also valid:

$$[K] \cdot \vec{\lambda} = \vec{D} \text{ ,}$$

where $[K]$ is the kriging matrix, $\vec{\lambda}$ is a vector with the weighting factors λ_1 to λ_n. The vector $\vec{D}$ contains the covariances of the sample points i in relation to the block B.:

$$\vec{D} = \begin{bmatrix} \sigma_{B,1} \\ \sigma_{B,2} \\ \vdots \\ \sigma_{B,n} \\ 1 \end{bmatrix}.$$

This is the vector for kriging without a mean. The final value 1 is omitted for kriging with a known mean.

The application of the block kriging technique can be explained by an example from the same gold deposit as was discussed in Sections 13.4.5.2.2 and 13.4.5.2.3. The same configuration of the drill holes is also used, as shown in Fig. 75, and the reference block B has a central drill hole 1. This hole, together with the four adjacent drill holes, is included in this kriging exercise.

Kriging with an unknown mean is carried out by using the matrix system that, analogous to the Eq. (10) in Section 13.4.5.2.2, is now written as follows:

$$
\begin{bmatrix}
\sigma_{11} & \sigma_{12} & \sigma_{13} & \sigma_{14} & \sigma_{15} & 1 \\
\sigma_{21} & \sigma_{22} & \sigma_{23} & \sigma_{24} & \sigma_{25} & 1 \\
\sigma_{31} & \sigma_{32} & \sigma_{33} & \sigma_{34} & \sigma_{35} & 1 \\
\sigma_{41} & \sigma_{42} & \sigma_{43} & \sigma_{44} & \sigma_{45} & 1 \\
\sigma_{51} & \sigma_{52} & \sigma_{53} & \sigma_{54} & \sigma_{55} & 1 \\
1 & 1 & 1 & 1 & 1 & 0
\end{bmatrix}
\cdot
\begin{bmatrix}
\lambda_1 \\ \lambda_2 \\ \lambda_3 \\ \lambda_4 \\ \lambda_5 \\ \mu
\end{bmatrix}
=
\begin{bmatrix}
\sigma_{B,1} \\ \sigma_{B,2} \\ \sigma_{B,3} \\ \sigma_{B,4} \\ \sigma_{B,5} \\ 1
\end{bmatrix}
\tag{16}
$$

All the values for σ_{ii} and σ_{ij} were determined for the kriging matrix in Section 13.4.5.2.2, where the symmetry of the λ-vector was also taken into consideration. The results can be derived from Eq. (11) in Section 13.4.5.2.2. Vector D, which includes the covariances between the drill holes 1 to 5 and the reference block B, is the only unknown that must be determined anew.

1st Step: Regarding the symmetry of the drill grid, it can be seen that drill holes 2, 3, 4 and 5 have the same location with respect to reference block B. Therefore the following must be true:

$$\sigma_{B,2} = \sigma_{B,3} = \sigma_{B,4} = \sigma_{B,5} .$$

Thus two covariance values have to be determined: $\sigma_{B,1}$ and $\sigma_{B,2}$.

2nd Step: The covariance between a point, in this case a drill hole, and a block can be derived from the variogram by replacing the block by numerous points within the block (theoretically an infinite number of points, in practice 9 to 16). The covariances are then determined from the variogram for the distance h_i between the drill site and each of the simulation points within the block, and these values are then averaged (Fig. 76).

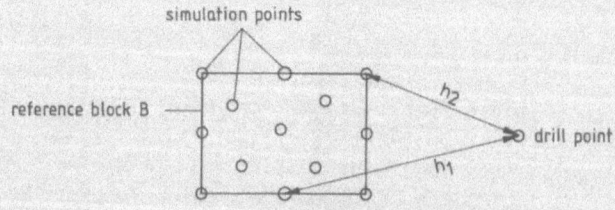

$h_1, h_2 .... h_i$ = distances for the determination of the covariances from the variogramm

Fig. 76. Method of calculating the covariance between a point and a block

Fig.77a-b. Method of calculating the covariances between the reference block and drill hole 1 (a) and drill hole 2 (b) for the gold deposit of Fig. 71

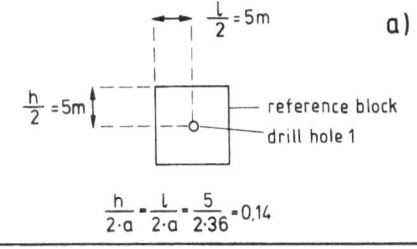

$$\frac{h}{2\cdot a}\cdot\frac{L}{2\cdot a}=\frac{5}{2\cdot 36}=0.14$$

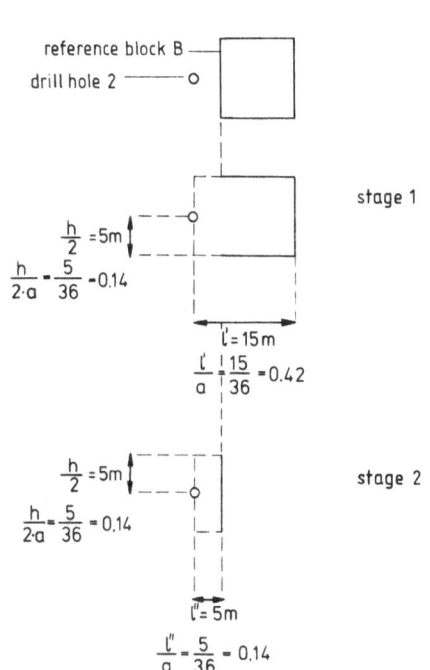

Tables can be used for simple geometric situations. The function $(1-H)^{33}$ is valid for rectangular blocks, and the values are reproduced as Appendix Table 7. As was the case for the F-function in Section 13.3.4.3.2, Fig. 66, the dimensions h and l of the block, and the distances of the point under consideration from the edges of the block, must be normalized by the range a of the variogram.

The variogram in Fig. 72 is valid for the gold deposit under consideration, and the range is a = 36 m.

---

[33] The function H itself is related to γ values, and not to the covariances. However, since the following is true (see Sect. 13.2.1 and Fig. 57):

$\gamma_{(x,V)} = C_0 + C_1 \cdot H(1/a, H/a)$ and

$\sigma_{(x,V)} = C_0 + C_1 - [C_0 + C_1 \cdot H(h/a,1/a)]$, therefore

$\sigma_{(x,V)} = C_1[1 - H(h/a,1/a)]$.

Consider first the drill hole 1 in the centre of the reference block B (Fig. 77a). The block is square, and so l = h.

The distance of drill hole 1 to the edges of the block is l/2. This is normalised with the range a, and therefore:

$$\frac{l}{2 \cdot a} = \frac{h}{2 \cdot a} = \frac{10}{2 \cdot 36} = 0.14 \ .$$

The function (1-H) is interpolated for this value (l = 0.14, h = 0.14) from Appendix Table 12 (an exemplary calculation for a step-wise interpolation is provided by example 2 in Section 9.3.5):

(1-H) for (0.14/0.14) = 0.826 .

The covariance $\sigma_{B,1}$ is then:

$$\sigma_{B,1} = C_1 \cdot (1\text{-}H) \ .$$

From the variogram in Fig. 72, the sill value $C_1 = 310$, so that the covariance is:

$$\sigma_{B,1} = 310 \cdot 0.826 = 256 \ .$$

3rd Step: The covariance between drill hole 2 and the reference block B is now examined. This is also determined by using the function (1-H). Since the drill hole is sited outside the reference block (Fig. 77b), the calculation is carried out in stages by examining the differences. Initially the function (1-H) is determined for the influence of drill hole 2 in the area h/2,l' (stage 1, Fig. 77b), and then for the influence in the area h/2,l" (stage 2, Fig. 77b). These two values are weighted by the difference of the two areas and subtracted. Therefore the weighting of the value derived in stage 1 is three times that of the value derived in stage 2.

Stage 1:
As is shown in Fig. 77b, the (1-H) value for (0.42/0.14) has to be determined. It is again derived from Appendix Table 12.
(1-H) for (0.42/0.14) = 0.664 .

Stage 2:
The (1-H) value for (0.14/0.14) was already determined in step 2, as

(1-H) for (0.14/0.14) = 0.826 .

Stage 3:
The weighted differences are generated:

$$\Delta(1\text{-}H) = \frac{1.5 \cdot 0.664 - 0.5 \cdot 0.826}{1.5 - 0.5} = 0.583 \ .$$

This value is now used to calculate the covariance between drill hole 2 and the reference block B:

$$\sigma_{B.2} = C_1 \cdot \Delta(1\text{-}H) = 310 \cdot 0.583 = 181 .$$

4th Step: All the necessary values are now available for the matrix system:

$$[K] \cdot \vec{\lambda} = \vec{D} .$$

This can be written as follows:

$$
\begin{bmatrix}
480 & 185 & 185 & 185 & 185 & 1 \\
185 & 480 & 125 & 65 & 125 & 1 \\
185 & 125 & 480 & 125 & 65 & 1 \\
185 & 65 & 125 & 480 & 125 & 1 \\
185 & 125 & 65 & 125 & 480 & 1 \\
1 & 1 & 1 & 1 & 1 & 1
\end{bmatrix}
\begin{bmatrix}
\lambda_1 \\ \lambda_2 \\ \lambda_2 \\ \lambda_2 \\ \lambda_2 \\ \mu
\end{bmatrix}
=
\begin{bmatrix}
256 \\ 181 \\ 181 \\ 181 \\ 181 \\ 1
\end{bmatrix}
\tag{17}
$$

This matrix system can now be solved by a procedure similar to that outlined in Section 13.4.5.2.2 [following the computational procedure from equation system (11) to (13) above], and this yields:

$$480 \cdot \lambda_1 + 740 \cdot \lambda_2 + \mu = 256 , \tag{18}$$

$$185 \cdot \lambda_1 + 795 \cdot \lambda_2 + \mu = 181 ,$$

$$\lambda_1 + 4 \cdot \lambda_2 = 1 .$$

This system of three equations with three unknowns can be either solved by the procedures in Section 13.4.5.2.2, or by the method outlined in Appendix Table 11 for determining eigenvalues. The computational procedures will not be repeated here. The result is then:

$$\lambda_1 = 0.287 ,$$

$$\lambda_2 = 0.178, \text{ and}$$

$$\mu = -13.798 .$$

5th Step: The accumulation value for the block, $\bar{x}_B$, can now be calculated with the weighting factors by applying the same method as outlined for the 7th step in Section 13.4.5.2.2. The accumulation value of drill hole 1 is weighted by λ_1, and the accumulation values for drill holes 2, 3, 4 and 5 are weighted by λ_2 (Fig. 75b):

$$\overline{x}_B = 0.287 \cdot 8.3 + 0.178(5.2 + 17.2 + 10.5 + 7.2)$$

$$\overline{x}_B = 2.38 + 7.14 = 9.52 \text{ g Au/t} \cdot \text{m} \ .$$

13.4.5.4
Summary Remarks on the Calculated Weighting Factors

In summary, Table 42 compares the results obtained for the reference point P by three methods and the result for the reference block B by block kriging. The three methods for determining the results for the reference point P used:

- weighting factors derived directly from the variogram (Sect. 13.4.4),
- weighting factors (kriging factors) derived by kriging without a mean (Sect. 13.4.5.2.2), and
- weighting factors (kriging factors) derived by kriging with a known mean (Chap. 13.4.5.2.3).

Table 42. Comparison of the weighting factors calculated by different methods

Method	Weighting factors directly from variogram	Point kriging method		Block kriging method
		Kriging without mean $\overline{x}$ of deposit	Kriging with mean $\overline{x}$ of deposit	Kriging without mean $\overline{x}$ of deposit
Weighting factor for central drill hole 1	0.30	0.448	0.450	0.287
Weighting factor for bordering drill holes 2, 3, 4 and 5	0.18	0.138	0.127	0.178
Weighting factor for mean $\overline{x}$ of deposit	-	-	0.042	-
Weighted value of reference point P	9.71	9.25	9.36	9.52

It can be seen that the two point kriging methods for determining the value at point P yielded values that are practically identical. However, this depends on the mean $\bar{x}$.

The method of deriving the weighting factors directly from the variogram, in which the covariances are used, yields a value that is about 4% higher than that derived by point kriging. This is usually an acceptable accuracy.

A comparison of the values obtained by block kriging with those obtained from point kriging, however, clearly shows that the central value in the block is given a significantly lower weighting while the outlying drill holes are given a higher weighting. The reason for this is obvious from the computational procedures for determining the covariances between the drill holes and the block (Fig. 76). The more marginal areas of the block are located at almost the same distance from the central drill hole as they are to the outlying holes. The weighting factors must therefore be more balanced as compared to a procedure based on a point in the centre of the block.

The best way to check which method yields the best results are cross validation techniques (see, e.g. Isaaks and Srivastava 1989). Such techniques theoretically could be applied to the point estimates in Table 42. Cross-validation is a technique by which data values are dropped one at a time and the estimates computed without this value. The comparison of two sample series with different sample character as described in Chapter 11.4 (diamond drill core samples vs. percussion drilling samples, see Fig. 43) via point kriging could be considered in practical terms as cross-validation.

If there are numerous samples, cross-validation techniques can be used to see if the method applied or the model fitted to the experimental variogram is acceptable or can be improved. The differences between the true and the calculated values, the so-called residuals, have to be examined. The centre, the skewness, and the spread of the statistical distribution of the residuals should be as close to zero as possible. In the early exploration stages there are, however, rarely enough data to do a meaningful cross-validation computation.

Nevertheless, the question remains how good geostatistical, especially kriging, estimations are in comparison to other methods and to the really achieved grade results in mining. In Table 43 a few case histories from the literature are compiled, indicating the relatively good performance of kriging estimations.
Champigny and Armstrong (1993) conducted a survey about geostatistics used in the estimation of gold deposits; 79% of the companies polled found a good agreement with production figures, 11% variable agreement and 5% no agreement.

In the introduction of the kriging chapter (Sect. 13.4.5.1), it was pointed out that kriging supplies the best linear unbiased estimator with minimal estimation variance. If one looks again at Fig. 48 (Chap. 12), which shows the correlation between the grades of samples with the corresponding mining blocks and the effect of the application of a cut-off grade, kriging has the effect of making the ellipse flatter, as in Fig. 48c; but also with kriging the ellipse still remains and therefore the sectors II and IV of wrongly assigned blocks have not disappeared totally. This means that a bias – although reduced by the application of the kriging method – exists towards a too optimistic average grade estimate. To reduce this bias further or even eliminate it, special kriging methods have to be applied, like disjunctive kriging. Disjunctive kriging is the most demanding kriging procedure, requiring computer facilities in any case. Such a method is applied only in the final evaluation stage, for example for a feasibility study, but not in an early exploration phase.

Table 43. Comparison of mined (true) and estimated grades

Mine/case	True grade (=mined grade)	Estimating by kriging	Other estimates		Reference
Louvem Mine, Quebec, Canada	2.04	2.04 (kriging with block envelope) 2.07/2.24 kriging with geological envelope	a) 2.51 b) 3.04 c) 2.62 (conventional methods with dilution)		Vallée et al. (1977)
Pima Cu mine, Arizona, USA	0 (relative)	- 0.03 (overestimation)	- 0.03 (overest.) (ELIP=modified distance weighting), - 0.05 (overest.) inverse distance squared weighting, - 0.06 (overest.) (polygons)		Knudsen et al. (1978)
Similkameen Cu mine, British Columbia, Canada	100 (relative)	101 (relative)	116 (relative) (polygons), 120 (relative) (modified distance weighting)		Raymond (1979)
Walker Lake/Nevada V sample set (geochem. data)	436.5 ppm	444.5 ppm (ordinary kriging)	488 ppm (polygons)		Isaaks and Srivastava (1989)
Cobar Mine, New South Wales, Australia	6.0-6.5% Cu 3.5-4.0% Cu	6.61% Cu 4.29% Cu (indicator kriging)	11.10% 6.46% (classic estimates on sections)	10.37% 5.84% (3D-Model method)	Carswell and Schofield (1993)
Broken Hill, Cu, -Pb, -Zn, -Ag deposit, Aggeneys, South Africa	100 Cu (relative) 100 Pb (relative) 100 Zn (relative) 100 Ag (relative)	98 Cu 107 Pb 100.5 Zn 104.5 Ag	120 Cu 112 Pb 113 Zn 114 Ag (polygon, overest.)		Nowak (1994)

It must be pointed out clearly that geostatistical results can only be as good as the model fitted to the orebody-relevant geological and structural parameters. Many industry practitioners are aware of case histories with various kriging methods applied (seldom published), which resulted in significant overestimation of the grades that were finally achieved in real life mining.

13.4.5.5
Calculation of the Kriging Variance
The kriging variance can now be calculated from the results of the kriging weighting factors in Section 13.4.5.3. The kriging variance in this case is the block variance for the accumulation value, which was determined by the kriging factors λ_i. The equation for the kriging variance σ_K^2 is:

a) for kriging with a known mean:

$$\sigma_K^2 = \sigma_B^2 - \sum_{i=1}^{n} \lambda_i \cdot \sigma_{B,i} \; ; \tag{19}$$

b) for kriging without a mean:

$$\sigma_K^2 = \sigma_B^2 - \sum_{i=1}^{n} \lambda_i \cdot \sigma_{B,i} - \mu \; , \tag{20}$$

where σ_B^2 is the block variance, λ_i are the kriging weighting factors, $\sigma_{B,i}$ are the individual elements in the vector $\underline{D}$ [see the Eq. system (16) in Sect. 13.4.5.3], and μ is the Lagrange multiplier.

σ_B^2 can be determined by the F-function in Fig. 66:

$$\sigma_B^2 = C_1 [1 - F(h/a, l/a)] \; . \tag{21}$$

The dimensions of the block B ($h = l = 10$ m, see Fig. 77a) must be normalised with respect to the range of the variogram in Fig. 72, specifically $a = 36$ m:

$l/a = h/a = 10/36 = 0.28$.

From the diagram in Fig. 66:

$F(0.28/0.28) = 0.22$.

Since $C_1 = 310$ from the variogram in Fig. 66, then:

$\sigma_B^2 = 310 \, (1 - 0.22) = 241$.

The kriging variance can now be calculated by Eq. (20) above using the values for λ_1, λ_2 and μ that were determined in Section 13.4.5.3:

$$\sigma_K^2 = \sigma_B^2 - \sum_{l=1}^{n} \lambda_i \sigma_{B,i} - \mu \, , \tag{20}$$

$$\sigma_K^2 = \sigma_B^2 - \left(\lambda_1 \cdot \sigma_{B,1} + 4\lambda_2 \cdot \sigma_{B,2} \right) - \mu \, ,$$

$$\sigma_K^2 = 241 - \left(0.287 \cdot 256 + 4 \cdot 0.178 \cdot 181 \right) + 13.8 \, ,$$

$$\sigma_K^2 = 241 - 202 + 13.8 = 52.8 \, .$$

The kriging standard deviation is therefore:

$$\sigma_K = \pm 7.27 \text{ g Au/t} \cdot \text{m} \, .$$

If the error for the block is now to be calculated for a 90% level of confidence, then the standard deviation must be multiplied by the factor 1.65[34], so that:

$$\text{ki}_{90} = \sigma_K \cdot 1.65 \, , \tag{22}$$

$$\text{ki}_{90} = \pm 7.27 \cdot 1.65 = \pm 12.00 \, .$$

The mean value $\bar{x}$ for the kriged block was 9.52 g Au/t . m, and therefore the relative error is:

$$\frac{\text{ki}_{90}}{\bar{x}} = \frac{\sigma_K \cdot 1.65}{\bar{x}} \, , \tag{23}$$

$$\frac{\text{ki}_{90}}{\bar{x}} = \pm \frac{12.00}{9.52} = \pm 1.26, \text{ or } +126 \%, \text{ in practical terms 100 \%.}$$

This is clearly a very high error for a single block, although it is a typical example of irregularities in gold-bearing quartz vein deposits.

---

[34] The factor 1.65 is derived from the area of 90% under the normal distribution corresponding to 90 % confidence limits (see discussion in Chap. 4). The value can be derived from Appendix Table 2. Since the normalized normal distribution is symmetrical about zero and the area is 1, then the x-value for $\emptyset(x) = 0.05$ and 0.95 is obtained from Appendix Table 2, and these are -1.65 and +1.65 respectively. The table for the Student's t-test, Appendix Table 4, can also be used since the Student's t-distribution approaches a normal distribution for a large number of values (cf. Chap. 7.1). For a large number of samples (column 1) and 90 % confidence limits (column 2) a figure of 1.65, is also obtained.

As mentioned in the introductory Section 13.4.5.1, the kriging weighting factors are determined by the kriging method in such a way that the estimation variance is a minimum. The kriging variance σ_k^2 is therefore the smallest possible estimation variance with a linear estimator for estimating the grade of blocks.

13.4.6
The Screen Effect

The screen effect was already introduced in Section 13.4.4, in which the weighting factors were derived directly from the variogram. The drill or sample points closest to the reference point or block act as a screen for the outlying drill or sample points. As a result the weighting factors for the outlying points are so small compared to the closer points that they can be ignored. If the more distant drill holes, and not only the closest holes, had also been taken into consideration for the kriging example in Section 13.4.5, then this effect would have been observed.

The screen effect is a function of the nugget effect, C_0. If there is absolutely no regional interdependence, so that the sill value $C_1 = 0$ (cf. Fig. 57) and the nugget effect C_0 is equal to the total sill value or variance s^2, then each drill hole can be regarded as a random sample. There is no justification in assigning any drill hole with a higher value, even if it lies in the centre of the block. From this, it can be deduced that all the weighting factors must be equal, or $1/n$ where n is the number of drill holes under consideration. The consequence of this is that the screen effect is zero. Conversely, the screen effect is a maximum if the nugget effect $C_0 = 0$. Most deposits have a nugget effect C_0 between 10 and 50% of the total sill value $C_1 + C_0$ (cf. Fig. 57), and the screen effect is therefore so well developed that the closest drill holes normally suffice for calculations in the early phases of an evaluation. In other words, drill holes (or other samples) that are located behind a closer drill hole (or sample) can be ignored.

13.4.7
Calculation of Variance by the de Wijs Variogram

The de Wijs type of variogram lacks the horizontal component, which characterizes the transitive type of variogram (Fig. 57), and thus has a continuously rising gradient (de Wijs, 1951, 1953; Krige, 1981). The lag, h, is plotted logarithmically, and this results in a linear relationship:

$$\gamma_{(h)} = 3\alpha \cdot \ln h + b . \tag{24}$$

3α is therefore the gradient of this line. (3α is also known as the absolute or intrinsic dispersion coefficient.) In the previous chapter about the screen effect, it was shown that only the closest blocks or drill holes need to be taken into consideration in an early evaluation. If these blocks are located within the range a of the spherical model, then the variogram (the curved section within the range a) can usually be satisfactorily plotted as a straight line on semilogarithmic paper.

As a result, variances and standard deviations can be easily calculated using the Eq. (24) above. The blocks or zones of a mineral deposit are plotted as linear equivalents. The linear equivalent L is:

$$L = sl + b + m/2, \text{ where } sl \geq b \geq m,\tag{25}$$

so that sl is the longest dimension of the block or zone of the mineral deposit under consideration and is usually parallel to the strike, b is the intermediate dimension and is usually parallel to the dip, and m is the shortest dimension and is usually the thickness. If a block with volume v and linear equivalent l is compared to a larger block or zone of the mineral deposit with volume V and linear equivalent L, then the following relationship is valid:

$$\sigma^2(v/V) = 3\alpha \ln(L/l),\tag{26}$$

where $\sigma^2(v/V)$ is the variance of the block v within the zone of mineralisation V. The application of the de Wijs variogram is explained by an example.

Example: The drives on the levels and sublevels of a gold mine are sampled at 1 m intervals with channel samples. During the first phase, trial stopes will be increased in height from the current 2.5 m by an additional 4.5 m to a new height of 7 m. Each 5 m comprises one block; 2.50 m can be drilled, blasted and removed in one shift. The stope is manned for two shifts. During mining, the analytical turn-around time for grade-control samples is 24 h, so that each mining advance of 24 h must be regarded as one unit, or 5 m as one block.

The variogram is calculated from the channel samples at 1-m spacing, and is plotted on semilogarithmic paper as in Fig. 78. The interpolation is linear according to the de Wijs type of variogram model.

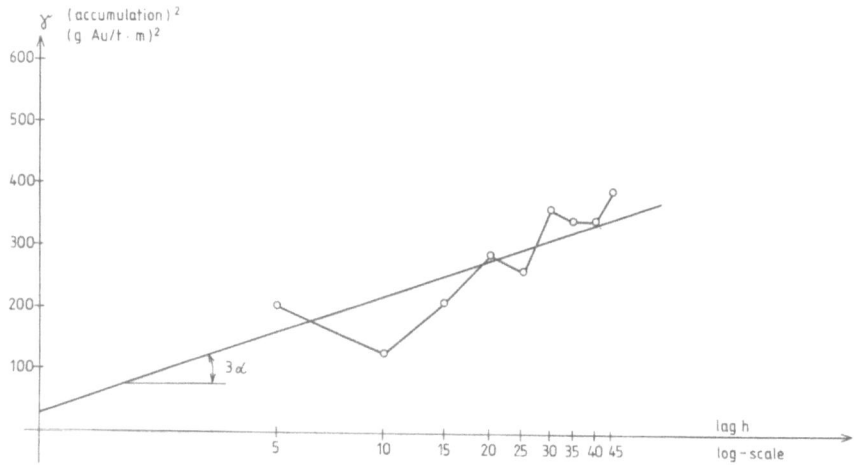

Fig. 78. Variogram of the de Wijs type for a gold deposit

The problem to be addressed is: what variation during 24 h, or for each 5 m advance, can be expected in comparison to the monthly average? An average of 20 days a month are working days, so that the monthly advance in the trial mine is 100 m.

1st Step: The de Wijs type of variogram is mathematically described by Eq. (24) above, and 3α must now be determined:

$$Y_{(h)} = 3\alpha \cdot \ln h + b. \tag{24}$$

Any two points are selected and compared, for example:

$$h = 5: \quad Y_{(5)} = 160 \, ,$$

$$h = 10: \quad Y_{(10)} = 220 \, .$$

Therefore according to Eq. (24):

$$Y_{(5)} = 3\alpha \cdot \ln 5 + b = 3\alpha \cdot 1.61 + b = 160 \, ,$$

$$Y_{(10)} = 3\alpha \cdot \ln 10 + b = 3\alpha \cdot 2.30 + b = 220 \, .$$

These two equations are now subtracted so that b is eliminated:

$$0.69 \cdot 3\alpha = 60$$

$$\text{or} \quad 3\alpha = 86.96 \approx 87 \, .$$

2nd Step: The linear equivalent of the daily mining block is now determined. The mining length is 5 m, the height is 4.5 m and the average thickness is 1.8 m. Therefore, according to Eq. (25) above:

$$l = sl + b + m/2 \, ,$$

$$l = 5 + 4.5 + 1.8/2 = 10.4 \, .$$

The linear equivalent for the volume that is mined in one month is L, where:

$$L = sl + b + m/2 = 100 + 4.5 + 1.8/2 = 105.4 \, .$$

3rd Step: The variance of each individual block within the monthly block is calculated according to the Eq. (26) above:

$$\sigma^2_{(v/V)} = 3\alpha \cdot \ln(L/l) \, , \tag{26}$$

$$\sigma^2_{(v/V)} = 87 \cdot \ln\frac{105.4}{10.4} = 87 \cdot \ln 10.13 = 87 \cdot 2.32 \, ,$$

$$\sigma^2_{(v/V)} = 201.49$$

and the standard deviation

$$\sigma_{(v/V)} = \pm 14.2 \, \text{g Au/t} \cdot \text{m}.$$

4th Step: The standard deviation is now calculated for the grades. The equation for this has already been introduced in Section 13.3.4.4, Eq. (11), as the relative estimation variance:

$$\frac{\sigma^2_G}{(\overline{G})^2} = \frac{\sigma^2_{GT}}{(\overline{GT})^2} + \frac{\sigma^2_T}{\overline{T}^2} - 2r \cdot \frac{\sigma_{GT} \cdot \sigma_T}{\overline{GT} \cdot \overline{T}} \, . \tag{27}$$

where:

$\dfrac{\sigma^2_G}{\overline{G}^2}$ the relative variance of the grades,

$\dfrac{\sigma^2_{GT}}{(\overline{GT})^2}$ the relative variance of the accumulation values, that in this case is:

$$\frac{\sigma^2_{GT}}{\left(\overline{GT}\right)^2} = \frac{\sigma^2(v/V)}{\left(\overline{GT}\right)^2} \, ,$$

GT   the accumulation value,

T   the thickness,

$\dfrac{\sigma^2_T}{\overline{T}^2}$ the relative variance of the thickness, and

r   the correlation coefficient between the thickness T and the accumulation value GT [Appendix Table 8, Eq. (4)].

The mean of the accumulation value $\overline{GT}$ be 12.5 g Au/t $\cdot$ m, so that the relative variance of the accumulation value is now:

$$\frac{\sigma^2_{GT}}{(GT)^2} = \frac{\sigma^2_{(v/V)}}{(GT)^2} = \frac{201.49}{(12.5)^2} = \frac{201.49}{156.25} = 1.29 .$$

The procedure for estimating the relative variance of the grades can be considerably reduced since the minimum mining thickness is 1.4 m, therefore in practice most of the vein has a constant thickness of 1.4 m. For practical purposes, the grade variation of the daily ore production is of interest, and not a theoretical value for the unmined and often very narrow vein.

Because the for practical purposes constant thickness of 1.4 m extends over large areas, the variance of the thickness is small compared to the variance of the grades. In addition, because of the constant thickness, the correlation between thicknesses and grades is low, and r is small. As a result, for a rough estimate, the last two components of Eq. (27) above can be ignored and the equation rewritten as follows:

$$\frac{\sigma^2_G}{G^2} \approx \frac{\sigma^2_{GT}}{(GT)^2} = \frac{\sigma^2_{(v/V)}}{(GT)^2} = 1.29 .$$

The relative standard deviation for the grades is then the square root of the above:

$$\frac{\sigma_G}{G} = \pm 1.14, \text{ or } \pm 114\% .$$

13.4.8
Extrapolation with Geostatistical Parameters

A very difficult task is extrapolation. The need to extrapolate occurs in exploration frequently, e.g. in cases where there are no drill holes to extrapolate beyond a drill hole with an ore intercept. Often then the ore boundary is extrapolated half the distance of the drilling grid. This, of course, is very arbitrary. Sometimes rules of thumb are applied using geostatistical parameters:

a) extrapolation to a distance which is equivalent to half of the range a;

b) by calculating the kriging variances or standard deviations respectively for outside blocks or points and extrapolating to a distance where the kriging standard deviation increases by 50% of the value of the kriging standard deviation of the marginal blocks with drill hole information. Such a case is shown in Fig. 79a.

A final remark concerning extrapolation: for extrapolation, frequently isoline maps are used. One has to be aware that for constructing the isolines mathematical models are used. Whereas in areas with data points the differences of the isolines based on different mathematical models for interpolation are in most cases minor, such differences in the area of extrapolation often become very significant. The same coal deposit on the Philippines which was used in the extrapolation example in Fig. 79a is used for different isoline maps displaying the thickness in Fig. 79b, c

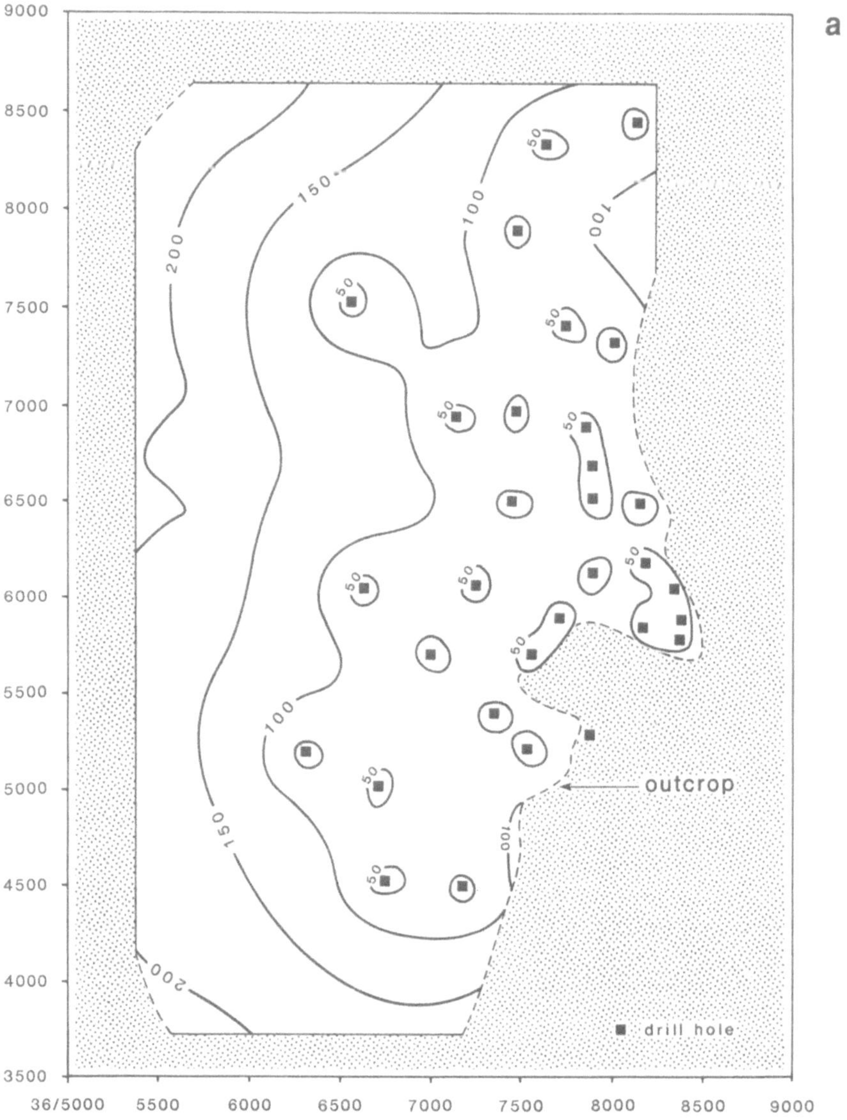

Fig. 79a. Isolines of relative kriging standard deviations of a coal deposit on the Philippines. b-d Isolines of thickness of a coal deposit on the Philippines. b Interpolation: inverse distance weighting. c Interpolation: negative exponentially weighting. d Interpolation: distance weighting least squares

and d. The three examples in Fig. 79b, c, d show the effects of different interpolation models.

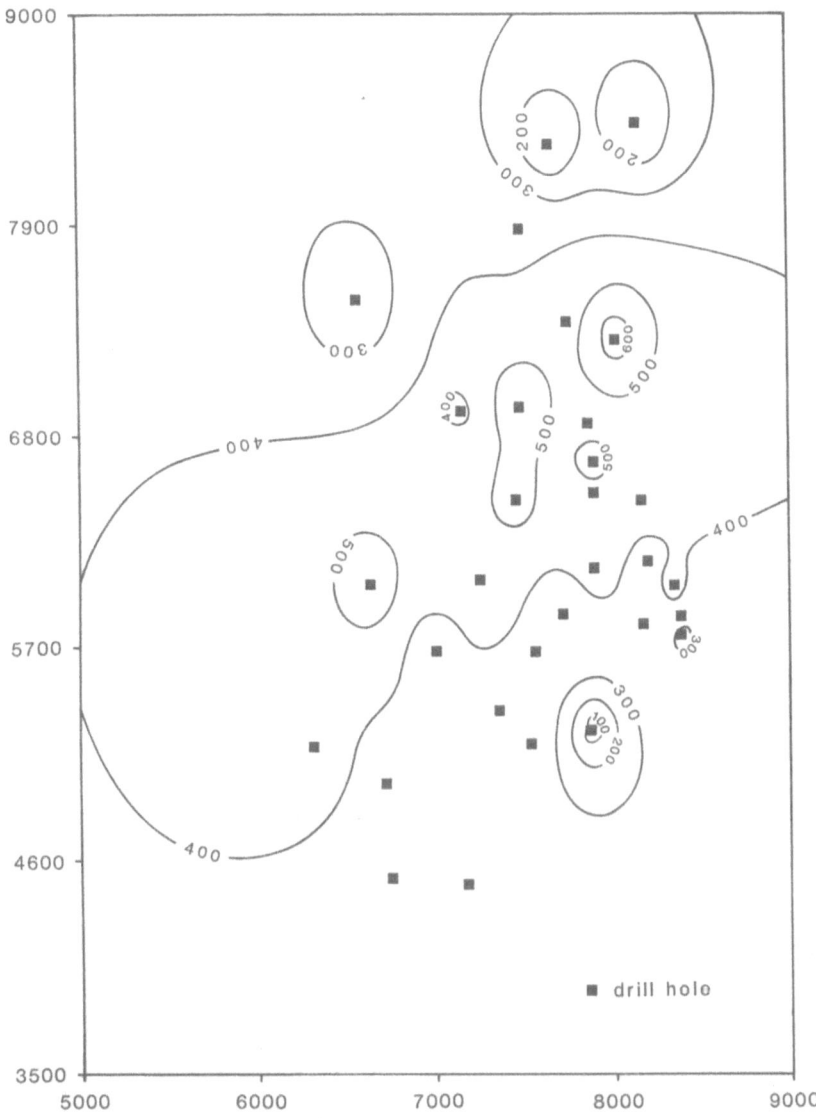

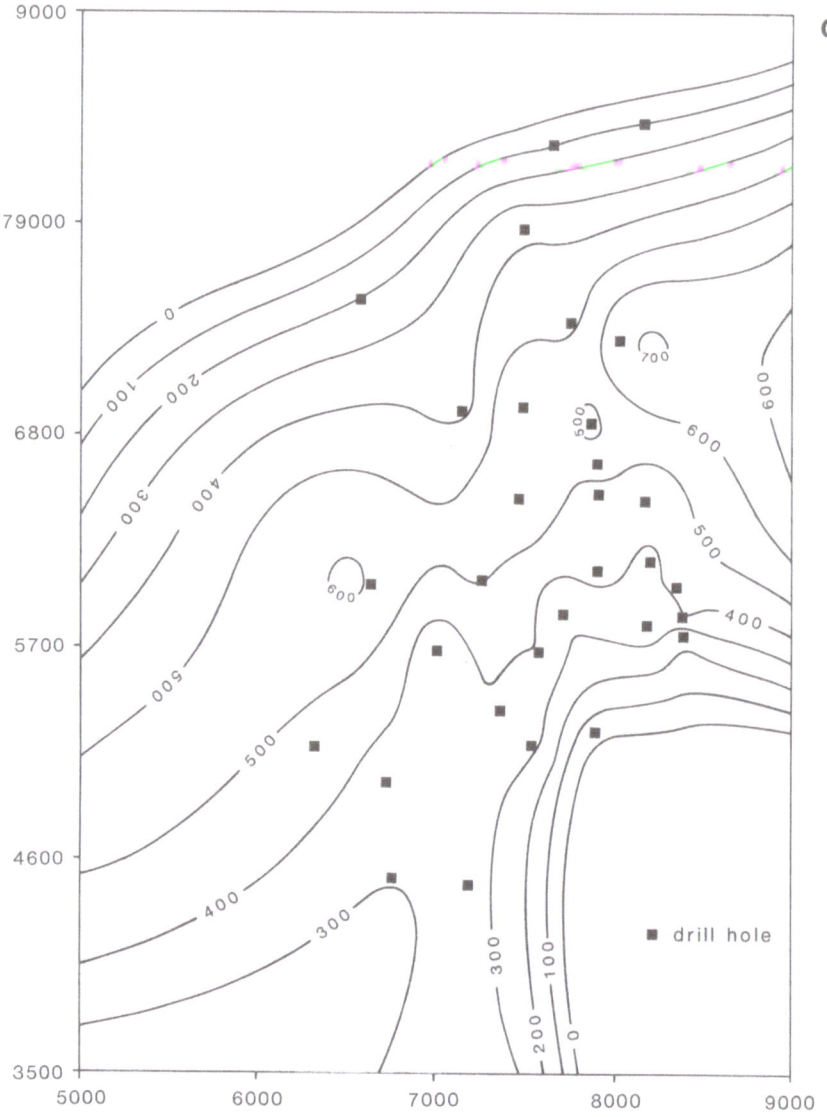

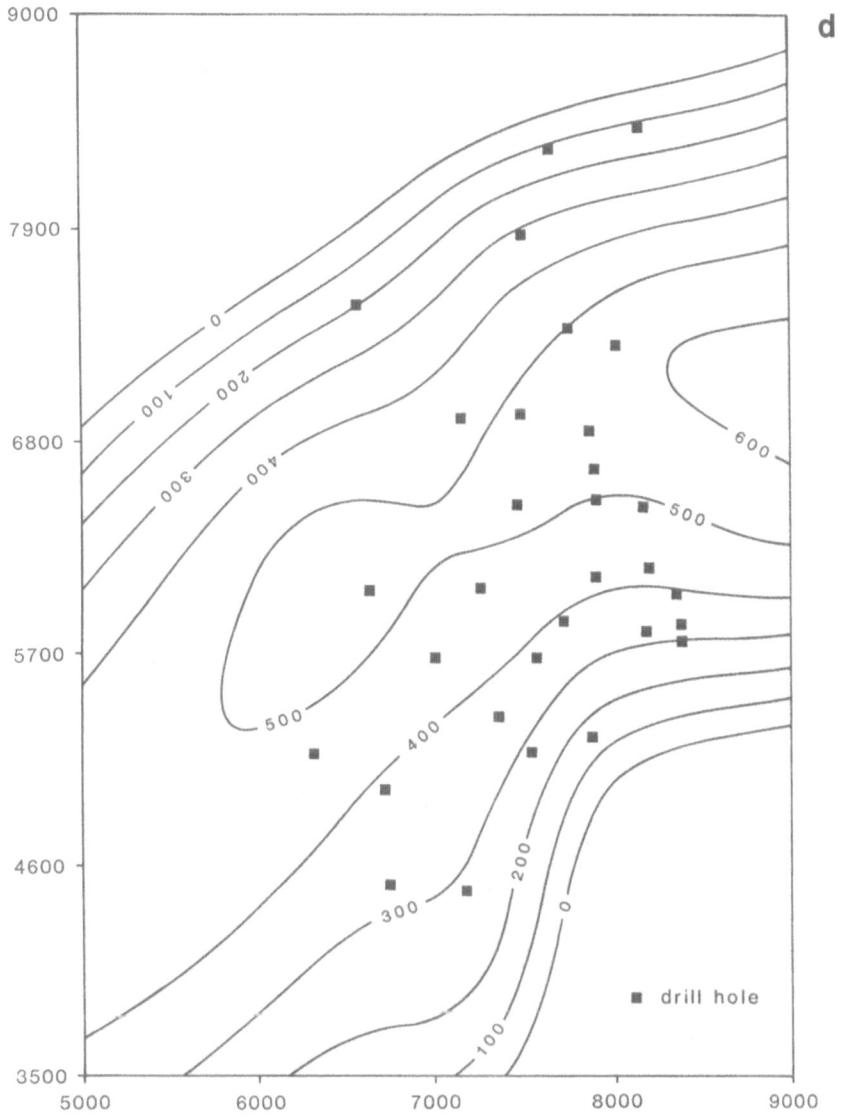

14
Further Statistical Considerations for Evaluating Mineral Deposits

It has already been mentioned that, if possible, other mines in the vicinity of the mineral deposit under examination should be studied. Most mines have established a number of statistical factors derived over years of experience, and these factors should be taken into consideration. It is naturally assumed that "our" mining company would do a better job of mining the deposit as compared to the current owner, but that is rarely the case during the start-up phase. However, with respect to the time value of money in calculating the economic feasibility (Economic Evaluations in Exploration, Chap. 10), this early start-up phase is extremely important for the economic performance of the whole project.

Some helpful values that can be obtained from other mines include, for example, the extraction quota; what proportion of the calculated reserves are actually mineable?

Another very important statistical parameter, in particular for very variable deposits, is the contained metal production relative to 1 m of development.

Hester (1982) published a table entitled "ounces mined for each foot of development advance in some vein gold mines", which was based mainly on figures from Canadian gold mines. These figures are reproduced in ounces/m as Appendix Table 13. From this table it can be seen that about 30 oz/m development is a requirement for an attractive gold mine. Such rules of thumb can also be recalculated for other metals by using the Net Smelter Return (NSR, Economic Evaluations in Exploration, Chap. 7) and the metal price relative to the gold price.

15
Bias in Reserve Calculations

After calculating the reserves, or any other calculations or estimates, it is always recommended to check carefully for any bias that might have become incorporated into the computations. One is often unaware that this might have happened. In the example of a lateritic nickel deposit, as described in Chapter 4.1 (Economic Evaluations in Exploration), a bias to higher nickel grades was caused by incorrect weighting.

The problem of a possible bias caused by core loss was discussed in Chapter 11.5.1.2, and a possible bias caused by incorrectly understanding the problems of comparing bulk samples to drill core samples was mentioned in Chapter 11.3.1.

Example: Structurally controlled uranium mineralization is drilled. There are two types of structure that dip at about 45° in opposite directions, so that on a large scale the mineralization can be described as a stockwork. The interpolation illustrated as a section in Fig. 80 was made from the drill results, and this was the basis for the ore reserve calculations. What bias has not been recognized (probably unknowingly) by the geologist?

Answer: The mineralization is intersected six times in four holes: A 1, A 17, A 18 and A 25. Although the geological model on which the interpretation is based is correct (there are two types of structure that dip at 45° in opposite directions as a corresponding system), in four out of six cases, or 66.7% of all intersections (intersections I to IV in Fig. 80), the geologist has defined two structures on the basis of one mineralized intersection.

Assume that the average spacing between the structures has been correctly interpreted, then an analysis of the section in Fig. 80 indicates that about 8% of the area of a dipping vein structure is comprised of the mineralised shear zone intersections or "nodes". Thus for each intersection of the structure, there is only a probability of $p = 0.08$ (or 8 %) of drilling through such a "node".

A precise analysis could be made by using the binomial distribution, similar to the example in Chapter 10. The two mutually exclusive events are: intersection of a "node" zone with a probability of $p = 0.08$, and missing it with $q = 0.92$. Clearly, even without doing the necessary calculations, it is extremely improbable that the three drill holes A 1, A 17 and A 18 would all intersect "nodes".

There is definitely a bias in the interpretation, and in practice it has resulted in doubling the reserves. In a realistic, but not unduly pessimistic, evaluation, only one structure should ever be derived from one intersection.

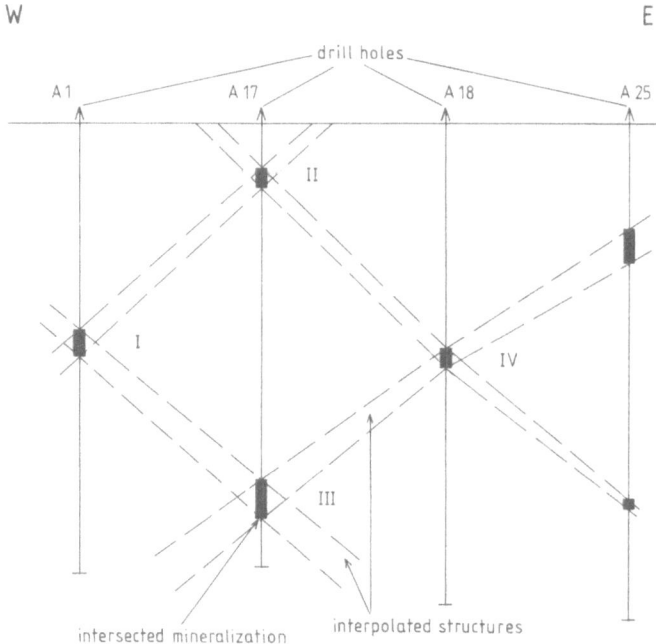

Fig. 80. Section through a structure controlled uranium mineralization

In some cases, the blocks used in a reserve calculation might not be approximately equal in size, which was the basis for the example of geostatistical error calculations in Chapter 13.3 (Fig. 63), but with blocks of different sizes (possibly caused by a very irregular drilling grid; an example being the definition of blocks on sections and plans in Economic Evaluations in Exploration, Chaps. 3.2 and 3.3). In such circumstances, it is always sensible to check if there is a positive correlation between the size and the grades of the block after the evaluation is completed. If there is a positive correlation, and the highest grades do indeed occur in the largest blocks so that disproportionately large tonnages of metal content are associated with only a few drill holes, then this would constitute a strong suspicion of bias.

Example: During the first phase of exploration a Cu-Ni deposit is drilled with 18 holes intersecting the mineralization. A partner is required in order to finance, by means of a disproportionately high farm-in premium (Economic Evaluations in Exploration, Chap. 12), the next detailed, and therefore expensive, exploration phase.

The previous drilling results are listed in Table 44. The Cu grades are lower than the Ni grades, and have been converted into Ni-equivalent values (cf. Economic Evaluations in Exploration, Chap. 5.3). The unweighted grade of the drill intersection is assigned to each respective block.

1st Step: Before beginning the calculation (such as a regression analysis using the equations from Appendix Table 8), the results are plotted graphically (Fig. 81).

2nd Step: An examination of the diagram in Fig. 81 shows there is a relatively compact area with grades between 1.2 and 2.4% Ni-equivalent and tonnages between 140 000 t and 220 000 t. Blocks 2, 10, 11 and 12 lie well outside this area.

It must now be examined if the high tonnages are related to thickness, or only to a greater spacing between the drill holes, which is equivalent to a lower level of information. If the latter is the case, and there is an inverse relationship between grades and amount of information, then for an initial evaluation the tonnages of the blocks should be reduced to the figure corresponding to that at the edge of the compact "tenor" field, which is about 22 000 t. The remainder represents a possible bonus or "upside potential".

If this is not the case, and there is a positive correlation between the grades and thickness, then this correlation is at least related to the geology and is therefore a much more reliable basis for an evaluation. It is even better if the blocks are in contact with each other, representing a high-grade and compact zone of mineralization, rather than being separated from each other as single blocks.

Table 44. Tonnages and grades of blocks of a Cu-Ni deposit

Block no.	Tonnage (t)	Ni equivalent (% Ni)	Metal content (t) (tonnage x Ni equivalent)
1	173 000	1.8	3114
2	348 000	2.8	9744
3	189 000	1.9	3591
4	198 000	1.6	3168
5	154 000	2.1	3234
6	183 000	2,0	3660
7	156 000	1.9	2964
8	188 000	2.4	4512
9	214 000	2.5	5350
10	286 000	3.7 (2.5 corr.)	10 582 (7150 corr.)
11	334 000	2.5	8350
12	275 000	2.2	6050
13	204 000	1.8	3672
14	168 000	1.9	3192
15	138 000	1.5	2070
16	148 000	1.4	2072
17	175 000	1.2	2100
18	184 000	2.4	4416
	3 715 000 t		81 841 (78 409 corr.)

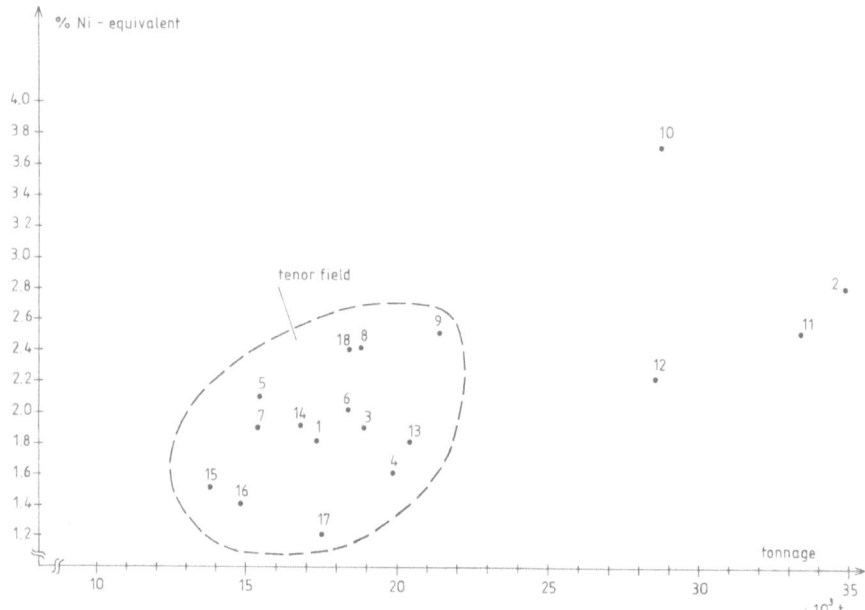

Fig. 81. Results of the first exploration phase of a Cu-Ni deposit; plot of tonnages of blocks against Ni-equivalent grades

If the blocks are isolated from each other, then weighting factors must be calculated by including the values from the adjacent drill holes as shown in Chapter 13.4, and the blocks re-evaluated.

Even if blocks 2, 10, 11 and 12 are in contact with each other, there is still a danger of overestimating those values that are greater than the average for the deposit, as was described in Chapter 11.3.1. This is true not only for the grades, but also for the thicknesses.

Since both the thicknesses and the grades in blocks 2, 10, 11 and 12 are greater than the average for the deposit, the risk of an overestimation is particularly serious during evaluation.

3rd Step: The percentage of the total metal content in the deposit that is represented by the four holes in blocks 2, 10, 11 and 12 can now be calculated. It is 34.700 t Ni, and thus 42% of the total calculated metal content is based on four drill holes, or only 22% of the drill holes. In block 10 alone, 13% of the metal content is based on one single hole, or 6% of the total number of holes. In other words, block 10 has double the weighting as compared to the average.

It is not realistic to be excessively cautious and reduce all these holes to the average. There is often intense competition for farming into mineral projects, and a hypercautious company would rarely be able to enter a project. However, it is clearly essential to be as realistic as possible. As mentioned above, it is possible to apply weighting factors in the recalculation of the block grades and thicknesses, or, better still, the GT accumulation values. Another practical method would be to group the four blocks 2, 10, 11 and 12 in Fig. 81 closer together by reducing the grade of block 10 to the average of blocks 2, 11 and 12, which is 2.5%. This would result in a reduction of the overall average grade from 2.2% to 2.1% Ni-equivalent.

Other reductions can also be made, for example the tonnage of the two blocks 2 and 11 could be reduced towards those of blocks 10 and 12, which are at the 30 000 t limit, and then new tonnages and grades recalculated. Ultimately, there are usually several cases, such as the simple situation and several cases with higher tonnages and grades, and the economic implications of each can then be evaluated by a sensitivity analysis (Economic Evaluations in Exploration, Chap. 10.6).

Part B Exploration Statistics

16
Introduction

As reliable an estimate as possible of the mean or average grade of an ore deposit was the fundamental cornerstone of mineral deposit statistics in Part A. This is, however, much less important for exploration statistics, where the main interest is recognising possible anomalous values, which could indicate mineralization. In other words, a signal must be identified in the geological or analytical noise.

The central part of a frequency distribution of analytical values is of critical importance for the evaluation of mineral deposits, whereas in geochemical exploration the values in the upper "tail" of the frequency distribution are critical to the recognition of a possible signal.

The anomalies must first be discovered, and before that the exploration areas must be selected. A large area is investigated in the first phase of an exploration programme. From this, various smaller areas are selected for reconnaissance and then detailed studies. The persistent concentration of exploration effort into increasingly smaller areas is described by Wellmer and Greinwald (1982).

The geologist uses various criteria, which are usually derived from literature studies and reconnaissance field visits, for the selection of the exploration target area. For example, the occurrence of acid volcanics, the proximity to volcanic centres, the thickness of the cover, the success of previous exploration in the area etc., are all criteria in the search for volcanogenic Cu-Zn deposits. The areas are usually selected intuitively, although some geologists will at least attempt to make a more rational decision by weighting the different factors by a point system.

Nowadays, computer programs have been developed for quantifying geological, geochemical and geophysical parameters by multivariate statistics, and on this basis a parameter of relative potential can be calculated. In a study of the Abitibi Belt in Canada, which is host to the world class Kidd Creek Zn-Cu deposits at Timmins as well as the Noranda deposits, cells with a surface area of 10 km^2 and a known metal potential were correlated with ten geological and geophysical parameters such as, for example, the proportion of acid volcanics or the aeromagnetic value in the centre of the cell. The correlation matrix was then transferred to areas with no known deposits, and a map was generated that showed the relative potential (Agterberg 1975). The Montcalm Cu-Ni deposit was subsequently discovered in an area of maximum potential, although the area selection prior to the discovery was not carried out with multivariate statistics but by the above-mentioned "manual" methods (Wellmer 1983e).

Multivariate statistics are very laborious and require computer programs, and will therefore not be considered here. Interested readers are referred to *Mathematical Geology*, Vol. 15, No. 1 (1983), in which there are numerous articles on and further references to this topic. In addition other sources of information include *Quantitative Analysis of Mineral and Energy Resources* by Chung et al. (1988).

Assuming that an exploration target area has been selected and that geochemical and/or geophysical exploration is about to commence, a search strategy for the discovery of anomalies must be developed before beginning the field work. Stream sediment geochemistry is usually applied for reconnaissance geochemistry, and subsequently, more detailed, geochemistry and geophysics is then based on a systematic exploration grid.

17
Defining an Exploration Grid

Geological Considerations

An area has been selected for further detailed exploration on the basis of preliminary geological investigations and economic criteria (Economic Evaluations in Exploration, Chap. 10). The definition of the grid for the subsequent geochemical and geophysical exploration represents the first step into exploration statistics. Initially, however, before any statistics are taken into consideration, all the geological information must be fully compiled and evaluated.

The geological strike of the target is particularly important. If stratiform mineralization is the target, then the regional strike is clearly of interest, and if structure-controlled mineralization is the target, then the strike of comparable structures in the region is obviously important. An idea of the minimum strike length can be derived by evaluating similar mineralization and estimating the required minimum tonnage.

The spacing between the grid lines should not be greater than this minimum strike length. If the spacing of the grid lines is S and the minimum strike length is L (Fig. 82a) then, presuming that the lines are perpendicular to the strike of the exploration target, the probability of intersecting the exploration target with at least one line is P:

$$P = \frac{L}{S}^{35}.$$ (1)

The next problem is the distance between measuring points on the lines. In geophysical surveys, the expected type of anomaly is of major importance. One or two measurements can represent an interesting magnetic anomaly. For electromagnetic surveys, a longer line is required before any conclusions can be drawn. It is always best to obtain exploration case histories for similar mineral deposits. The Bulletin of the Canadian Institute of Mining and Metallurgy, the Proceedings and the Bulletin of the Australasian Institute of Mining and Metallurgy, or the Transactions of the Institute of Mining and Metallurgy in London, as well as the international journals Geoexploration and the Journal of

---

[35] Mathematically, P can be greater than 1 in this, and also for some of the equations in Section 17.2.1. However, the maximum value for the probability P is 1, which represents certainty. A calculated value of, for example, P = 2 means that two lines intersect the target body. Slichter (1960) uses the term "completeness of search" for this situation.

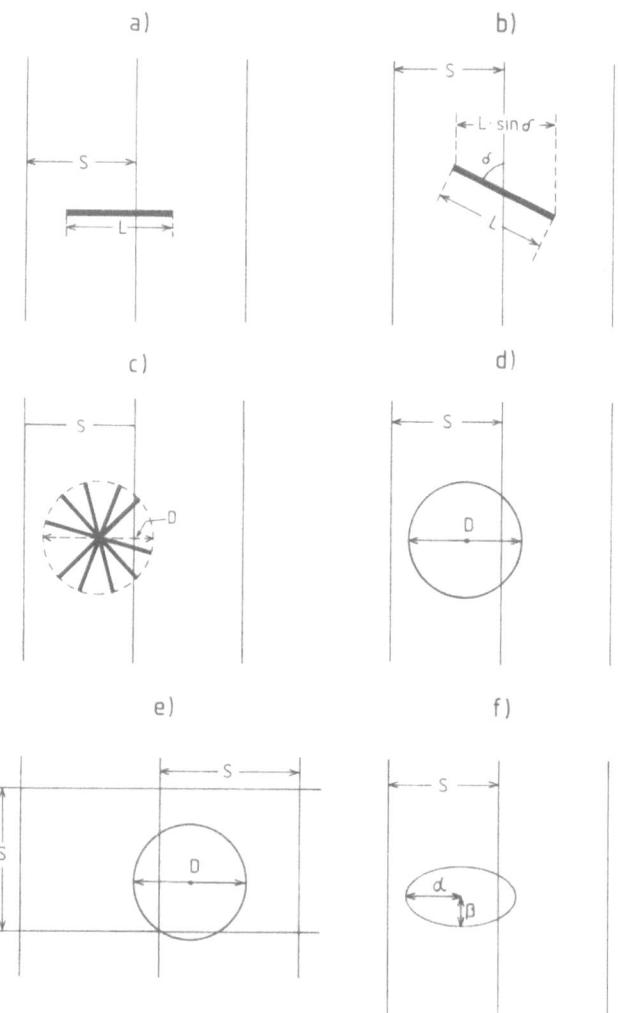

Fig. 82a-f. Targets with different geometries in a grid with lines at a spacing S

Geochemical Exploration are good sources of exploration case histories for metalliferous deposits.

A survey point spacing of between 10 to 15 m has become established for geophysical surveys, although a 5-m spacing is used occasionally for detailed magnetic surveys.

In geochemical surveys it is important to determine the expected outcrop width (apparent thickness) as well as whether the target mineralization is expected to be marked by a geochemical halo or not. This is not always the case as is illustrated, for example, by a geochemical profile perpendicular across the steeply dipping

Meggen Zn-Pb-pyrite deposit that yielded a geochemical anomaly only 15 to 20 m wide (Friedrich et al. 1972).

The sample spacing should be planned so that at least one value is located within the expected anomalous zone. If anomalous values do occur, then the grid can be tightened for subsequent work. The sample spacing along geochemical soil survey lines is usually between 10 and 25 m.

17.2
Statistical Considerations

Some styles of mineralization have no clearly defined preferred orientation, for example porphyry copper deposits. Furthermore, in areas of glacial overburden the strike can be estimated only very approximately from the available regional geological data or from geophysical maps such as aeromagnetic maps. A sudden change in the strike, which is a particularly common observation near mineral deposits, can cause further complications. Some types of ore deposit occur preferentially in the noses of folds, where the economic concentrations of the mineralization accumulate.

Mineralization is observed to plunge in many deposits, and this represents a further factor for consideration. The plunge can result in a significant reduction in the strike length of the deposit at the outcrop, which is fundamentally important for a geochemical anomaly. Several simple formulae are provided for the completeness of the search, P, whereby a target body is intersected by a survey line or is covered by a survey grid (Agocs, 1955; Slichter, 1955; Sinclair, 1975). The spacing between the survey lines is always S (Fig. 82). It is assumed in Section 17.2.1 that measurements are made continuously along the lines, such as is the case for airborne surveys. On the other hand, in Section 17.2.2 it is assumed that the measurements are made only at specific intervals.

17.2.1
Spacing Between Survey Lines

The spacing between the survey lines is denoted by S (Fig. 82).

1. Practically linear targets, with an outcrop length L.
    a) Grid lines perpendicular to the strike (Fig. 82a):

$$P = \frac{L}{S} \text{ (see equation (1) above)}[36] \tag{1}$$

---

[36] See also the footnote 35

b) Grid lines form an angle δ with the strike direction (Fig. 82b). The critical length is then the projection of the strike perpendicular to the grid, which is $L \cdot \sin \delta$, and therefore P is:

$$P = \frac{L \cdot \sin \delta}{S} . \tag{2}$$

For various angles δ the factor P is plotted in Appendix Table 14 A. In an area where the strike can be determined within specific limits, a minimum and maximum probability can then be calculated with this formula.

c) Randomly oriented targets (Fig, 82c):

   i) line spacing S is greater than the strike length L:

   ii) $P = \dfrac{2L}{\pi \cdot S}$;[37] $\tag{3}$

   iii) line spacing S is less than the strike length L:

   iv) $P = \dfrac{2L}{\pi \cdot S}(1 - \sin \Theta_0) + \dfrac{2 \Theta_0}{\pi}$, $\tag{4}$

where $\cos \theta_0 = S/L$.

If $S/L = 1$, then Eq. (4) above is reduced to Eq. (3), or further to $P = \dfrac{2}{\pi}$.

For a randomly oriented target with different S/L or L/S ratios (S>L and S<L), the factor P is plotted in Appendix Table 14 B and C.

2. Rectangular Targets

If the target strikes perpendicular to the orientation of the survey lines, then the width of the target B is of no significance, and Eq. (1) above is valid.
If the target is randomly oriented, then:

$$P = \frac{2(L + B)}{\pi \cdot S} . \tag{5}$$

---

[37] This is also known as Buffon´s needle problem. Interestingly, the French mathematician P.S. Laplace used it to determine the number π (Beckmann 1971).

3. Circular targets (Fig 82d)
   c) Grid with survey lines in only one orientation.
   If the diameter is D, then

$$P = \frac{D}{S} \, . \tag{6}$$

The orientation of the grid is obviously of no significance for circular targets (such as stockwork mineralization).

d) Grids with survey lines perpendicular to each other (Fig. 82e).
If the target is a circular body then, since it has no preferred orientation, two independent grids with survey lines that are perpendicular to each other might be used (Fig. 82e). The probability of at least one line cross-cutting the target is then:

$$P_1 = 2\frac{D}{S} - \frac{D^2}{S^2} \quad \text{if } D \leq S \, . \tag{7}$$

The probability of cross-cutting the target with two lines is then:

$$P_2 = \frac{D^2}{S^2} \quad \text{if } D \leq S \, . \tag{8}$$

The factor P as a function of the ratio D/S is plotted in Appendix Table 14 D for both cases.

4. The general formula for targets with any natural shape (for survey lines in only one orientation) is:

$$P = \frac{A}{\pi \cdot S} \, , \tag{9}$$

where A is the target area.

5. Most geochemical anomalies can be approximated to ellipses (Slichter 1960). The circumference of an ellipse is approximated by:

$$A \approx \pi(\alpha + \beta) \, , \,^{38} \tag{10a}$$

where α and β are the lengths of the half-axes of the ellipse (Fig. 82f).

---

[38] The exact formular for the circumference A is a geometric sequence

$$A = \pi(\alpha + \beta)\left(1 + \frac{\lambda^2}{4} + \frac{\lambda^4}{64} \ldots..\right) \quad \text{with } \lambda = \frac{\alpha - \beta}{\alpha + \beta} \, . \text{ If the ratio } \frac{\alpha}{\beta} = 4 \, A \text{ is already 20\% higher than}$$

approximated in Eq. (10a) above, i.e. P in Eq. (10b) is 20 % higher. The more elongated this elliptical target is, the higher the correction factor due to the geometric sequence. For a ratio $\alpha:\beta$ of 25 the correction would be 30%.

a) If the elliptical target is randomly oriented, then:

$$P \approx \frac{\pi(\alpha+\beta)}{\pi \cdot S} = \frac{\alpha+\beta}{S} \quad . \tag{10b}$$

b) If the long axis of the ellipse is perpendicular to the grid orientation, then the following equation is similar to Eq. (1) (Fig. 82f):

$$P = \frac{2\alpha}{S} \quad . \tag{11}$$

17.2.2
Spacing Between Lines and Between Survey Points on the Lines

The spacing between the survey lines is denoted by S, and the spacing between the survey points on the lines is denoted by T.

1. Circular Targets
   Assuming that the survey measurements on the lines are not made continuously (or quasi continuously), but only at specific points on the grid, then the factor P = 1 for the "completeness of the search" will never be attained, even if the line spacing S is equal to the diameter D (Fig. 83a).

Fig.83a-e. Circular and elliptical targets in a grid with lines at a spacing S and the spacing of survey points on the lines T

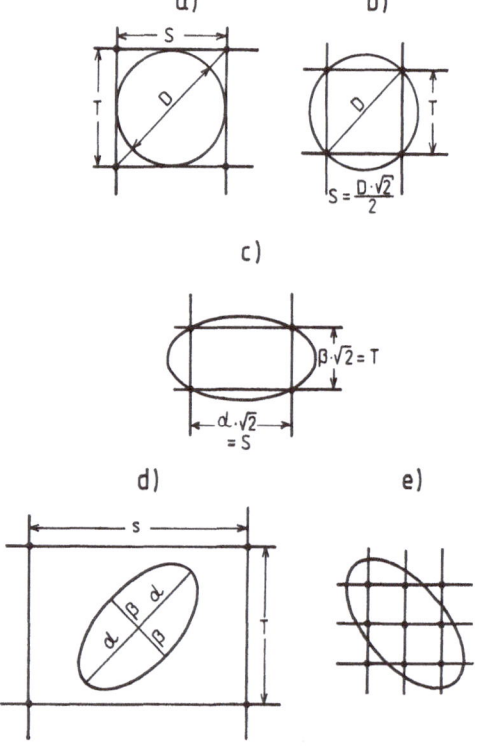

The "completeness of the search" P=1 is attained only if $S = \dfrac{D \cdot \sqrt{2}}{2}$ (Fig. 83b).

It is obvious that an isometric grid would be ideal for a circular target.

2. Elliptical Targets

b) The optimal orientation of the grid for an elliptical target is illustrated in Fig. 83c, where the grid is oriented parallel to the half-axes and the spacing between the lines S and the survey points T is:

$$S = \alpha \cdot \sqrt{2} , \tag{12}$$

$$T = \beta \cdot \sqrt{2} .$$

c) If a target of any shape can be contained within one cell of the grid, then the probability of one survey point being located on the target is:

$$P = \frac{F_Z}{F_R} , \tag{13}$$

where F_Z is the area of the target and F_R is the area of the cells between the grid lines, then:

$$F_R = S \cdot T \text{ (see above)} .$$

Since the area F_e of the ellipse is:

$$F_e = \pi \cdot \alpha \cdot \beta ,$$

then Eq. (11) for this situation (elliptical target contained within a grid cell, Fig. 83d) is written as:

$$P = \frac{\pi \cdot \alpha \cdot \beta}{S \cdot T} . \tag{14}$$

d) If a grid cell can be contained completely within the elliptical target, then the probability is clearly $p = 1$, and it is certain that the target body would be discovered by the exploration grid (Fig. 83e).

e) The calculation of the "completeness of search" is very tedious for those situations that are somewhere between the end-members depicted in Fig. 83d and 83e. The factors for the "completeness of search" itself are a function of the size of the ellipse, the size of the grid cells, the ratio of the half-axes of the ellipse, and the orientation of the ellipse. Savinskii (1965) and de Geoffroy and Wignall (1985) have calculated tables for this purpose.

These formulae will be briefly explained by an example.

Example: Carbonate-hosted Pb-Zn deposits are the target of an exploration program. The bedding dips at 45°, so that the regional trend of the anomaly can be expected to be parallel to the strike of the potentially mineralized carbonate horizon. The geochemical sampling grid is planned accordingly. An economic estimate based on similar deposits in the district suggests that the minimum strike length will be 250 m. The evaluation of geochemical case histories shows that the maximum expected width of the anomaly is about 40 m. What is the shape of the ideal grid?

Answer: The anomaly is considered to be elliptical. With respect to this ellipse, the minimum strike length of 250 m = 2α and the width of the anomaly of 30 m = 2β (cf. Fig. 83d). Therefore, according to Eq. (12) above and Fig. 83d, the ideal spacing between the lines is:

$$S = \alpha \cdot \sqrt{2} = \frac{250}{2} \cdot \sqrt{2} = 177 \text{ m},$$

and the ideal spacing between the survey points is:

$$T = \beta \cdot \sqrt{2} = \frac{30}{2} \cdot \sqrt{2} = 21 \text{ m}.$$

For practical reasons, the line spacing is selected to be 170 m, or even better 150 m, and the spacing between the points to be 20 m.

18
Determining Anomalies from Geochemical Exploration Data

In mineral exploration (soil or rock analyses) the frequency distributions of geochemical analytical data are usually so skewed that they are described statistically by the lognormal distribution (Chap. 9). As already mentioned, the high and therefore potentially anomalous values are usually the objectives of exploration geochemistry. These values contrast clearly compared to the remainder, and can therefore provide the first indications for the presence of economic mineralization.

The statistical problem is to identify the anomalous values and partition them from the normal values. The degree of accuracy of these calculations need not be so great as that for ore deposit statistics. Furthermore, in geochemical exploration, greater tolerance is acceptable in the analytical accuracy, especially if this results in a reduction in analytical costs. Although a 20% underestimate of the grades of a mineral deposit is often economically fatal, a similar fluctuation of the results in geochemical exploration has scarcely any effect on the success of the program.

18.1
Preparation of the Data Set

The number of individual values within a geochemical exploration data set is so great that, for practical reasons, the data are nearly always condensed into classes before being manually assessed. The appropriate procedures were described in Chapter 2.2.

Although about 10 classes are usually selected for mineral deposit data, between 12 to 20 classes are required for geochemical data. Experience indicates that the number of classes should be about $\sqrt{n}$, where n is the number of individual values.

Example: There are n = 210 geochemical values. The recommended number of classes is therefore:

$$\sqrt{210} = 14.5,$$

or either 14 or 15 classes.

Although the data are condensed into classes for practical reasons, it must be made clear that it is basically paradoxical to do this for geochemical data. If the evaluation is undertaken by computer, then the data set is not usually subdivided into classes, since this results in the homogenization of all the data within each class interval, and this problem has already been mentioned in Chapter 2.2. Section 18.2.2 will emphasize that the subdivision into class intervals can result in a loss of the detailed information that might otherwise be derived from the data set as frequency or cumulative frequency distribution curves.

The statistical parameters are usually determined graphically on probability paper. Chapter 9.2.2 (see Appendices B1 and B2 at the back of this book) discussed the use of logarithmic probability paper for the treatment of mineral deposit data. Appendices C1 and C2 at the back of this book contain examples of the logarithmic probability grid, which is rather more suitable for treating geochemical data. (Again, as mentioned in Chap. 5.1 for the probability grid of the normal distribution and in Chap. 9.22 for the lognormal distribution, in some English speaking countries the x- and y- axis are often reversed. For readers already more used to the reversed version of the probability grid, such a version is enclosed as enclosure C2 at the back of this book.)

There are two methods, each of which is favoured by certain geochemists, for plotting the cumulative frequency of geochemical data. As shown in Chapters 5.1.3 and 9.2.2, if the frequency accumulation commences with the lowest values, then the highest values for a cumulative frequency of 100% do not appear on the probability paper. However, since the higher values are of interest in geochemical exploration, most geochemists commence their accumulation with the highest values so that they are fully represented on the probability paper. This results in a cumulative frequency curve that runs from the lower right to upper left on the paper (e.g. Fig. 87), which is the opposite trend to that shown in, for example, Fig. 14 or Fig. 24. Some geochemists, however, accumulate the values by starting with the lowest values, as was the case in these examples, because the resulting frequency curve is more suitable for discriminating between several populations (e.g. James 1986).

Neither of the above methods is "correct" or "wrong", they are both aids to making a decision based on geochemical and geological information. The method that is selected should be the most suitable for identifying the fundamental features of the distribution. The following sections are based on this principle.

18.2
Defining Anomalous Values and Populations

18.2.1
Low Number of Anomalous Values

In the following examples it is always assumed that the high values are not statistically insignificant outliers (see example in Chap. 11.2.2), or that the population is a mixed distribution. Instead, the following methods can generate the first indications for possible anomalies. A suspected anomaly can be confirmed if, for example, anomalous values are spatially associated and occur as clusters.

Although the following discussions are based on classical statistical techniques, which treat geochemical values as statistically independent events, in reality the variables are often spatially interrelated (cf. discussion in Chap. 13.1). The explorationist realizes this, and makes use of this knowledge. For example, five isolated and probably anomalous values are considered to be less important as compared to five similar, or even slightly lower, values that cluster together into a group.

18.2.1.1
Evaluation Using the Median and Standard Deviation

18.2.1.1.1
Fundamentals

If only a few high values can be expected from the target style of mineralisation, or from a particular type of sampling (e.g. wide-spaced reconnaissance geochemistry), then the potentially anomalous values are determined by using the statistical parameters of the total population. This necessitates the definition of the threshold value, above which every value is considered to be anomalous.

The problems related to a large number of background values and only a few possibly anomalous values occur typically for stream sediment sampling, in which individual high values could indicate a mineralized area. A relatively thin, stratiform mineralization without any halo, or with only a very restricted one, would be a further example of this problem. (The Pb, Zn and Cu anomalies at the Meggen deposit represent a good example; Friedrich et al. 1972, p. 212.)

The so-called background is used to identify the threshold value. According to Hawkes and Webb (1962) and Lepeltier (1969), the background is defined as the median m (Chap. 3.4), which is the value that divides a sample size or population into two equal halves. On the probability grids in Appendices A1 (and A2) and C1 (and C2) at the back of this book, the median can be directly derived from the 50% line, or it is easily determined from a table of values.

It has already been demonstrated in Chapters 3.4 and 9.2.1 that the median is the same as the arithmetic mean $\bar{x}$ if the distribution is symmetrical, such as the normal distribution (Fig. 11), and it is the same as the geometric mean g if the distribution is lognormal (Fig. 22). The median can also be easily determined for other distributions. Since the definition of the median is independent of the type of distribution, it can also be calculated if some of the samples lie beneath the detection limit. Furthermore, the median is not affected by isolated high values. It is therefore more suitable for the definition of the background as compared, for example, to the arithmetic mean, which is very strongly influenced by occasional high values, as was shown in Chapter 9.3.2.

The background is normally understood to be not one single value, but a range of values, and therefore it should be strictly referred to as the background values. The median is then the mean of this range of values (Fig. 84). The range of background values can also be compared to geological noise (cf. Chap. 19.1).

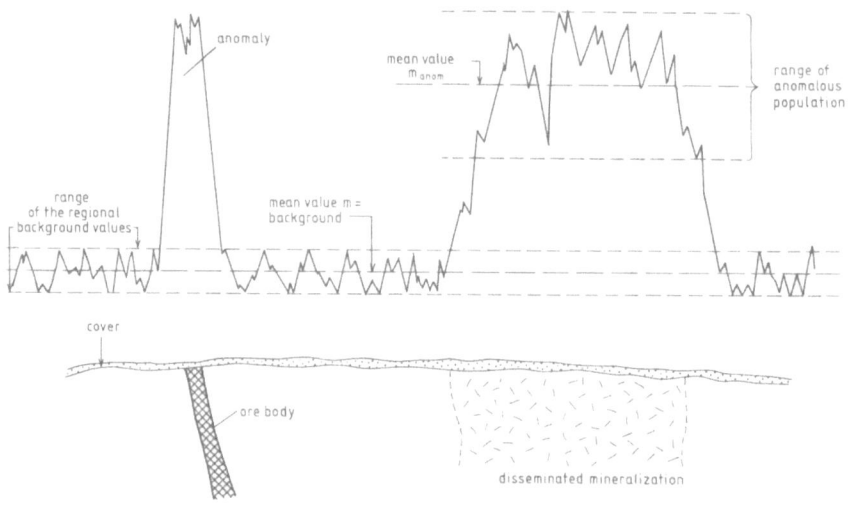

Fig. 84. Sketch for the definition of background and anomaly

The standard deviation s of the sample size, or s of the population, is another commonly used statistical dimension. In Chapter 5.1.3 for the normal distribution and Chapter 9.2.3.1 for the lognormal distribution it has already been demonstrated how the values s and s can be derived respectively from between the lines 16 and 50%, and between the lines 50 and 84% on the probability grid.

During the assessment of geochemically anomalous values, it is always suspected that the distributions no longer have normal or lognormal distributions at the upper end of the curve. It is therefore recommended always to derive the standard deviation s from the lower intervals between the 16 and 50% lines.

The threshold is then defined as

$$th = m + 2\,s \tag{1}$$

(or $th = m + 2\,s$, if the whole population is under consideration).

For a normal or a lognormal distribution this threshold is therefore at 97.7% of all the values, as is illustrated in Fig. 85. Only the upper "tail" of the distribution, or half of the difference of 100 - 95.5%, or 2.25%, is of interest.

On this basis, Hawkes and Webb (1962) suggested that:

1. all values greater than m + 2s are considered to be *possibly anomalous*, and
2. all values greater than m + 3s are considered to be *probably anomalous*.

18.2.1.1.2
Distribution Tests

In the previous section the normal and lognormal distributions were emphasized. The probability grid for the normal distribution was described in Chapter 5.1.2 and is reproduced as Appendix A1 (and A2), and that for the lognormal distribution was described in Chapter 9.2.2 and is reproduced as Appendix C1 (and C2) at the back of this book. Most geochemists use the probability grid as a method of deciding which is the most suitable distribution by determining if the Hazen line for the lower values is approximately straight (the values from the upper tail are not used since they are expected to be anomalous).

Some geochemists (e.g. Cameron 1983) use a coefficient of skewness, which was mentioned in Chapter 8.2, and a coefficient of kurtosis as a test. According to Eq. (1) in Chapter 8.2, the coefficient of skewness Sf is:

$$Sf = \frac{\sum\limits_{i=1}^{n} (x_i - \bar{x})^3}{n \ s^3}, \tag{2}$$

or for data in classes:

$$Sf = \frac{\sum\limits_{i=1}^{n} \left[f_i (x_i - \bar{x})^3 \right]}{\sum\limits_{i=1}^{n} f_i \ s^3}.$$

The kurtosis Kur is:

$$Kur = \frac{\sum\limits_{i=1}^{n} (x_i - \bar{x})^4}{n \ s^4} - 3, \tag{3a}$$

or for data is classes:

$$Kur = \frac{\sum\limits_{i=1}^{n} \left[f_i (x_i - \bar{x})^4 \right]}{\sum\limits_{i=1}^{n} f_i \ s^4} - 3. \tag{3b}$$

The individual values are x_i, and $\bar{x}$ is the arithmetic mean $\bar{x} = \frac{1}{n} \cdot \sum_{i=1}^{n} x_i$, or if the data is subdivided into classes, then x_i is the middle of the class interval and f_i is the frequency of the class, s is the standard deviation and n is the total number of values. If the coefficient of skewness Sf = 0, then the distribution is symmetrical. If the kurtosis Kur = 0, then the distribution of the data is approximately normal.

Example: As part of an exploration programme for volcanogenic Cu-Zn deposits, electromagnetic airborne anomalies were followed up with ground geophysical surveys. In addition, soil geochemical surveys were also carried out in the areas where geophysical anomalies had been identified. The soil geochemical survey over an electromagnetic anomaly with interesting geophysical characteristics yielded the following Zn values, which have already been subdivided into class intervals:

The coefficients of skewness, Sf, and kurtosis, Kur, are calculated by the Eqs. (2) and (3b) above, respectively. The computational procedure is not shown here in detail (see Chap. 3.2.2.2 for the calculation of data subdivided into classes and Chap. 8.2 for the calculation of the skewness).

The results are then:

Sf = +0.19, which indicates a slight skewness to the right (Chap. 8.2), and Kur = 0.57.

The kurtosis is small enough to enable the use of the normal distribution.

Table 45. Zn values of a geochemical soil sample survey

Zn grade (ppm)	Class mean	Relative frequency f_i (%)	Relative cumulative frequency
10- 20	15	2.2	2.2
21- 30	25	5.1	7.3
31- 40	35	9.0	16.3
41- 50	45	12.6	28.9
51- 60	55	19.6	48.5
61- 70	65	17.2	65.7
71- 80	75	16.6	82.3
81- 90	85	9.8	92.1
91-100	95	4.9	97.0
101-110	105	2.5	99.5
111-120	115	0.3	99.8
•			
•			
•			
161-170	165	0.2	100.0

The weakness of this method is obvious. If the highest class in which the anomalous values are suspected was not 161 to 170 ppm Zn, but 195 to 200 ppm Zn, then the coefficient of skewness Sf would be 0.44, which is still an acceptable value, but the coefficient of kurtosis Kur would be 2.33, which is too high.

The graphical technique using the probability grids, as was described above, is therefore the more suitable method.

18.2.1.1.3
Examples of Determining the Threshold Values

The determination of the threshold is illustrated by two examples, one mathematical and the other graphical.

1st example: The Zn geochemical values in Table 45 are used again for this example.

The values in the 161 to 170 ppm Zn class interval are suspected to be anomalous. What is the threshold?

1st Step: Since the higher values are suspected of being anomalous, the case for the normal population is tested from the lowest values upwards and the cumulative frequency percent values are calculated (column 4 in Table 45).

2nd Step: The median, m, is determined. It lies at the 50% relative cumulative frequency, which is within the 61-70 ppm Zn class interval although closer to 61 than to 70 ppm. The precise value for m is determined by interpolation:

The difference between 50% and the next lowest cumulative frequency (51-60 ppm Zn) of 48.5% is 1.5%, and the difference between 65.7% and 48.5% is 17.2%. The interval from 61 to 70 is therefore subdivided by the ratio:

$$\frac{x}{10} = \frac{1.5}{17.2}, \text{ and thus } x = 0.87.$$

Therefore $m = 61 + 0.87 = 61.87$, or 62 ppm.

3rd Step: The standard deviation is calculated from the equation for the variance of values in class intervals [Chap. 3.2.2.2, Eq. (1)]:

$$s^2 = \frac{\sum_{i=1}^{n} f_i x_i^2 - (\sum_{i=1}^{n} f_i)\,\bar{x}^2}{\sum_{i=1}^{n} f_i - 1},$$

where x_i are the middle values of the class intervals, f_i are the frequencies of each of the classes.

Table 46. Hg values of a geochemical rock sample survey

Hg grade (ppb)	Relative frequency f_i (%)	Relative cumulative frequency
-10	5.1	5.1
11- 15	5.7	10.8
16- 25	6.5	17.3
26- 35	7.9	25.2
36- 50	5.8	31.0
51- 75	12.2	43.2
76- 110	9.6	52.8
111- 170	8.2	61.0
171- 250	13.9	74.9
251- 380	6.4	81.3
381- 550	5.2	86.5
551- 850	5.2	91.7
851-1300	3.6	95.3
1301-2000	2.8	98.1
2001-3000	0.7	98.8
3001-4500	0.4	99.2
4501-7000	0.8	100.00

Using the data in Table 45, this equation yields a variance of:

$$s^2 = 444.4 \, ,$$

and therefore the standard deviation is

$$s = \sqrt{444.4} \doteq 21.1 \, .$$

The Sheppard correction is not applied to this calculation since this procedure would be unnecessarily accurate.

4th Step: The threshold is derived from:

$$th = m + 2s = 62 + 2 \cdot 21.1 = 104.2, \text{ or } 104 \text{ ppm Zn,}$$

and the values in the interval 161 to 170 ppm are probably anomalous since they are greater than m + 3s:

$$thprob = m + 3s = 62 + 3 \cdot 21.1 = 125.3, \text{ or } 125 \text{ ppm Zn.}$$

2nd Example: Rock geochemical sampling was undertaken during exploration for epithermal gold deposits. The Hg values in ppb are listed in the following Table 46. They have already been subdivided into approximately equal, logarithmic class intervals.
Which values should be considered as anomalous?

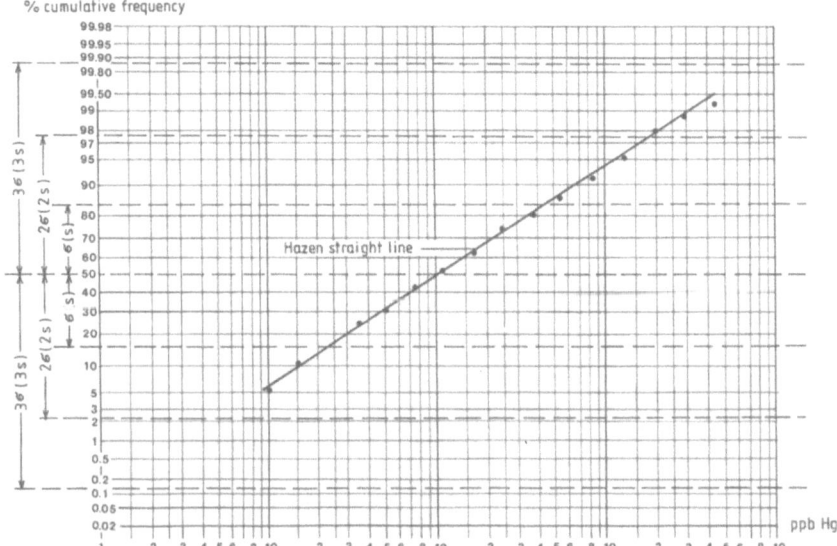

Fig. 85. Cumulative frequency distribution of Hg geochemical data of Table 46

The problem is solved graphically on the logarithmic probability grid (Appendix C1). Since a few high, and possibly anomalous, values will be distinguished from the remainder of the normal population, the cumulative frequency is therefore calculated from the lower values upwards and the relative cumulative frequency is also calculated from the base upwards (column 3 in Table 46). The cumulative frequencies are then plotted on the logarithmic probability grid (Fig. 85). Since the accumulation is from the lower values upwards, the cumulative frequency is plotted against the upper limit of the class interval because the graph indicates the percent of the population or sample size that is less than or equal to the upper limit of the class interval (cf. Chap. 5.1.3).

The threshold can be immediately derived from the m + 2 s line, or the 97.7% line, where it can be seen to be 2000 ppb Hg, or 2 ppm Hg. All values greater than this are considered to be possibly anomalous. The limit for the probably anomalous samples is read from the 99.87% line at 10 000 ppb, or 10 ppm Hg. There are no values that satisfy this condition.

If the lognormal distribution is being used, then the threshold th should preferably always be determined on the probability grid in order to avoid errors. In Eq. (1) from Chapter 18.2.1.1.1

$$th = m + 2\ s.$$

s is obviously the standard deviation of the logarithms of the analytical values x_i. Clearly, s below and above m is very different for non-logarithmic values (see the calculation of the variance and standard deviation for lognormal distributions in Chap. 9.2.3.1)

18.2.1.2
Rough Estimates Using the Median Only

It has already been discussed in the previous section that the median is a relatively stable statistical figure. In Chapter 9.3.3, this was also demonstrated for the lognormal distribution, in which the median is the geometric mean g.

The background is often recognized quite quickly in the early stages of an exploration program. It usually has a relatively small variation over large areas because it is related to specific lithologies. However, the standard deviation is commonly more variable.

In order to decide which areas should be sampled in more detail in the next exploration phase, a rule of thumb is applied that roughly classifies the values relative to the background. If the background is, for example, 60 ppm Zn and there is a value of 250 ppm Zn, then that value is greater than four times the background. Another value of 100 ppm Cu relative to a background of 30 ppm Cu would be three times the background. Hence, the Zn value would be examined more closely. Indeed, this is also a common method during the evaluation of geophysical anomalies. It is often said that a local maximum must be at least double as high as the background in order for it to be recognized as an anomaly, for example in airborne electromagnetic survey or gravity surveys.

18.2.2
Numerous Anomalous Values

In cases where regional geochemical sampling is carried out in the search for better deposits, such as porphyry copper deposits with large halos, the number of potentially anomalous values is relatively so high that they can be considered to comprise their own population (Fig. 84 right) The problem is to discriminate this anomalous population from the normal population. Sometimes there are not only two, but several populations. The methods for discriminating between populations were developed first for zoology in order to separate populations with different age groups in a fishing haul (Cassie, 1954). The fundamental publications of Sinclair (1974, 1976) describe the application to geochemistry.

18.2.2.1
The Identification of Populations

The effect of the mean standard deviation and population size, all of which are basic parameters, on the cumulative frequency curve will be examined in order to provide a better understanding of the procedure discriminating between populations. Consider two equal-sized, normally distributed, populations of geochemical lead analyses (Fig. 86). Population A has a mean of 150 ppm, and population B of 500 ppm Pb. Since the higher values are required for the identification of anomalous populations, the class accumulation is calculated downwards from the highest values, so that the cumulative frequency curve runs downwards from upper left to lower right (Sect. 18.1).

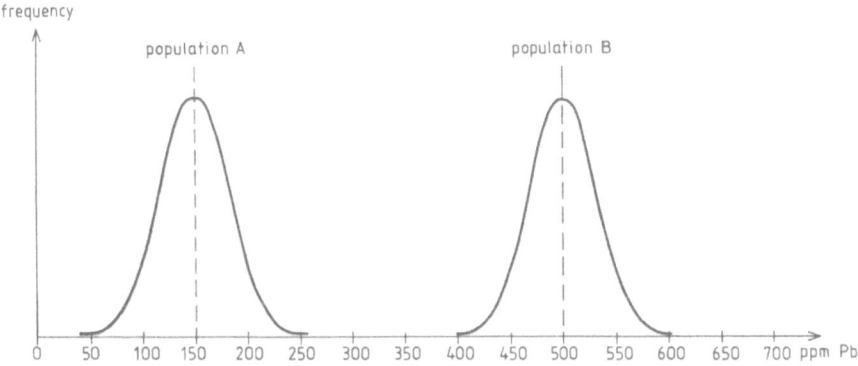

Fig. 86. Frequency distribution of two populations of geochemical Pb analyses

The cumulative frequency lines for populations A and B, or the Hazen lines, are plotted on the probability grid as in Fig. 87.

In the next step, populations A and B are considered as one population together, so that it is a mixed population. Since less that 0.2% of the values in the population actually overlap, each population can be examined on its own and the cumulative frequencies for the subpopulations A and B calculated with respect to the total mixed population. This implies that the cumulative frequencies of the

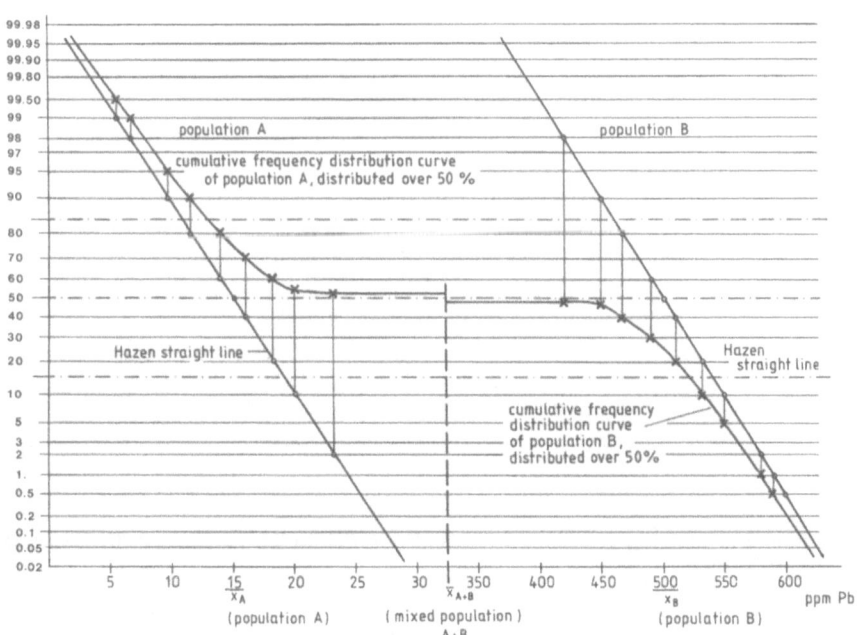

Fig. 87. Cumulative frequency curves of the populations of Pb analyses of Fig. 86

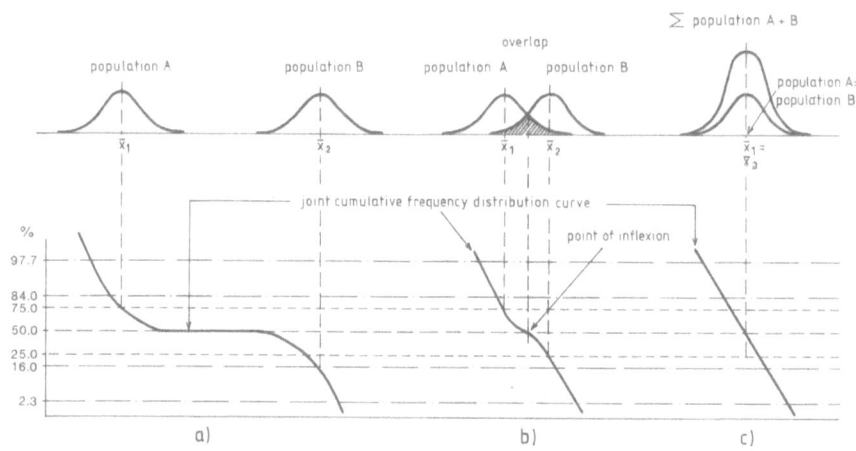

Fig. 88. Creating a mixed population by stepwise joining two populations

subpopulation B with the mean B must be halved (subpopulations A and B are of the same size, see above).

The same is true for subpopulation A, although half of the cumulative frequency value must now be added to 50% since the first 50% of the cumulative frequency is already filled by the subpopulation B.

In Fig. 87 it is clear that these cumulative curves are only approximately linear in their lower (for subpopulation B) and their upper (for sub-population A) sections. Since the normal distribution theoretically extends over the range from -¥ to +¥, and thus the cumulative frequency value is theoretically first attained at -¥, or 0 ppm Pb, then the cumulative frequency curve for the subpopulation B approaches the 50% line asymptotically from below. Similarly, the cumulative frequency curve for the subpopulation A approaches the 50% asymptotically from above. This results in the characteristic shape of the curve in Fig. 87.

Both curves are then combined for the mixed population A + B at the mean of the mixed population, which is (150 + 500)/2 ppm Pb = 325 ppm Pb. This is schematically shown for the mixed population A + B in Fig. 88.

If the mixed population is now changed so that the subpopulations approach each other and therefore have a higher degree of overlap (Fig. 88b), then the cumulative frequency curves for subpopulations A and B approach each other. The flat section in the middle with a shallow gradient and low rate of change of curvature becomes shorter.

If the two subpopulations A and B coincide with each other (Fig. 88c), then there is no longer a mixed population but a total population with a normal distribution and the same standard deviation as the original sub-populations A and B. The cumulative frequency curve is therefore a continuous straight line, or the Hazen line. If there are different populations that not only have different mean values, but also different standard deviations, then the corresponding Hazen lines have different gradients (cf. Chap. 5.1.2).

It can be deduced from this that different populations can be identified and discriminated from each other by the following criteria:

a) the cumulative frequency curve has various gradients;

b) flexured sections with significantly lower gradients or inflexion points occur on the curve.

The accumulated frequencies of each population can be derived directly from the inflexion points on the cumulative frequency curve of the mixed population (Sinclair 1976). For example, if the inflexion points occurs at 68%, then 68% of the values are classified with one population and 32% with the other population, assuming there are two subpopulations. If there are several subpopulations, then the percent values between the inflexion points yield the relative proportion of the values that belong to the individual populations.

Other geochemists use histograms in which the subsidiary maxima can sometimes be better identified, and these are then treated as individual populations (e.g. Müller 1986). However, it must be realized that not every population is marked by a maximum on the histogram and, conversely, not every subsidiary maximum must necessarily be related to a specific population. This often occurs if the sample size is not sufficiently representative of the population. At the same time, it must be noted that a subpopulation can also be so disguised that it is no longer identified on the cumulative frequency curve.

The correct separation of a greater number of populations is usually necessary if the data were not already discriminated according to areas with different geological and geochemical characteristics. Explorationists should always attempt to separate data from specific geochemical and geological environments. There are then normally two populations and it can be assumed that each sample can be classified with one or other of the populations. This should not be compared with mixed samples such as stream sediment samples, which are derived from a variety of source areas such as granites, basic volcanics etc. In practice, it is critical to successful exploration that the population with the highest, potentially anomalous, values can be identified and defined.

18.2.2.2
Simple Separation of two Populations

This method exploits the approximation that the mixed population is divided into two subpopulations at the inflexion point of the cumulative frequency curve. The characteristic values for the first subpopulation are derived from all the values to the left of the inflexion point, and those for the second population are derived from the values to the right of the inflexion point. This is a simplification, whereby no consideration is given to the possibility that values from the area where the subpopulations overlap might actually belong to the other subpopulation.

The disadvantage of this truncation is potentially the elimination of extensive overlaps. The means are thus shifted and the standard deviations are smaller. From the slope of the cumulative frequency curve of all the samples at the point of

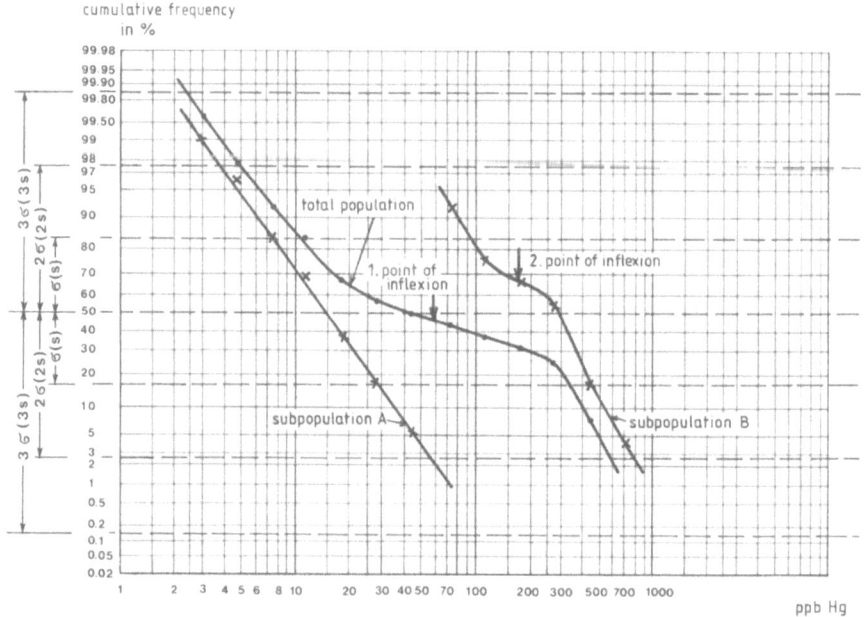

Fig. 89. Cumulative frequency curves of the mixed population of Hg values in Table 47a and b

inflexion, the degree of overlap can be estimated. If there is practically no overlap at all, the curve must be horizontal at the point of inflexion (see Fig. 88). The less pronounced the break in slope, the more overlay is present. Since the slope is rather flat at the point of inflection in our example in Fig. 89, it is justified to use the simplified method of truncation.

The procedure is best explained by an example.

Example: 201 rock samples were collected during a rock geochemical program and they were analysed for Hg as well as other elements. In Table 47a the values are already classified into logarithmically equal intervals.

In the next step, the cumulative frequency is calculated commencing with the highest values (4th column). These values are now plotted on logarithmic probability paper (Fig. 89). The lower limits to the class intervals must be used for plotting the cumulative frequency since the accumulation is calculated downwards from the high values. The class interval limit shows the percentage of values that are equal to or greater than a defined value. For example, the cumulative frequency of 7.5% in the penultimate row of Table 47a indicates that 7.5% of the values are greater than 467.8 ppb Hg, and therefore the value of 467.8 ppm must be plotted against 7.5% on the cumulative frequency axis.

A cumulative frequency curve is interpolated after plotting the cumulative frequency percents (Fig. 89) and the inflexion point or limiting value is determined by eye as being about 60 ppb Hg. The value of 60 ppb Hg does not mark a class limit, and the total number of values (n = 13) in the interval for 46.8 to 74.1 ppb Hg must be appropriately subdivided through the interval. The difference between 46.8

Table 47a. Hg values of a geochemical rock sample survey

Hg grade (ppb)	Absolute frequency	Relative frequency f_i (%)	Relative cumulative frequency Σf_i
2.0- 2.9	1	0.5	100.0
3.0- 4.7	3	1.5	99.5
4.8- 7.4	12	6.0	98.0
7.5- 11.7	16	7.9	92.0
11.8- 18.6	32	15.9	84.1
18.7- 29.5	24	11.9	68.2
29.6- 46.7	12	6.0	56.3
46.8- 74.1	13	6.5	50.3
74.2- 117.5	16	7.9	43.8
117.6- 186.2	8	4.0	35.9
186.3- 295.1	13	6.5	31.9
295.2- 467.7	36	17.9	25.4
467.8- 741.3	11	5.5	7.5
741.4-1174.9	4	2.0	2.0

and 74.1 is 27.3, and the difference between 46.8 and 60 is 13.2. Therefore the 13 samples are split according to the relationship

$$\frac{x}{13} = \frac{13.2}{27.3} \text{, or } x = 6.29 \text{, rounded to 6.}$$

Six values must be assigned to the lower population and seven to the higher population.

The whole population is therefore divided into the following subpopulations A and B (Table 47b).

The subpopulations A and B are now plotted separately on the probability diagram in Fig. 89. The plots of the two subpopulations A and B on the logarithmic probability grid in Fig. 89 show:

a) The values in the subpopulation A are satisfactorily located on a Hazen line. It is therefore assumed that this population has been separated from the total population.

b) The values of the subpopulation B, on the other hand, form another S-shaped cumulative frequency curve, which implies there are still two (or more) populations. The inflexion point of this curve is at 180 ppb Hg.

The above procedure must now be repeated for subpopulation B.

Table 47b. The Hg values of Table 47a partitioned into two populations

Hg-grade (ppb)	Absolute frequency n	Relative frequency f_i (%)	Relative cumulative frequency $\sum f_i$
Population A			
2.0- 2.9	1	1.0	100.0
3.0- 4.7	3	2.8	99.0
4.8- 7.4	12	11.3	96.2
7.5-11.7	16	15.1	84.9
11.8-18.6	32	30.2	69.8
18.7-29.5	24	22.6	39.6
29.6-46.7	12	11.3	17.0
6.8-60.0	6	5.7	5.7
$\sum n = 106$			
Population B			
60.1- 74.1	7	7.4	100.0
74.2- 117.5	16	16.8	92.6
117.6- 186.2	8	8.4	75.8
186.3- 295.1	13	13.7	67.4
295.2- 467.7	36	37.9	53.7
467.8- 741.3	11	11.6	15.8
741.4-1174.9	4	4.6	4.2
$\sum n = 95$			

If several inflexion points are identified on the cumulative frequency curve of the total population, then it can be immediately subdivided into several sub-populations at each of the inflexion points by the procedure described above.

18.2.2.3
More Detailed Discrimination Between Two Populations

The next method that will be described is based on a manuscript by Rehder (1988). Again, it is best explained by an example.

Laterite samples are collected in an area of lateritic weathering and are analysed for As, which is a pathfinder element used in gold exploration. The As analytical results are plotted on a logarithmic frequency grid as in Fig. 90 (for the sake of simplicity, only nine classes are used). The relative cumulative frequency values, expressed as fractions of 1 and denoted here as Q, are listed in Table 48 (column 2).

The shape of the cumulative frequency curve in Fig. 90 is flexured, and the presence of a mixed population, or two populations, is therefore considered likely. The inflexion point w is determined visually at w = 0.75. The cumulative frequency values of the two populations are denoted by Q_1 and Q_2.

The following is true for the total mixed population:

$$Q = w \; Q_1 + (1 - w) \; Q_2 \; . \tag{4}$$

The cumulative frequency curve in Fig. 90 is now examined from right to left. If only values from the first population are present, then:

$$Q_1 = \frac{Q}{w}. \tag{5}$$

This value can obviously have a maximum of only 1.

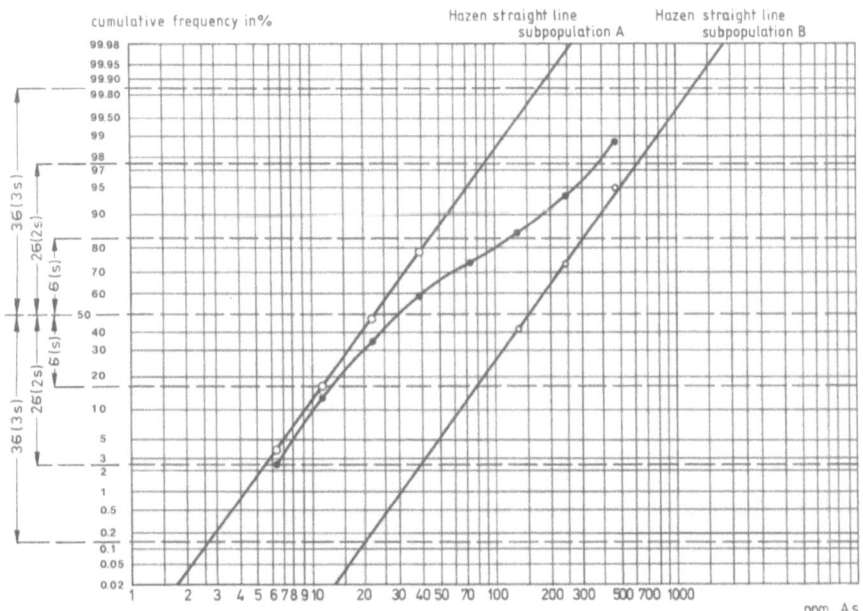

Fig. 90. Cumulative frequency curve for As values from a gold exploration survey

Conversely, outside the area of population 1 where there are only values from population 2, then the following is valid:

$$Q_2 = \frac{Q - w}{1 - w} \qquad (6)$$

must be true. The minimum for this value is 0.

The values Q_1 and Q_2 are now calculated with respect to the maximum and minimum conditions noted above, and are listed in columns 3 and 4 in Table 48. The heading for column 3 is denoted $Q_1 \min(1, \frac{Q}{w})$, which means that either 1 or $\frac{Q}{w}$, whichever is the smaller, is listed in column 3. The equivalent procedure is used in column 4 for $Q_{2\max}$.

If these values Q_1 and Q_2 are now plotted on the cumulative frequency diagram in Fig. 90, it can be seen that the points lie close to two Hazen straight lines, and that the populations overlap each other extensively. This method avoids the critical truncation of values into either one or the other population, as was the case in Section 18.2.2.2.

Table 48. As values from a geochemical laterite sample survey

Column 1	Column 2	Column 3	Column 4
Class of As values (ppm)	Cumulative frequency values Q	$Q_{1\min}\left(1, \dfrac{Q}{W}\right)$	$Q_{2\max}\left(0, \dfrac{Q-W}{1-W}\right)$
≤ 6	0.028	0.037	0
6.1- 11.0	0.14	0.187	0
11.1- 20.2	0.36	0.48	0
20.3- 37.1	0.60	0.8	0
37.2- 68.0	0.75	1	0
68.1-124.8	0.86	1	0.44
124.9-229.0	0.94	1	0.76
229.1-420.0	0.988	1	0.952
≥ 420.1	1.000	1	1.000

18.2.2.4
Determining the Threshold for Anomalous Populations

Once the anomalous population has been separated from the normal population, the question of defining the threshold arises. Theoretically, both populations, including the anomalous one, extend from $-¥$ to $+¥$. In this case there are no conventional methods for defining the threshold, although there are for the populations with only a few high values, as was described in Section 18.2.1.1.1. The inflexion point on the cumulative frequency curve can be chosen and, if the populations were separated by the simple procedure described in Section 18.2.2.2, this would then be the logical threshold value.

Sinclair (1976) recommends that the threshold should be at a cumulative frequency value of about 1 or 2% of the normal population or 98 or 99% of the anomalous population. If the populations are relatively clearly separated from each other, then the 1% threshold is selected. If there is extensive overlap of the populations, then a 2% threshold is selected in order not to reject too many values of the anomalous populations.

The term "anomalous" is always relative: anomalous with respect to something that can be considered as normal. It is therefore fair to identify particularly anomalous values, and distinguish them from the rest of the anomalous population. Effectively, the anomalous population is treated as the local background and the individual high values become the anomalies. It is essential that the geological aspects are considered in the interpretation, and not just the pure statistics. A broad geochemical halo around a relative thin mineralization would be an example. Very high values can occur over this mineralization and, with respect to the anomalous population over the halo, they appear to be even more anomalous. In cases such as this, the threshold th is defined by the median m and the standard deviation s:

$$th = m + 2\ s$$

according to Eq. (1) in Section 18.2.1.1.1. The median and standard deviation can be derived from the Hazen line for the anomalous populations on the probability grid, and the threshold can be read directly from $m + 2\ s$. For the example of As geochemistry in Fig. 90, the value for $m + 2\ s = th$ is about 610 ppm As.

18.2.2.5
Appraising Other Distributions Obtained During Geochemical Exploration

It is quite commonly found that the cumulative frequency curve has a marked nick-point within the area of the high values on the lognormal probability grid (Fig. 91). This implies that the distribution of the logarithmic values is still skewed to the right. There is absolutely no geological reason why geochemical distributions must be lognormal (or normal), although it has often been observed that geochemical populations are best described by the lognormal distribution. At the same time, it is

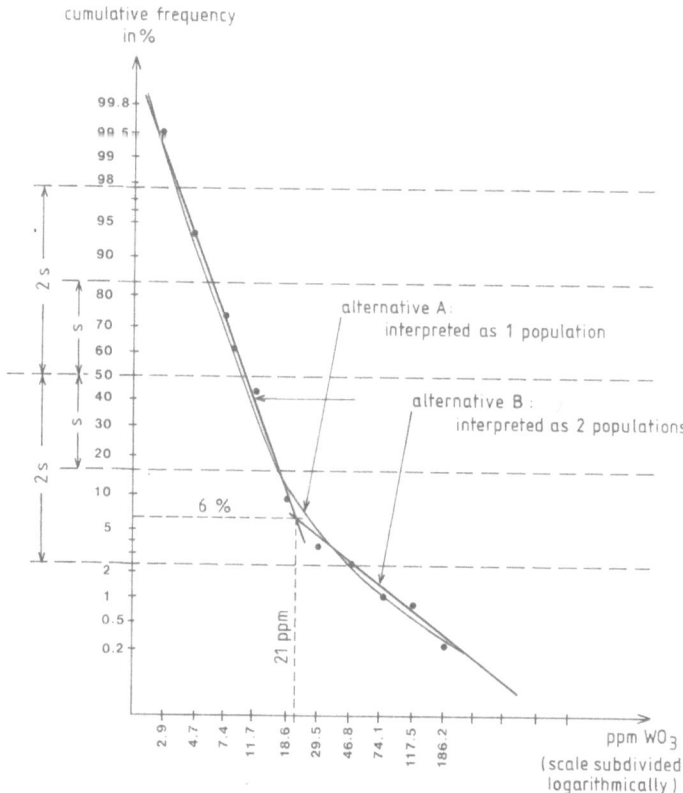

Fig. 91. Cumulative frequency curve for WO_3 analyses from a heavy mineral exploration survey

quite reasonable to expect that some populations will have logarithmic values tending to high values, even the distribution of logarithmic values will be skewed to the right.

If all the values are considered to belong to one population, then the threshold and thus the anomalous values are defined by Eq. (1) in Section 18.2.1.1.1.

If the bent tail is considered to be a population on its own, then the nickpoint separates the populations according to the procedure described in Section 18.2.2.2. On Fig. 91 this would be at 6%.

18.3

Determining the Relative Geochemical Contrast

Before carrying out a geochemical exploration program, the most suitable technique is ascertained by an orientation survey over known mineral occurrences. For example, in Cu, Pb and Zn exploration the various chemical extraction techniques are tested in order to discover if the less expensive cold extraction method is as satisfactory as the other methods of analyzing for the total Pb, Zn and Cu content.

The relative geochemical contrast between anomalies and background is not just the difference between the mean, but the standard deviation must also be used to estimate the variability. According to Dixon and Massey (1957), the contrast t can be determined by a Student's t-test, and thus different methods can be quantitatively evaluated.

The appropriate formula is:

$$t = \frac{\bar{x}_a - \bar{x}_b}{s_p \sqrt{\dfrac{1}{n_a} + \dfrac{1}{n_b}}} \; . \tag{7}$$

$\bar{x}_a$ and $\bar{x}_b$ are the means for the anomalous and background values, n_a and n_b are the number of anomalous and background values, and sp is the common standard deviation. Formula (7) has certain similarities to formula (1) in Chapter 11.2.2, in which two sequences of analyses were examined for significant differences by the Student's t-test.

The common standard deviation sp is calculated as follows:

$$s_p = \sqrt{\frac{(n_a - 1)\, s_a^2 + (n_b - 1)\, s_b^2}{n_a + n_b - 2}} \; . \tag{8}$$

s_a^2 and s_b^2 are the variances of the anomalous and the background values.

This procedure can be explained, with reference to Rose and Keith (1976), by an example.

Example: Prior to undertaking an exploration program for sandstone-hosted uranium deposits, stream sediment samples were collected for an orientation survey from one area with known uranium mineralization (anomalous) and another

Table 49. Determination of the geochemical contrast in a geochemical uranium survey

Geochemical method	Mean of logarithmic value = geometric mean non-logarithmic		Logarithmic mean = α		Standard deviation of logarithmic values β		No. of sample values		Geochemical contrast
	Anomal	Background	Anomal	Background	Anomal	Background	Anomal	Background	t
Total U (ppm U)	3.9	3.2	1.36	1.16	1.85	0.17	35	45	0.72
HNO₃ extract. U (ppm U)	1.7	0.6	0.53	-0.51	0.65	0.47	35	45	8.25

area with no known uranium mineralization (background). Different extraction techniques were compared and in Table 49 the results for total analysis are compared with the HNO_3 extraction method. Logarithmic values have been used so that the mean a and the variance b^2 of the log-transformed values are substituted in Eq. (7) and (8) above (cf. Chap. 1.2.3.1).

As an example, the contrast, t, is now calculated for the first row, which is for the total uranium content. The common standard deviation sp is calculated from Eq. (8) above:

$$sp = \sqrt{\frac{(n_a - 1)\, s_a^2 + (n_b - 1)\, s_b^2}{n_a + n_b - 2}} \,, \tag{8}$$

$$sp = \sqrt{\frac{(35 - 1)\, 1.85^2 + (45 - 1)\, 0.17^2}{35 + 45 - 2}} = \sqrt{\frac{34 \cdot 3.42 + 44 \cdot 0.03}{78}} \,,$$

$$sp = \sqrt{\frac{117.60}{78}} = \sqrt{1.51} = \pm 1.23 \,.$$

Equation (7) is solved by using this value for sp,:

$$t = \frac{\overline{x}_a - \overline{x}_b}{s_p \sqrt{\dfrac{1}{n_a} + \dfrac{1}{n_b}}} \tag{7}$$

$$t = \frac{1.36 - 1.16}{1.23 \sqrt{\dfrac{1}{35} + \dfrac{1}{45}}} = \frac{0.20}{1.23 \sqrt{0.029 + 0.022}}$$

$$t = \frac{0.20}{1.23 \cdot 0.225} = 0.723 \,.$$

If this calculation is now carried out for the 2nd row in Table 49 of the HNO_3 extractable uranium, then it is found that t = 8.25.

The HNO_3 method is therefore clearly the superior technique, and is also the only method by which a significant contrast could be identified. If one plots the value t = 0.72 from the first method in the diagram in Fig. 31, then it is clear than no significant contrast can be related to this value.

19
Other Methods for Defining Anomalies

19.1
Filter Methods

Filtering techniques are used for quasi-continuous measurements, such as during airborne geophysical surveys, in order to identify not only the well-defined and obvious anomalies but also those anomalies that are relatively weak with respect to the background noise. In addition, filter techniques can be, and are, used for geochemical exploration since the geochemical analytical data are usually spatially interdependent, or regionalised variables (Chap. 13.1).

19.1.1
Filtering with Moving Averages

The moving average was already introduced in Economic Evaluations in Exploration, Chapter 2.2, for calculating projected metal prices. The moving average determines the mean for a defined number of values, and the averaging interval moves successively from value to value.

Example: Five values are selected from a magnetic survey and they are assessed with a moving average of three values.

This sort of calculation for the moving average is already a filter since the high frequency noise that is manifested by occasional peaks is subdued by averaging with three or more other points.

Strong regional gradients, together with the noise, make it difficult to recognize weak anomalies, and this is schematically shown in Fig. 92. The anomalies are even less clear on the isoline maps, since the superimposed strong gradient does not yield any well-defined highs and lows, and the isolines are either only bunched more closely together or are more widely separated (Fig. 92 right). After taking into

Table 50. Calculation of moving averages for a magnetic survey

Survey data in γ	1st moving average	2nd moving average	3rd moving average
75	(75+		
83	83+ }88	(83+	
107	107):3	107+ }97	(107+
100		100):3	100+ }98
87			87):3

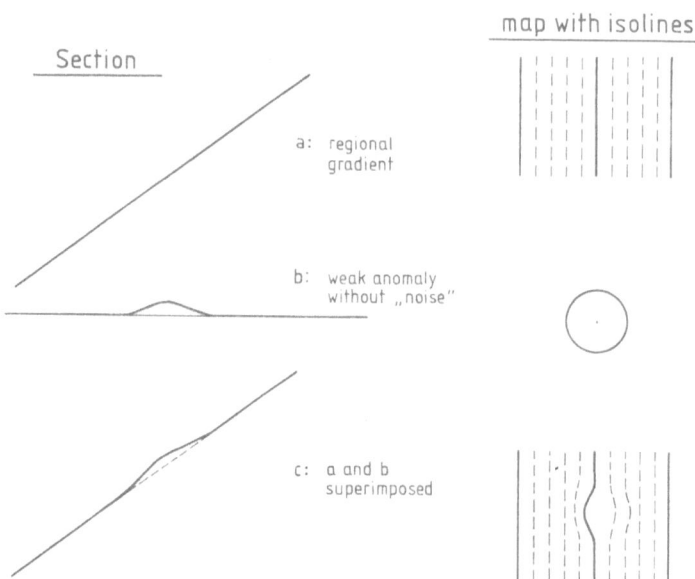

Fig. 92. Weak anomaly superimposed on regional gradient

consideration that the isolines are never regular, but are always irregular, then it is clearly difficult to recognize such an anomaly at all.

Moving averages with different averaging intervals are subtracted from each other in order to eliminate both the effects of high frequency noise as well as the regional gradients.

A regional gradient, which extends over a significantly larger area as compared to the broadest moving average interval, is unaffected by all such moving averages, and is thus eliminated by the subtraction of two moving averages. On the other hand, short frequency anomalies will be reflected differently by each of the two filters, and therefore the subtraction of the two filters results in a filtered anomaly without the regional gradient.

An example for this effect of differences between moving averages is provided by the subtraction of one moving average derived from an interval with seven values (eliminating the regional gradients) from another moving average derived from an interval with three values (reduces the noise). This is demonstrated by an idealized survey profile in Table 51 and on Fig. 93.

The lower half of Fig. 93 shows how the regional gradient, noise peaks and anomalies are summed together. The upper half of Fig. 93 shows the filtered anomaly, from which the regional gradient has been removed. The slight skewness of the anomaly is due to the effect of the noise peak (one value) on the filter compilation.

It should also be briefly mentioned that geostatistical methods, such as universal kriging or kriging with generalized covariances (e.g. Akin and Siemes 1988), can also be used to eliminate the effects of regional gradients (known as drifts in geostatistical terminology).

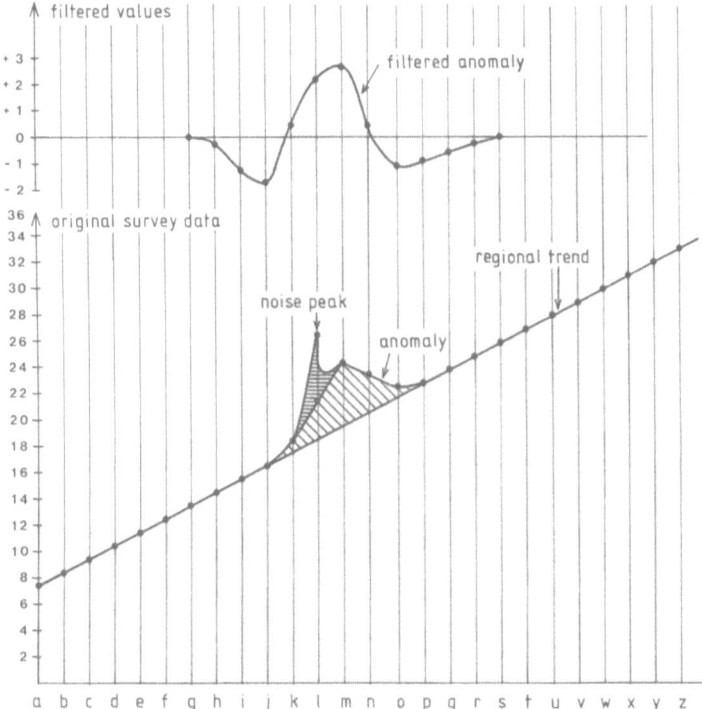

Fig. 93. Filtering of the survey data of Table 51

19.1.2
The Fraser Filter

The Fraser filter technique is commonly applied in geophysical exploration, and it was developed by the Canadian geophysicist D.C. Fraser (1971, 1981) for the processing of angles of declination derived from electromagnetic surveys, particularly VLF surveys[39] . The Fraser filter is now also used for other types of surveys.

---

[39] VLF = Very Low Frequency. This geophysical technique is based on transmitters with the frequency range between 15 and 25 kHz. Either the angle of declination is measured, whereby the cross-over point marks the location of the conductor, or the resistivity is measured, whereby the apparent resistivity is determined from the orthogonal components of the electrical and magnetic fields. Because the field apparatus is relatively light-weight, and because it can be operated quickly by one person, this technique is suitable for identifying airborne anomalies on the ground. The electromagnetic surveys' relatively high survey frequency restricts the technique to areas with relatively high resistivities (Paterson and Ronka 1971).

The so-called cross-overs are critical in the measurement of the angle of declination during geophysical surveys, as it is the point at which the direction of the declination angle reverses. The profile derived from this type of survey has a characteristic curved shape. The Fraser filter changes these curves into highs and lows, and thereby reduces the high frequency noise. Figure 94 demonstrates how this method results in a cross-over being transformed into a "high" or peak.

The Fraser filter technique is virtually a weighted moving average over four points. The first two points are summed together, the next two points are then subtracted from this figure. This corresponds to a multiplication of the four points by the weighting factors +1, +1, -1, -1. A normal moving average procedure is then carried out as described in Section 19.1.1.

Table 51. Examples for a filter calculation

Filter

Survey station	Survey value	Moving average for 3 values = GM_3	Moving average for 7 values = GM_7	Filter value $(GM_3 - GM_7)$
a	7			
b	8	(8)		
c	9	(9)		
d	10	10	10	0
e	11	11	11	0
f	12	12	12	0
g	13	13	13	0
h	14	14	14.14	- 0.14
i	15	15	16.29	- 1.29
j	16	16.33	18	- 1.67
k	18	20	19.43	+ 0.57
l	26	22.67	20.57	+ 2.10
m	24	24.33	21.57	+ 2.76
n	23	23	22.57	+ 0.43
o	22	22.33	22.43	- 1.10
p	22	22.33	23.29	- 0.94
q	23	23	23.57	- 0.57
r	24	24	24.14	- 0.14
s	25	25	25	0
t	26	26	26	0
u	27	27	27	0
v	28	28	28	0
w	29	29	29	0
x	30	(30)		
y	31	(31)		
z	32			

Fig.94. Example for the
application of the Fraser filter

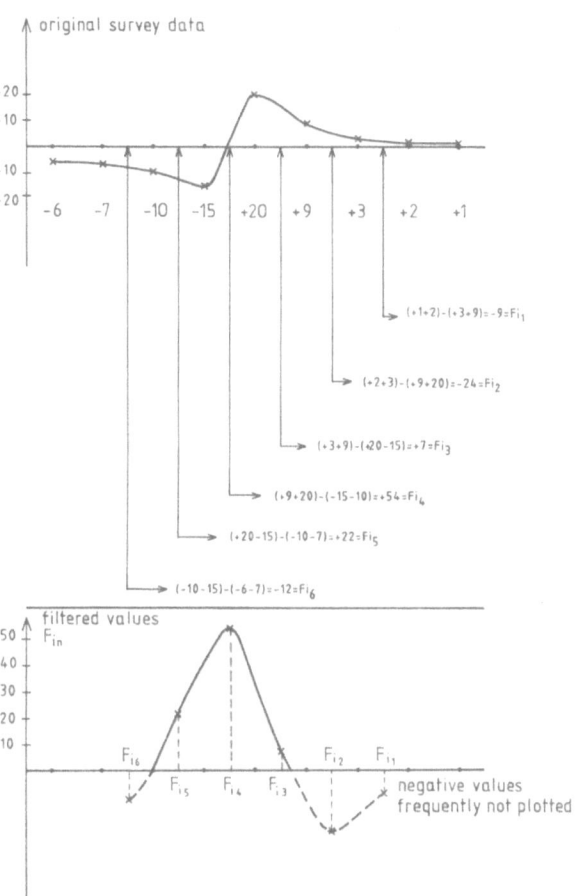

Example: There are four survey points with the results: 2, 17, -5, -3. The Fraser filter value is then $Fi = (2+17) - (-5-3) = +27$.

In effect, this is the difference between two moving averages with a two point interval, at a spacing of two points. The point Fi (Fig. 94) is assigned to the middle point of the filter and is therefore located between points 2 and 3. A constant background is eliminated as a result of the subtraction.

The negative values are often not included in isoline maps since they occur only on the flanks of the peaks.

This filter effect, in which inflexion points are converted into peaks, can also be applied even if there are no cross-overs. In Fig. 95 it is assumed that all values are overprinted by a regional background of 40. The weighting procedure, twice with +1 and twice with -1, eliminates the background, and this is clearly shown for the central point of the anomaly in Fig. 95. The filtered values for both cases in Fig. 94 and 95 are therefore identical (lower parts of Figs. 94 and 95).

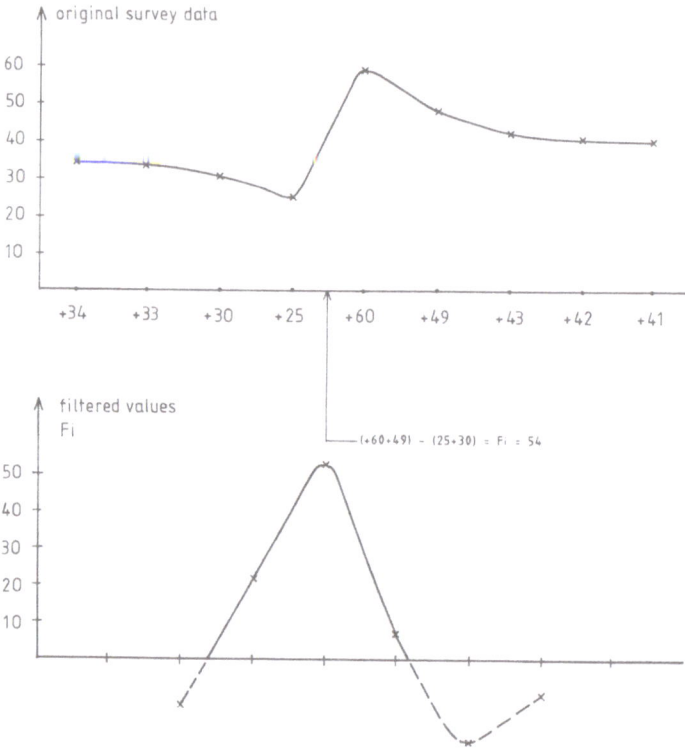

Fig. 95. Fraser filter, applied to the anomaly in Fig. 94, to which a regional background has been added

19.2
Addition and Multiplication Methods

The stacking procedure is used in seismics. The recordings of repeated readings are added together (stacked). The highs and lows related to noise are irregular, and are therefore eliminated by the repeated additions. On the other hand, true as well as weak anomalies are always reflected by a peak at the same point and, as a result of the continuous addition of the repeated recordings, the anomalies become better defined.

Virtually the same procedure is used in geochemistry when comparing anomaly maps of different elements. An explorationist will normally consider that a weak geochemical anomaly, which is reflected by the elements Cu, Pb and Zn and correlates with a magnetic anomaly, is more significant than a single isolated Zn anomaly, even if it is a higher value. Geochemical values for several elements, superimposed in order to enhance possible anomalies, can also be carried out mathematically by multiplying or adding the values.

If there are two or three elements, then they are multiplied. If there are four or more elements, as is commonly the case for base metal exploration when combinations such as Cu, Pb, Zn and Ag are used, then for practical reasons the values are added.

Table 52. Normalization of Zn and Ag values of a geochemical survey

Sample point	Zn (ppm)	Ag (ppm)	Zn normalised = x · th (th=42)	Ag normalized = y · th (th=2)
17	48	2	1.14 · th	1 · th
18	45	9	1.17 · th	4.5 · th
19	51	3	1.21 · th	1.5 · th

During the addition it is important to note that the elements should be normalized with respect to their background[40] . The background is statistically determined by the procedures described in Chapter 18.2.1.1.1.

Each geochemical value is then treated as a product of the factor x multiplied by the background.

Example: The following Zn and Ag values in ppm were determined for samples from the adjacent sample sites 17, 18 and 19:

The background for Zn is th = 42 ppm Zn and for Ag th = 2 ppm Ag, and the standard deviation for Zn is s = 20 ppm Zn and for Ag is s = 2 ppm Ag. The Ag value at sample site 18 is therefore probably anomalous since it is greater than th + 3 s (cf. Chap. 18.2.1.1.1).

If Zn and Ag are now added without any corrections, then the following values would be derived:

sample site 17:   Zn+Ag = 50,
sample site 18:   Zn+Ag = 53,
sample site 19:   Zn+Ag = 54.

Since the range of the background values for Zn is as broad (cf. Fig. 84) as the probable Ag anomaly, then the Ag anomaly is obliterated by the addition, which results in an effect exactly the opposite to that required: a smoothing rather than an enhancement of the data. In order to avoid this, the Ag and Zn values are normalized with respect to the background by dividing them by the background value th so that they are expressed as x · th (columns 4 and 5 in Table 52).

These values are now added and are used as the basis for further examination of the data:

sample site 17:   1.14 + 1   = 2.24,
sample site 18:   1.17 + 4.5 = 5.67,
sample site 19:   1.21 + 1.5 = 2.71.

---

[40] To be completely correct, the anomalous values should be normalized with respect to the background and the standard deviation, since both parameters are included in the definition of anomalous values (cf. Chap. 18.2.1.1.1). The computational procedure would then be similar to that for the normalization in the χ^2 test (Chap. 5.2).

As an example, Fig. 96 illustrates the multiplied values from a section of the map of a Cu and Au geochemical investigation program which was carried out during the exploration for complex, volcanogenic Cu-Zn-Pb-Ag-Au deposits. A feeder zone was intersected in subsequent drilling. It was found to contain practically no zinc although it is weakly reflected by Cu and Au on the surface.

A comparison of the right-hand map of the multiplied values in Fig. 96 indicates that these anomalies are clearer than the individual anomalies for Cu or Au.

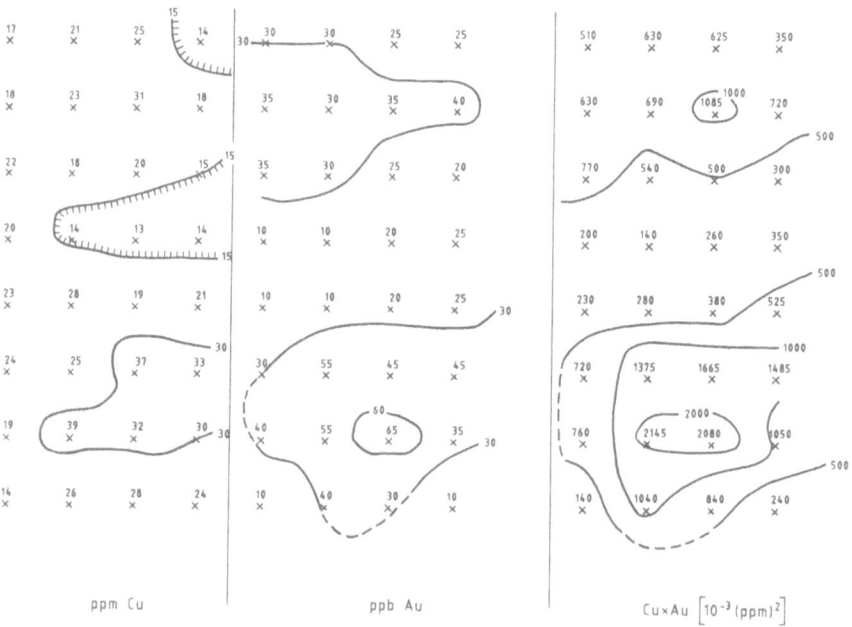

Fig. 96. Part of a geochemical map with Cu and Au values and with factors Cu x-times Au

20
Defining a Drill Grid

20.1
Basic Considerations

20.1.1
Probability of Intersecting a Blind Target

During exploration, one usually drills targets that are defined by geochemical or geophysical anomalies, or by a geological concept, although in mineral exploration it is more usual to drill geochemical and geophysical anomalies. However, in exceptional circumstances, a blind target that has been defined by a geological concept will be tested by drilling.

Fig.97. Completeness of search as a function of the ratio S/F_e

(spacing of drilling S in a square grid and area of target F_e).

(After Slichter 1960)

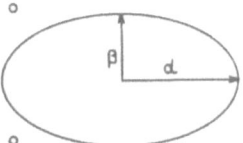

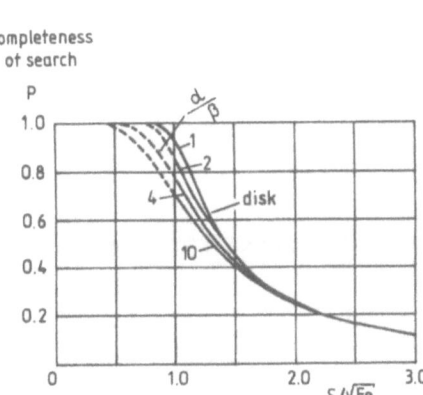

In such cases, it is important to know what the probability is of the drill hole intersecting a model orebody. Orebodies are often expressed as ellipses, and Slichter (1960) published a relevant diagram (Fig. 97). Fe is the area of the ore deposit ellipse, that in extreme cases tends to a circle:

$$Fe = \pi \cdot \alpha \cdot \beta \ . \tag{1}$$

S is the spacing between the drill holes on a square drill grid. The calculation of the probability of intersection is similar to the problem in Chapter 17.2.2 of defining the spacing between survey lines and the location of measuring sites on these lines. Before commencing exploration, normally an economic modelling study is carried out on the basis of the minimum size of the expected mineral deposit. As a result, the minimum dimensions of the deposit are already established and the parameter Fe in the diagram (Fig. 97) can be estimated. It is obviously important that the half-axes of the idealised ore deposit ellipse are defined on a plane on which the drill grid is a square, and this plane is usually perpendicular to the drill axis.

Example: Exploration is being carried out for carbonate-hosted, Mississippi Valley-type, Pb-Zn mineralisation. The deposits mostly have an irregular shape with no preferred axes.

The target deposit is modelled as a circle with radius r (Fig. 98). The carbonate horizon dips at $\delta = 80°$. It is assumed that the Pb-Zn deposit is approximately stratabound. The carbonate horizon is drilled with a 60° angled hole ($\xi = 60°$). The drill grid is square with respect to the plane perpendicular to the 60° drill hole, and the spacing between the holes is S = 100 m. What is the probability of intersection if the radius of an ideal, minimum-sized, ore deposit is r = 50 m?

Answer: The circular model ore deposit must be projected on the plane that is perpendicular to the 60° drill hole (Fig. 98). As a result, the circle becomes an ellipse. The longest axis α remains r, and the shortest axis β equals the projection of r onto the reference plane perpendicular to the drill hole. Therefore:

$$\beta = r \cdot \cos(\delta + \xi - 90°) \ ,$$

$$\beta = r \cdot \sin(\delta + \xi) \ .$$

Therefore in this case:

$$\alpha = r = 50 \text{ m, and}$$

$$\beta = r \cdot \sin(\delta + \xi) = 50 \cdot \sin(80° + 60°)$$

$$\beta = 50 \cdot 0.6428 = 32.14 \text{ m.}$$

Thus:

$$Fe = \pi \cdot \alpha \cdot \beta = \pi \cdot 50 \cdot 32{,}14 = 5048.5$$

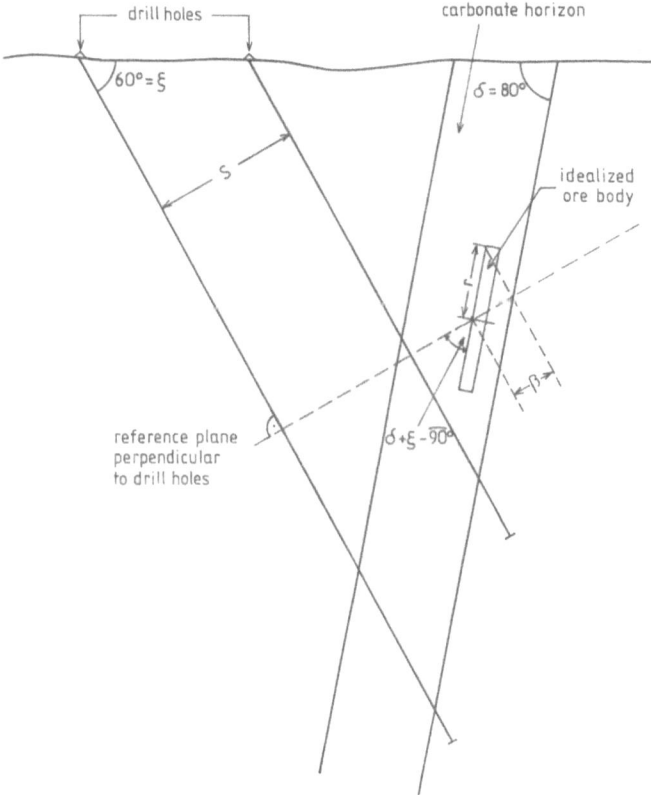

Fig. 98. Section through a Pb-Zn orebody in carbonates

and

$$\sqrt{F_e} = 71.05 \, .$$

The quotient on the x-axis in Fig. 97 is therefore:

$$\frac{S}{\sqrt{F_e}} = \frac{100}{71.05} = 1.41$$

and the ratio

$$\frac{\alpha}{\beta} = \frac{50}{32.14} = 1.56 \, .$$

Therefore, according to Fig. 97, the completeness of the search is approximately $P = 0.5$.

20.1.2
The Type of Grid

In exploration, the drill targets are usually either defined by geophysical or geochemical anomalies or, in some cases, geological concepts.

Once a discovery hole has been drilled, then the step out holes are still determined by geological, geochemical or geophysical criteria, such as the strike extension of the anomaly. However, if continued drilling proves successful, the question arises of what grid should be used for systematically drilling the potential ore deposit.

Irregular mineral deposits, which are deposits with a large coefficient of variation (Chap. 3.3 and Appendix Table 1), must be drilled on a close-spaced grid (gold veins, for example, are drilled at 10- to 15-m intervals). More regular deposits, which are deposits with a low coefficient of variation such as stratiform synsedimentary Pb-Zn deposits, are drilled at a wider spacing (50 to 100 m). In the USA, for example, the reserves of coal deposits within a radius of 400 m of an intersection are defined as "measured" (Wood et al. 1983). In Queensland and New South Wales in Australia, for coal deposits the distances would be even 500 m (Whitchurch et al. 1990). This more or less corresponds with "proven" in the German definition (Appendix Table 3; Wellmer 1983a).

Anisotropic features must also be taken into account. If pitches within the mineralisation are identified, then drill spacing will be closer in the direction of greater variability, which is perpendicular to the pitches, as compared to the orientation parallel to the pitches.

If variograms of mineral deposits similar to the type that is being drilled are available, and the range a (Chap. 13.2.1 and Fig. 57) is approximately known, then the drill spacing should initially be about a/2. Then, once about 15 to 20 values are available, it is possible to generate a variogram. The optimal drill spacing for defining reserves according to the standard procedures can then be determined geostatistically (Sect. 20.2).

The grid pattern for drilling a mineral deposit must also be defined. The usual pattern is a square or rectangular grid. However, with respect to the efficiency of a systematic and homogeneous coverage, an equilateral triangular grid pattern is preferable to the square or rectangular patterns. This is demonstrated by simple geometrical principles (Fig. 99):

1st Step: An area should be drilled so that no point is further than r from a drill hole, and that the circles of influence with radius r intersect in the centre of the grid cell, regardless of whether the cell is square (Fig. 99a) or equilateral triangular (Fig. 99b). The size of the grid cell can be derived from these conditions.

Case 1: Rectangular square grid, Fig. 99a
From the right-angled triangle ABC, r is defined by:

$$r^2 = \left(\frac{b}{2}\right)^2 + \left(\frac{b}{2}\right)^2 = \frac{b^2}{2} \ ,$$

$$b = r \cdot \sqrt{2} \ .$$

Fig.99. Comparison of a square
and an equilateral triangular
grid

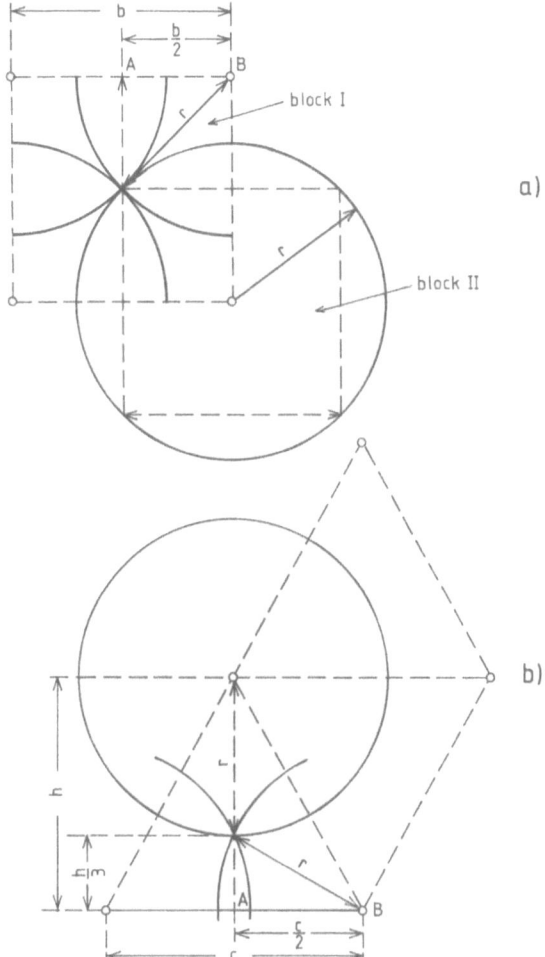

a)

b)

So that the area of a grid cell is:

$$F_v = b^2 = 2r^2 \ .$$

(1)

Case 2: Equilateral triangular grid, Fig. 99b. The perpendiculars of a triangle intersect at h/3. Since $r = \dfrac{2}{3} \cdot h$ for an equilateral triangle, then:

$$\frac{h}{3} = \frac{r}{2} \ .$$

(2)

Therefore for the triangle ABC

$$r^2 = \frac{c^2}{4} + \left(\frac{h}{3}\right)^2$$

and therefore with (2):

$$r^2 = \frac{c^2}{4} + \frac{r^2}{4} \;,$$

$$c^2 = 3r^2$$

$$c = r \cdot \sqrt{3} \;.$$

Since the area of an equilateral triangle is

$$F = \frac{c^2}{4} \cdot \sqrt{3} \;,$$

then the area of the triangular grid cell is:

$$Fd = \frac{c^2}{4} \cdot \sqrt{3} = \frac{r^2 \cdot 3 \cdot \sqrt{3}}{4} = 1.3 r^2 \;. \tag{3}$$

2nd Step: It must now be decided how many drill holes should be drilled within each grid cell.

Case 1: Square grid, Fig. 99a
It is obvious from Fig. 99a that the grid cells are identical regardless of whether the drill holes are sited at their corners (Block I) or in the centre (Block II). From this it is clear there must be one drill hole in each cell.
Case 2: Equilateral triangle grid, Fig. 99b.
Since the sum of the angles in a triangle is only 180°, in practice every drill hole cuts a 60° section out of the circle of influence, and therefore only 0.5 holes need be drilled in one cell.

Table 53. Attributable area for a drill hole in a rectangular and triangular drilling grid

	Area	Attributed drill holes	Area/drill holes
Rectangular cell	$2\ r^2$	1	$2\ r^2$
Triangular cell	$1.3\ r^2$	0.5	$2.6\ r^2$

3rd Step: The above can be summarised as follows.

Table 53 shows that the triangular grid pattern covers the area 30% more efficiently than the square pattern.

The advantages and disadvantages of these two grid patterns have been discussed at length in a series of articles in the *Bulletin of the Australasian Institute of Mining and Metallurgy* (Armstrong 1983; Rudenno 1985).

20.2
Geostatistical Methods for Determining the Drill Spacing

As discussed in Chapter 13, all geostatistical calculations are based on the variogram, and a minimum data base from about 15 to 20 drill holes is required before the variogram can be generated. If this is available, then the following geostatistical methods for calculating the drill hole spacing can be applied.

20.2.1
Application of the Matheron Diagram

Matheron (1971) published a diagram for determining the extension variances, σ_e^2, of simple cases by using the spherical model (i.e. with the transitive type of variogram, Chap. 13.2.1). In Fig. 100, the curves are provided for the three most important cases,:

- sampling along a drive, with the sample site in the middle of the sample profile, and
- drilling in the middle of a square block;
- in addition, the case of circular blocks to classify coal reserves like in the USA, as mentioned above (Wood et al. 1983), is given. The extension variances for this case are from Günther and Kelter (1994).

The application of the diagram is the same as that for the diagram in Fig. 65 and 66 as described for the example in Chapter 13.3.5. For one block, the extension variance σ_e^2 is:

$$\sigma_{e_1}^2 = C_0 + M \cdot C_1 \ . \tag{1}$$

The factor M is derived from the diagram (Fig. 100). If n equal-sized blocks are taken into consideration, then the extension variance must be divided by n:

$$\sigma_{e_{(n)}}^2 = \frac{1}{n}(C_0 + M \cdot C_1) \ . \tag{2}$$

Example: A vein near a gold deposit, which was examined in Chapter 13.4.4 and 13.4.5.2.2, as well as in 13.4.5.2.3, will be explored further. Only reconnaissance holes have been drilled at a spacing of 70 m. The variogram calculation yields γ values

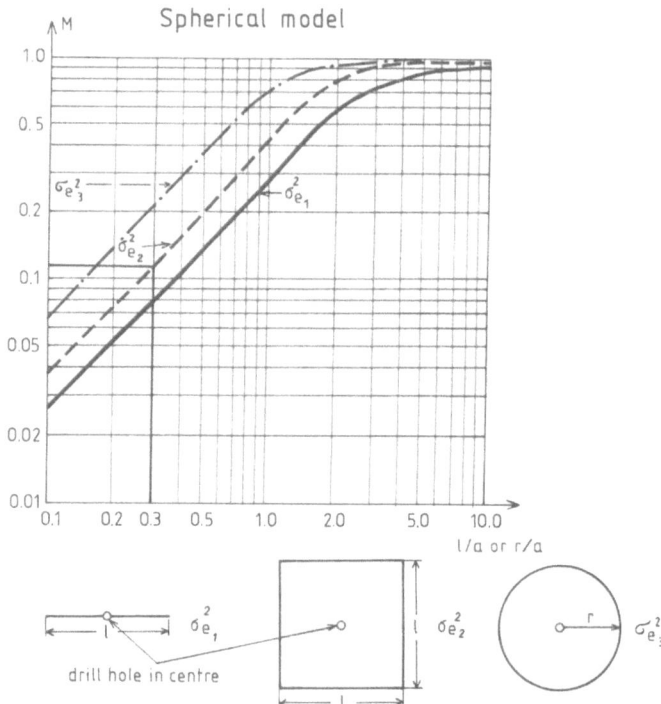

Fig. 100. Diagram for the determination of the extension variance based on the spherical model. e_1 Sampling in a drift, sample in the centre of a stretch of length l. e_2 Drill hole in the centre of a square block with edge dimension l. e_3 Drill hole in the centre of a circular block with radius r

only near the sill value, which is the variance s^2. This implies that all the values are still statistically independent (cf. Chap. 13.2.1). The previous drill spacing of 70 m is therefore greater than the range, a. To a first approximation, and as a working hypothesis, the a-priori information from the explored vein is taken into account. It is assumed that the relative variogram for the explored gold deposit (Fig. 72 shows the absolute variogram) with a range of 36 m is the same as that for the vein to be explored nearby.

The problem is to estimate how the extension variance, or the relative error, changes if the drill spacing is reduced to 35 m (fill-in drilling).

1st Step: The absolute parameters of the variogram in Fig. 72 are converted to relative values.

The nugget effect C_0 was 170, and the sill value $C_1 = 310$.

The average accumulation value for the deposit is 22.5 g Au/t m = $\overline{GT}$. The relative nugget effect and sill values are therefore:

$$C_{0_{rel}} = \frac{C_0}{(GT)^2} = \frac{170}{(22.5)^2} = 0.34 \, ,$$

$$C_{1_{rel}} = \frac{C_1}{(GT)^2} = \frac{310}{(22.5)^2} = 0.61 \, .$$

2nd Step: The ratio of the edges of the grid to the range a, which is 36 m, is required for the diagram in Fig. 100, so that

for 70-m drill spacing: $l/a = 70/36 = 1.94$,

for 35-m drill spacing: $l/a = 35/36 = 0.97$,

which is 1 for practical purposes. These values are then applied to the diagram in Fig. 100, and the following M factors are derived:

for 70-m drill spacing: $M = 0.8$, and

for 35-m drill spacing: $M = 0.42$.

3rd Step: These values are substituted in Eq. (2) above. The number of blocks is also required, and for the 70-m drill grid this number is 10. By halving the drill spacing, the number of drill holes is quadrupled, and is therefore 40.

1. This yields the following for a 70 m drill spacing:

$$\sigma^2_{e_{(70)}} = (C_0 + M \cdot C_1) \cdot \frac{1}{n} \, ,$$

$$\sigma^2_{e_{(70)}} = \frac{0.34 + 0.8 \cdot 0.61}{10} = \frac{0.83}{10} = 0.083 \, .$$

The standard deviation is therefore:

$$\sigma_{e_{(70)}} = 0.29.$$

2. This yields the following for a 35-m drill spacing:

$$\sigma^2_{e_{(35)}} = (C_0 + M \cdot C_1) \cdot \frac{1}{n} \, ,$$

$$\sigma^2_{e_{(35)}} = \frac{0.34 + 0.42 \cdot 0.61}{40} = \frac{0.60}{40} = 0.015 \, .$$

The standard deviation is therefore:

$$\sigma_{e_{(35)}} = 0.12$$

This shows that if the drill hole spacing is halved, then the standard deviation σ_e improves by about 40% of its original value.

This example illustrates the superiority of the geostatistical method as compared to classical statistical methods, in which the spatial interdependence is not allowed for. In a purely statistical calculation, the variance in the numerator would remain constant, and only the figure for n in the denominator would be quadrupled by halving the drill spacing, so that by changing from a 70-m spacing to a 35-m spacing

$$\frac{\sigma}{\sqrt{10}} \rightarrow \frac{\sigma}{\sqrt{40}}$$

the standard deviation is only halved.

If the classical statistical methods were used, then the variance in the numerator would also be greater, specifically C_0+C_1 (since the total sill value is the same as the variance; Chap. 13.2.1). Therefore, for a drill spacing of 35 m, the standard deviation would be:

$$\sqrt{\frac{C_0+C_1}{40}} = \sqrt{\frac{0.34+0.61}{40}} = \sqrt{\frac{0.95}{40}} = \sqrt{0.024} = \pm\,0.154\,.$$

4th Step: The relative error, after multiplication by the Student's t-factor [cf. Chap. 13.3.4.1, Eq. (3)], shows a slight improvement since the factor decreases with increasing values for n. The Student's t-factor is derived from Appendix Table 4 (for 90% confidence limits):

For a 70-m drill spacing: $t_{70}= 1.83$, since $n = 10$:

therefore the relative error is:

$$ki_{rel} = t \cdot \sigma_{e_{(70)}}$$

$$ki_{rel} = 1.83 \cdot 0.29 = 0.53\,.$$

For a 35-m drill spacing: $t_{35}= 1.68$, since $n = 40$:

therefore the relative error is:

$$ki_{rel} = t \cdot \sigma_{e_{(35)}}$$

$$ki_{rel} = 1.68 \cdot 0.12 = 0.20\,.$$

Therefore there is a relative improvement of 62% as compared to the value for a drill spacing of 70 m

20.2.2
Consideration of Rectangular Blocks

Because there are geologically related anisotropies, rectangular blocks, rather than square blocks, are used in most exploration programs. However, the problem of determining the optimum drill spacing remains.

It has been mentioned above that geostatistical methods can be used only if there is a variogram, and that requires a certain minimum amount of data. The basic structure of a drill grid is therefore already determined by this previous drilling since, for example, if the previous drilling was on a 100-m grid, then is not possible to transfer to a 40 x 40-m grid unless it is acceptable that an additional hole is drilled in every third block along strike and down dip. This is clearly not an optimal drill grid. Even if the drill grid is not predetermined, a practical spacing will normally be selected and nobody would select, for example, a 57.5 x 68.5-m grid.

Because of this, the type of grid that is selected will also depend on practical considerations. Furthermore, it is necessary to know the core reserves that will be mined in the first year of operation, since the dimensions of the definition block will depend on this (cf. Chap. 13.3.2 and Fig. 62). For every possible grid there is then a particular number, n, of blocks within the definition block.

Once these parameters have been determined, the calculations for the different grid patterns are carried out according to Chapter 13.3.4.3.1 for a regular grid (drill holes are in the centre of the pattern), or to Chapter 13.3.4.3.2 for the quasi-regular pattern of the random stratified grid (the drill holes are not in the centre of the pattern) if major, but unpredictable, deviations of the drill holes must be expected. The diagrams in Figs. 65 and 66 are fundamental to these calculations.

The calculation is the same as that in example in Chapter 13.3.5. The result can be graphically displayed as a function of the dimensions of the grid, or of the blocks (see Fig. 101 as an example of a spodumene deposit).

20.3
Defining a Random Grid Pattern

A systematic drilling grid that is designed to obtain as much information as possible about the geological anisotropy is always the most preferable. However, sometimes this ideal situation is just not practical. For example, a potential mineral deposit that has been sampled by only a few drill holes or trenches is offered for sale. Trends in the distribution of the mineralization cannot be positively identified, and it is obviously too early to apply geostatistical methods to the evaluation. The owner of the property will only permit a very short time for examining the mineralization, before a major option payment is due. During this short period, it is only possible to drill a few additional holes or dig a few trenches. These few drill holes or trenches could be distributed evenly over the potential mineral deposit, but several companies prefer random sampling, using a random grid.

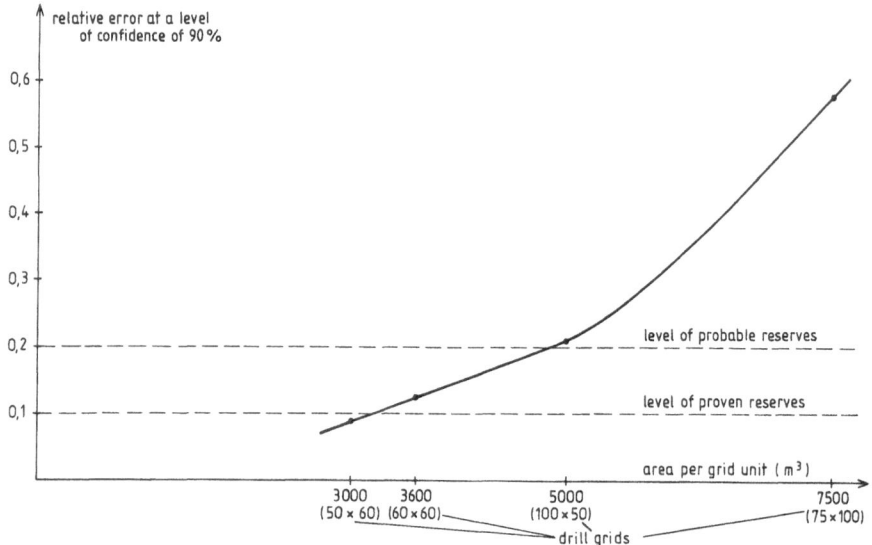

Fig. 101. Diagram for the determination of the optimal drill grid for a spodumene deposit

A random grid means that every point on the grid has the same chance of being a drill site, and thus the drill sites are more or less decided by the throw of a die. Random number tables are required for this, and they are provided as Appendix Table 15.

Figure 102 illustrates an example of defining a random grid. The area of interest is covered by a grid and the intersection points are numbered, and in this case they are all two-digit figures. Any row or column of the random number table (Appendix Table 15) is then selected. In this case, presume that the first row is selected.

Fig.102. Drill sites (circled) selected with random numbers in a regular drill pattern

0	1	0	2	0	3	0	4	0	5	0	6	7	0	8	0	9	10
11	12	13	14	15	16	17	18	19	20
21	22	23	24	25	26	27	28	29	30
31	32	33	34	35	36	37	38	39	40
41	42	43	44	45	46	47	48	49	50
51	52	53	54	55	56	57	58	59	60
61	62	63	64	65	66	67	68	69	70
71	72	73	74	75	76	77	78	79	80

The first five numbers are:

7559   2521   7011   5411   4602.

Two-digit numbers are required for the grid, and therefore two digits of the random numbers are combined. If a number is encountered that is not represented on the grid, or if the number has already occurred, then the next number is valid. Therefore the following nine random points on the grid are selected and are circled in Fig. 102:

Table 54. Selection of random drill locations using the random number table (Appendix Table 15)

Drill site	Coordinate	Drill site	Coordinate
1	75	6	11
2	59	7	54
3	25	8	11 is skipped, therefore 46
4	21	9	02
5	70		

20.4
Testing for Randomness

An irregular drill grid is often accompanied by the comment that it represents a quasi-random sample, or, in other words, the sampling is not affected by any trends. These irregularities are caused, for example, by lack of access in some areas because of difficult mountainous terrain or problems with the landowners.

If the variogram calculations indicated that the individual drill holes are not correlated, and presuming that the drill grid is truly random, then the population can be accepted as a representative sample so long as there are sufficient drill holes. This implies that a mean can be calculated from the drill holes. However, this is the case only if the grid is random.

Hazen (1967) proposed the next-neighbour test, based on Clarke (1956), as a test for randomness. He provides an example that is repeated here.

Figure 103 shows a drill grid that will be tested for randomness. The test is based on the spacing between the drill holes, and any of the drill holes can be treated as the starting point. The test examines if the starting hole is the nearest neighbour to its own first, second or third neighbour. The second and third neighbours of the starting hole can be taken into consideration only once in the same round, and cannot be treated as neighbours again until in the next round.

If the starting hole is also the next neighbour for its own next neighbour, then this neighbour is reflexive = R. If it is not the next neighbour, and another drill hole is closer than the starting hole, then this neighbour is non reflexive = NR.

Fig.103. Drill pattern to be
tested for randomness. (After
Hazen 1967)

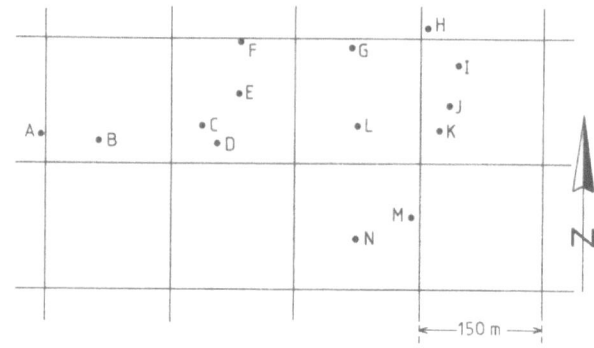

1st example: Drill hole B in Fig. 103 is the starting point.

1st neighbour: The next neighbour is A. Conversely, B is also the first neighbour for
A, and therefore the 1st neighbour is type R.

2nd Neighbour: The second closest neighbour is C. However, D is closer to C than
the starting hole B, and therefore the 2nd neighbour is type NR.

3rd neighbour: The third closest neighbour is D. C has already been used and
cannot be considered any more in this round. E is the next neighbour after point D,
and E is closer to D than B, and therefore the 3rd neighbour is also type NR.

2nd Example: Drill hole C in Fig. 103 is the starting point.

1st neighbour: D is the closest neighbour, and C is also the closest neighbour to D,
and therefore the 1st neighbour is type R.

2nd neighbour: E is the second-closest neighbour, and, conversely, C is also the
closest neighbour to E, since E is located closer to C than to F. The 2nd neighbour is
therefore also type R.

3rd neighbour: F is the third-closest neighbour to C. Since E cannot be used again in
this round, C is also the closest neighbour to F, since F is closer to C than it is to G.
Therefore the 3rd neighbour is also type R.

This test is now carried out successively for all the drill holes, and the results are
added together as in Table 55.

If the grid is a random grid, then the ratio V_R should be greater than

- 0.6215 for the 1st neighbours,
- 0.3863 for the 2nd neighbours and
- 0.2401 for the 3rd neighbours.

Table 55. Summary of the test of randomness (Hazen 1967)

Starting point drill hole	1st neighbour	2nd neighbour	3rd neighbour
A	R	NR	NR
B	R	NR	NR
C	R	R	R
D	R	R	R
E	NR	R	R
F	NR	NR	R
G	NR	R	NR
H	NR	R	NR
I	NR	R	R
J	R	R	NR
K	R	NR	NR
L	NR	NR	NR
M	R	NR	NR
N	R	NR	NR
Sum of all R	8	7	5
Ratio V_R (14 drill points totally)	$\dfrac{8}{14} = 0.571$	$\dfrac{7}{14} = 0.5$	$\dfrac{5}{14} = 0.357$

Examination of the results in Table 55 shows that these conditions are almost satisfied for the 1st neighbours, and clearly satisfied for the 2nd and 3rd neighbours. The whole grid, and therefore the samples, can thus be accepted as random.

21
Assessing the Exploration Risk

21.1
Introductory Comments

The techniques for assessing the exploration risk were developed mostly by the oil industry and are regularly used there. The probability factors are considerably less precise in metal exploration, and therefore a quantitative estimate of the exploration risk is in most cases extremely doubtful. For example, current experience indicates that the success rate of drilling wild cat holes in the oil industry is about 27%. There is no similarly high success rate for the metal exploration industry. A rule of thumb used in Canada for many years claimed that one deposit is found after about 45 000 line-km of airborne electromagnetic surveying, and this figure could be used to estimate the exploration risk (Sect. 21.4). Subsequently, over a period of many years, no mineral deposits were found by airborne electromagnetic surveying (Wellmer 1983d), and clearly this method had been applied to saturation level. New ore deposits were then discovered only after other exploration techniques were improved, and these techniques obviously have quite different success statistics.

However, in spite of the above, some methods for assessing the exploration risk will be briefly described. Sometimes, these methods can be used for a rough estimate, and therefore purely intuitive decision-making can be rendered a bit more rational.

21.2
The Expected Monetary Value (EMV) Method

The expected monetary value (EMV) method is regularly applied in the oil industry for decision-making (e.g. Harbaugh et al. 1977; Caldwell and Johnston 1985).

The EMV method essentially compares the monetary reward weighted by the probability of success, with the expenditure of risk capital weighted by the chance of a failure. The final total must be positive in order to justify a positive decision for the expenditure of risk capital (for example, the drilling of a hole). If p is the probability of success, then the probability of failure is:

$$q = 1\text{-}p .$$

The monetary success, or risk reward, is B, and the risk capital that is invested is Ri, then the EMV is:

$$EMV = B\cdot p - Ri\cdot(1-p) . \tag{1}$$

The final reward, B, is the monetary success in terms of the the cash value at the time of the drilling, in other words the net present value (NPV). The calculation of the NPV is described in Economic Evaluations in Exploration, Chapter 10.2.3.

Example: A basin analysis calculation suggests that, with probability of 50%, an offshore oil field contains sufficient oil to generate an NPV of $200 million with the commissioning of production.

The probability of success is assumed to be 27%, which is the average probability of success for wild cat drilling.

The offshore drilling is expected to cost $7 million. Is it justified to drill this hole?

Answer: There are two success probabilities in this problem:

a) the 50% probability, p_1, that oil worth an NPV of $200 million is actually proved; and
b) the 27% worldwide probability of success for wild cat drilling is p_2.
   Both probabilities are independent of each other, and both probabilities must be multiplied together in order to derive the overall probability.
   Consequently, Eq. (1) above for the EMV yields:

$$EMV = B \cdot p_1 \cdot p_2 - Ri \cdot (1 - p_2) \ .$$
$$EMV = 200 \cdot 0.5 \cdot 0.27 - 7 \cdot 0.73$$
$$EMV = 27 - 5.11 = \$21.89 \text{ million.}$$

The EMV is therefore positive, and it is justified to drill the hole.
This EMV method will be referred to again in the discussion on "decision trees" in Section 21.6.

21.3
The Expected Value of Each Discovery

As explained in the introduction to this chapter there are only a few, more or less reliable, probability figures for metal exploration. As a result, the average costs within a country per discovery of an economic ore deposit are often used (e.g. Levy 1982).

This way of considering the problem eliminates the necessity of using individual probabilities of success, but is suitable only for major and strategic exploration decisions.

The overall probability of success, pg, can be subdivided into three separate probabilities:

a) the probability, p_3, of a deposit occurring within the exploration area;
b) the probability, p_2, of discovering the deposit (p_3 and p_2 together correspond to the probability p_2 in the example in Sect. 21.2); and
c) the probability, p_1, of the economic value (this corresponds again to the probability p_1 in the example in Sect. 21.2).

All the above facets, each of which is associated with one of the three probabilities, are independent of each other. Therefore the probability of an economic success, pg, is:

$$P_g = P_1 \cdot P_2 \cdot P_3 .$$

The expected value (ev) per discovery is:

$$ev = P_g \cdot B - C . \tag{2}$$

pg is the probability of discovering an economic and productive mine. The average value for an economic discovery is used, and therefore according to the definition:

$$P_g = 1 .$$

B is again the NPV of an economic discovery and C is the cost per economic discovery. Since these costs, or exploration expenditures, extend over several years, the expenditures must be discounted for the interest rate (see Chap. 11.2.3.1 in Economic Evaluations in Exploration for calculating discounted money values), so that the risk reward B and the discovery costs C can be compared with the same time money values.

Example: Mackenzie and Bilodeau (1984) examined the Australian exploration scene and concluded that the discovery of a base metal deposit in the Australian Proterozoic had cost an average of A$ 58.9 million calculated in 1980 cash values. The average risk reward, discounted for interest to the moment when the decision was made to bring the deposit into production, was A$ 265 million (pretax, and also in 1980 cash values). If it is now assumed that an exploration company has decided to invest A$ 6 million per year in the search for base metal deposits, then the average cost per discovery is attained in nearly 10 years. There is always a certain time lapse between the discovery and the decision to invest in a mine, such as low metal prices, for example (e.g. Burgin 1976; Gries 1979), and it is assumed that 5 years pass before the investment decision is made.

The start of the exploration programme is taken as the reference point for this example, and therefore the following discounts for interest must be made:

a) The risk reward, B, must be discounted over 15 years (10 years exploration and 5 years lapse time between the discovery and the investment decision). Assuming an interest rate of 10%, so that i = 0.1, then the discount factor $q^{-n} = 0.239$ (Economic Evaluations in Exploration, Appendix Table 13).

b) The exploration expenditures are made annually at the same rate of A$ 6 million. This figure must be multiplied by the annuity present value factor for 10% (i = 0.1) and 10 years. This factor $b_n = 6.145$ (Economic Evaluations in Exploration, Appendix Table 14).

Therefore substituting these values in Eq. (2) above:

$$ev = p_g \cdot B - C \ . \tag{2}$$

for which:

$p_g = 1$,

$B = 265 \cdot 0.239$, and

$C = 6 \cdot 6.145$,

i.e. $ev = 1 \cdot 265 \cdot 0.239 - 6 \cdot 6.145$

$ev = 26.47$.

The expected value per discovery is therefore positive, and the exploration strategy for the search of base metal deposits in the Proterozoic of Australia can be recommended.

This type of study can be criticised since the probability p_g of making a discovery after expenditure of the average cost per discovery is not 1 ($p = 1$ implies certainty). Every explorationist knows that there are very few excellent prospects and anomalies, many moderate ones and even more marginal ones. There is obviously a great difference in the probability of success if the same exploration expenditures are spent only on drilling excellent prospects or only on marginal ones. Better approaches are described in Section 21.5.

The law of gambler's ruin must be taken into account.

21.4
Calculating the Exploration Success by the Law of Gambler's Ruin

The rule of gambling theory that is used in exploration for the calculation of probabilities of success, p_e, is the law of gambler's ruin:

$$p_e = 1 - e^{-p_s \cdot n} \ . \tag{3}$$

p_s is the probability of success of a single exploration project, and n is the number of exploration projects.

Since the expression $e^{-p_s \, n}$ will never be zero even for large values of n, but only approaches zero asymptotically, then the probability of success of the whole exploration project, p_e, can therefore never be 1, and thus there is no certainty. This is indeed found to be the real case in exploration.

Example: A rule of thumb used in Canada for many years in the exploration for volcanogenic base metal deposits in the Canadian Shield suggested that one deposit is found for every 45 000 line-km of airborne electromagnetic surveying. Obviously, no sensible exploration geologist is going to carry out 45 000 km of airborne electromagnetics just anywhere. An airborne survey area over about 100 to 500 km is selected in one of the potentially mineralized greenstone belts after careful research of all the available geological, geochemical and geophysical information as well as a ground reconnaissance. The spacing between the flight lines for this type of survey is nowadays about 200 m, so that 1 km on the ground is covered by 5 line-km. It is assumed that the average size of all the areas selected for the airborne survey is 400 km , which means that 2000 line-km will on average be flown in each area. The chance p_s of discovering a deposit in each survey area, according to the above probability of success, is:

$$p_s = \frac{2000}{45000} = 0.044.$$

This value can be substituted for p_s in Eq. (3) above.

If there are 23 airborne survey areas, so that $n = 23$ (and the total line-km is 45 000), then according to Eq. (3) above:

$$P_e = 1 - e^{-p_s \cdot n} , \qquad\qquad (3)$$

$$P_e = 1 - e^{-0.044 \cdot 23} = 1 - e^{-1.01} = 0.64 ,$$

and thus the probability is nowhere near 1, or certainty.

This simulated example from mineral exploration is actually quite realistic. Even if the best target areas have been selected, there is always a major degree of uncertainty as to whether there is actually a deposit within the area and, if there is, whether it will be discovered. Every area represents a new chance, just as in gambling. Each time the probability of discovering a deposit is unrelated to the probabilities in the previous exploration areas (as long as no major breakthrough has been made in the understanding of the model or geology that could positively influence the probability). The probability is therefore similar to throwing a die, whereby the probability of throwing a six is always 1/6, regardless of whether a six has been previously thrown 20 times or only once. Equation (3) above is suitable for calculating how many trials must be made in order to achieve a certain minimum probability. Each of the trials must be scientifically well based, or otherwise the value for the probability of success, p_s , of each individual project is no longer valid.

A minimum probability of 50% for the total project p_e is desirable (this implies that the chances of success are balanced with the chances of failure). If the chance of success is less than 50%, then the exploration programme should not even be started. p_e in Eq. (3) above is 0.5 for a 50% probability of success. The value for n is the unknown. Therefore:

$$P_e = 1 - e^{-P_s \cdot n} \text{ or} \tag{3}$$

$$e^{-P_s \cdot n} = 1 - P_e \,,$$

$$-p_s \cdot n = \ln(1 - p_e) \,;$$

$$n = -\frac{\ln(1 - p_e)}{P_s} \,. \tag{4}$$

If $p_e = 0.5$ and $p_s = 0.044$, then Eq. (4) above yields:

$$n = \mathrm{abs}\left[-\frac{\ln(1 - 0.5)}{0.044}\right] + 1 = \mathrm{abs}\left[\frac{+0.693}{0.044}\right] + 1 \,,$$

$$n = \mathrm{abs}[15.75] + 1 = 15 + 1 = 16 \,,$$

which means that 16 trials must be made, or 16 areas must be selected and flown with the airborne survey with the same degree of carefulness. It is obvious that persistence is an important ingredient in exploration programmes.

If the probability of success is to be greater, and approach certainty with, for example, an overall probability of success of 90% (pe = 0.9), then the number of tests that would be necessary are again derived from Eq. (4) above:

$$n = \mathrm{abs}\left[-\frac{\ln(1 - p_e)}{P_s}\right] + 1 \,, \tag{4}$$

$$n = \mathrm{abs}\left[-\frac{\ln(1 - 0.9)}{0.044}\right] \,,$$

$$n = \mathrm{abs}\left[\frac{2.30}{0.044}\right] + 1 = \mathrm{abs}[52,3] + 1 = 53 \,,$$

or 53 tests are required.

An additional law associated with mineral exploration is obvious from the results of the calculations for a 50% and then a 90% overall probability of success, and that is the law of diminishing returns. An overall probability of success of 0.5 is achieved after the first 16 tests, and the probability is only increased by 0.4 with the next 53-16 = 37 tests. This is expressed graphically in Fig. 104.

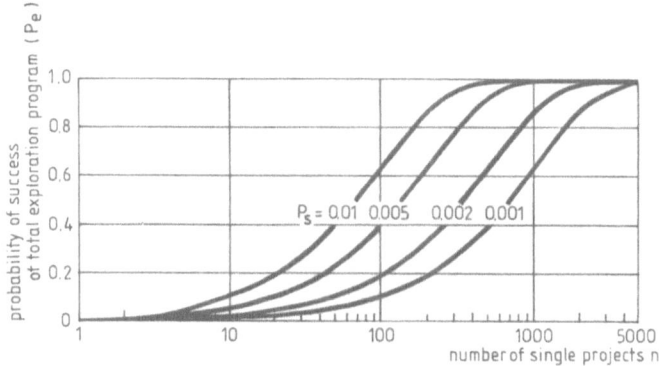

P_S = probability of success of a single exploration project

P_e = $1 - e^{-n \cdot P_s}$

P_e = 1 absolute success

P_e = 0 absolute failure

Fig. 104. Probability of success of total exploration programme as function of number of single projects. (After Sames and Wellmer 1981)

21.5
Calculation of the Minimum Exploration Budget

Mackenzie (1973) made the following proposal in order to define the minimum exploration budget A_{min}:

The probability of a single exploration project being successful is again ps. The chance of failure is then $1 - p_s$. In exploration, the chance of failure is known to be much greater than that of success. The chance that two projects are successively failures is $(1 - p_s) \cdot (1 - p_s)$, because the probabilities of success or failure in each of the exploration projects are independent of each other.

If there are n exploration projects, then the probability of a total failure is therefore $(1 - p_s)^n$. Alternatively, the probability of making at least one discovery is:

$$P_{min} = 1 - (1 - p_S)^n .$$ (5)

If Amin is the minimum exploration budget, and C is the average cost of a project, then:

$$n = A_{min}/C ,$$ (6)

substituting for n in equation (5):

$$P_{min} = 1 - (1 - p_S)^{A_{min}/C} ,$$ (7)

$$1 - P_{min} = (1 - p_S)^{A_{min}/C} ,$$

$$\ln(1-p_{min}) = \frac{A_{min}}{C}\,\ln(1-p_s)\ ,$$

$$A_{min} = C\,\frac{\ln(1-p_{min})}{\ln(1-p_s)}$$

Example: The costs for each individual exploration project are $400 000, the desired minimum probability of success is again 0.5, and the probability of each project being successful is again ps = 0.044, as in the example in Section 21.4.

Then according to Eq. (7) above for the minimum exploration budget Amin:

$$Amin = 400\,000\ \frac{\ln(1-0.5)}{\ln(1-0.044)} = 400\,000\ \frac{-0.693}{-0.045}\ ,$$

Amin = 400 000 15.4

so that practically 16 times the budget of a single project is required for a 50/50 chance of success, which is virtually the same result as in Section 21.4.

Relationship (3) in Section 21.4 can also be used to calculate the minimum exploration budget.

$$p_e = 1 - e^{-p_s\,n} \tag{3}$$

If the average cost of a discovery C_f is considered one exploration program, which was successful, then the probability $p_s = 1$.

For a given probability of a total exploration program, p_e, n is then the number by which the average discovery costs, C_f, have to be multiplied for the total program probability p_e.

Example: This calculation shows how many times the average discovery costs have to be budgeted for in a systematic exploration campaign to yield a 90% probability ($p_e = 0.9$) of actually making a discovery.

According to the above Eq. (3), the following relationship is true for n:

$$0.9 = 1 - e^{-1\,n},$$

$$0.1 = e^{-n}\quad\text{or}\quad e^n = 1/0.1 = 10$$

$$n = \ln 10 = 2.3\ .$$

This means that 2.3 times the average discovery cost must be included in a budget for a 90% discovery chance (see, for example, tables in Mackenzie and Woodall 1988).

21.6
Assessing Various Exploration Alternatives

21.6.1
Assessment with a Decision Diagram

The EMV method was introduced in Section 21.1 in order to make a yes/no decision (whether a hole should be drilled or not). This method can also be used if there are several exploration alternatives. A relationship similar to Eq. (2) in Section 21.3 is applied. The net present value of the reward, B, is multiplied with the probability of success, p, and this figure is then compared with the costs or the investment of risk capital, Ri. The costs are therefore not multiplied by the probability of failure. This value is called the expectancy value, EW. Therefore:

$$EW = p \cdot B\text{-}Ri. \quad (8)$$

A decision diagram or tree is constructed if a decision has to be made between several exploration possibilities. The diagram, or tree, can have any number of decision points (e.g. Hammond 1967; Harbaugh et al. 1977), and it contains both decision points and chance nodes (Fig. 105a). A yes/no decision must be made at the decision points. At the chance nodes, the decision is determined by the chances for either of both possibilities, each of which has a defined probability. The choice is therefore not determined by a personal decision.

Figure 105a shows an example from oil exploration after Hammond (1967). The money values shown on Fig. 105a are relative to a particular year, as was the case in Sections 21.2 and 21.3. It is assumed that the seismic investigations and the drilling all occur within 1 year. The expenditures for this are covered by the risk capital Ri, which can also be regarded as negative cash flow (Economic Evaluations in Exploration, Chap. 11.2.1). The costs for drilling are $1 000 000, and the costs for a seismic survey are $300 000. If the project is successful, then the net profits from the producing field must be discounted to the exploration year, which is the reference year, in order to derive the net present value (Economic Evaluations in Exploration, Chap. 11.2.3). The net present value of the profits of the producing field are estimated to be $4 000 000.

The calculation is made backwards, and the EW value is calculated with Eq. (8) above:

At decision point B (Fig. 105a and 105b):

The decision "yes drill" has an EW of

$$EW = -1\,000\,000 + 0.85 \cdot 4\,000\,000 = 2\,400\,000 \, \$$$

The decision "no drill" obviously always has an EW of zero since there are no costs, but at the same time there is no chance of finding anything.

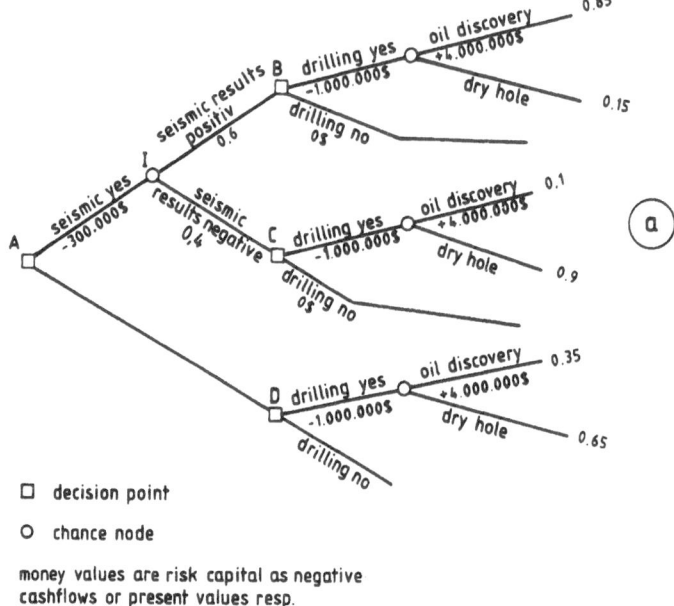

decision point

chance node

money values are risk capital as negative
cashflows or present values resp.

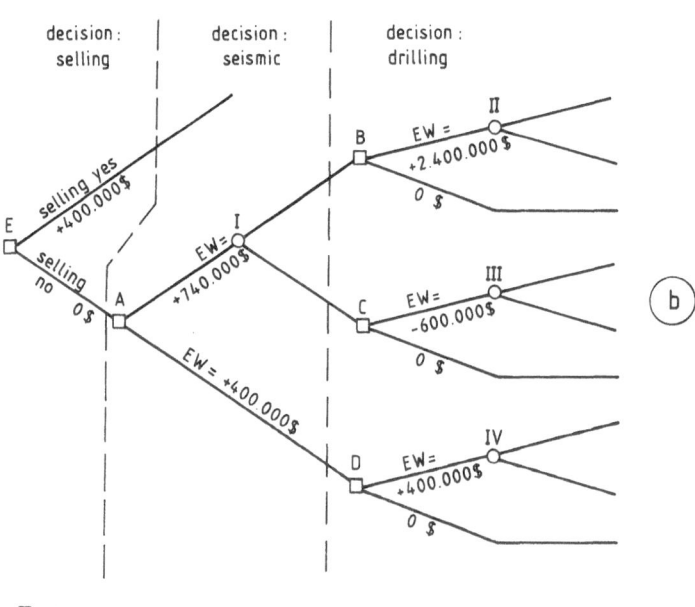

decision point
chance node

Fig. 105a-b. Decision tree for an oil exploration project. a Decision steps with probabilities, costs and present values of profits. b Expectancy value at decision points

Decision point C:
The EW for the decision "yes drill" is:

EW = -1 000 000 + 0.1 · 4 000 000 = -600 000 $

Decision point D:
The drilling to be decided at point D obviously has less chances of success than at point B, since the location of the drill site is made without previous seismic investigations.
The EW for the decision "yes drill" is therefore:

EW = -1 000 000 + 0.35 · 4 000 000 = +400 000 $

The decision of whether to drill a hole to test a negative seismic result can also be made at this point. The answer is no, since, as expected, the EW value for the decision point C is considerably smaller than for the decision point B, which marks a positive result from the seismic survey.

In the next step backwards in the calculation, it is examined if the seismic investigations should be carried out at all (decision point A). The EW value at decision point A for the decisions "seismics yes" and "drilling yes" is:

EW = -300 000 - 1 000 000 + 0.6 · 0.85 · 4 000 000 ,   (9)

EW = -1 300 000 + 2 040 000 = +740 000 $.

Since the probabilities at chance points I and II for a successful seismic investigation and a positive oil well, are independent of each other, they must be multiplied together in Eq. (9) above.

In this case the EW is greater (+740 000 $) as for the route via AD "seismic no" and "drilling yes", which has an EW of +400 000 $. Thus, a seismic investigation should definitely be undertaken.

The oil company is now suddenly offered $400 000 for the prospect (decision point E, Fig. 105b). Should the oil company accept or not? The two alternatives are:

- it receives the $400 000 with certainty (p = 1).
- the probability of discovering an oil field with a net present value of $4 million is p= 0.6 0.85 = 0.51, or 51%, and in order to accomplish this the company must with certainty (p = 1) expend $ 1.3 million on seismics and drilling.

A company that is permanently involved in several exploration projects, and intends to remain in the oil business, will refuse the offer. The reason is that the EW of selling is +$400 000, which is less than the EW of the alternative route via EAB – or "sell no", "seismic yes" and "drilling yes" with an EW of +$740 000.

An investor with no long term goals would probably respond according to the saying: a bird in the hand, or the sale for $400 000, is worth more than two in the bush. With an EW value of $740 000, but only a 51% chance of positive results from the exploration, the investor may therefore prefer to sell the property.

This example clearly demonstrates the real essence of exploration. If high rewards are expected from exploration in the form of major economic ore deposits, then high risks must be taken. However, even with such high investments, there is no way of being certain with a probability of success p=1.

21.6.2
Application in Mineral Exploration

It has already been noted that there are very few reliable probability factors in mineral exploration. They are mostly very subjective. However, one can apply the method of decision-making described in the previous section in order to derive a relative assessment of the probability factors, and therefore also to make a decision. Although geologists are also very much at odds about absolute probabilities, they are mostly in agreement about orders of magnitude of the relative probabilities (one decision path A is considered to be either less promising or else very much more promising as the other decision path B). This can be explained by an example:

Example: A helicopter-borne electromagnetic survey is undertaken at the start of a base metal exploration program. An interesting airborne anomaly is identified on the ground, and is surveyed with ground magnetics. The question of undertaking an additional IP survey is now raised. This IP survey should define the anomaly more precisely prior to testing it by drilling. The drilling costs, including mobilization costs, are estimated to be $15 000, and the IP survey would cost $10 000. The expected net present value of an average discovery is $NPV_L = \$20$ million, and the probability of success for drilling this type of anomaly is about 1%, so that $p_1 = 0.01$. The decision tree is displayed as Fig. 106.

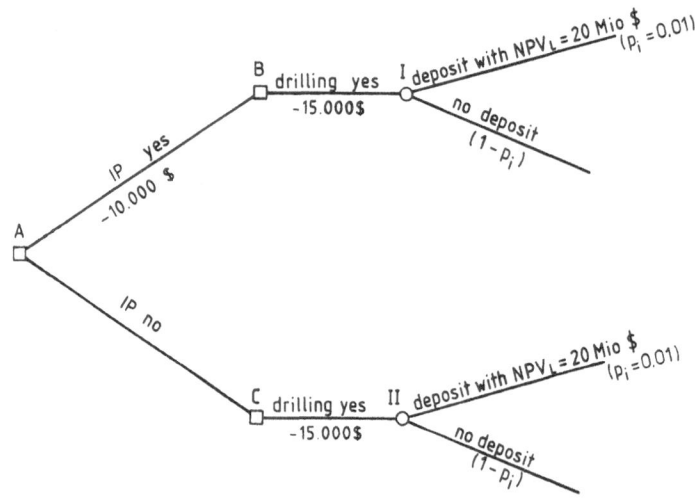

Fig. 106. Decision tree in base metal exploration

According to Eq. (8) in Section 21.6.1, the EW for the decision path AC, without the IP survey, is:

$$EW_{AC} = -15\ 000 + p_1 \cdot NPVL = -15\ 000 + 0.01 \cdot 20 \cdot 10^6$$

$$EW_{AC} = -15\ 000 + 200\ 000 = +\$185\ 000$$

The probability of success on the decision path AB, with the IP survey, changes to p_2. The relative EW is therefore:.

$$EW_{AB} = -15\ 000\ -10\ 000 + p_2 \cdot 20 \cdot 10^6$$

$$EW_{AB} = -25\ 000 + p_2 \cdot 20 \cdot 10^6\ \$.$$

In order that the costs of the IP survey are justified, the EW_{AB} value must be greater than, or at least similar to, the EW_{AC} value. This condition is now examined. The following must be tested:

$$EW_{AB} = EW_{AC}\ \text{or}$$

$$+185\ 000 = -25\ 000 + p_2 \cdot 20 \cdot 10^6$$

$$210\ 000 = p_2 \cdot 20 \cdot 10^6$$

$$p_2 = 0.0105\ (\text{or}\ 1.05\%)\ .$$

p_1 was 0.01 (or 1%). Therefore, if one is of the opinion that the chances of drilling into a mineral deposit are improved by at least 5%, then it is justified to carry out the IP survey.

The IP survey costs almost as much as a drill hole. Therefore it must now be decided whether, instead of the IP survey, it is better to drill two holes at marginally increased costs, rather than drilling only one hole, particularly since the mobilization charges have to be paid only once.

21.7
Assessment of Alternative Exploration Strategies Using Slichter's Method

Slichter (1960) introduced a factor of merit in order to evaluate alternative exploration strategies. The factor of merit, A_t, is the probability of a discovery P_e divided by the costs of a discovery, K_e:

$$A_t = P_e / K_e\ .\tag{1}$$

Example: It must be decided whether it is better to explore for a circular target with diameter D using a simple survey grid, for example by airborne geophysics, or with an orthogonal survey grid (cf. Fig. 82d and e).

In Chapter 17.2.1, Eq. (6), it was explained that, for this type of situation with a simple survey grid with parallel lines at a spacing S, the probability P_1 of discovery on one line is:

$$P_1 = D/S \text{ if } D < S .\tag{2}$$

If $D \geq S$ then the probability is obviously:

$$P_1 = 1.$$

The costs are directly proportional to the line-km flown. Therefore if the flight line spacing is halved, then the line-km and the costs K_e are doubled:

$$K_e = C_e/S ,\tag{3}$$

where C_e is a constant.

Consider the case of the orthogonal survey grid with survey lines at a spacing S, then the probability of discovery on one line (or better at least one line [Chap. 17.2.1, Eq. (7)] is:

$$P_1 = 2\frac{D}{S} - \frac{D^2}{S^2} \text{ if } D \pounds S .\tag{4}$$

If $D \geq S$, then the following is still true:

$$P_1 = 1.$$

On an orthogonal grid with survey lines at a spacing of S, double as many lines are flown as for the simple survey grid. Therefore the costs:

$$K_e = \frac{2\,C_e}{S} .\tag{5}$$

According to Eq. (1) above, the merit factor A_1 is:

Simple Grid:

$$A_{t1} = P_e/K_e ,$$

$$A_{t1} = \frac{D}{S}\frac{S}{C_e} = \frac{D}{C_e} \text{ if } D \ddagger S .$$

Fig.107. Assessment of two alternative search strategies. (After Slichter 1960)

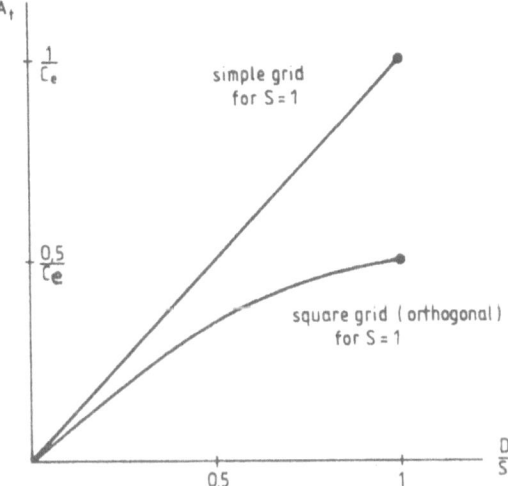

Orthogonal grid:

$$A_{12} = P_t/K_e \,,$$

$$A_{t2} = \frac{\left(\dfrac{2D}{S} - \dfrac{D^2}{S^2}\right) \cdot S}{2C_e} = \frac{2D - \dfrac{D^2}{S}}{2C_e} \quad \text{if } D \leq S \,.$$

If S is fixed at 1, then the relationships are derived up to $D/S = 1$ (i.e. D and S are equal), as they are shown in Fig. 107.

From this, it is clear that, if one wishes to cross the target with only one line, then a simple parallel grid is superior to the orthogonal grid. However, if one wishes to cross the target with two lines, which is only possible for the simple survey grid if $D>S$, then the converse is true (Slichter 1960).

Appendix Tables

Appendix Table 1. Examples of coefficients of variation of deposits based on drill hole data (total intersection summarized to one value)

Deposit		Coefficient of variation C
Name	Type	
North America	Various Cu porphyry deposits	0.5 - 0.8
Ramsbeck, Germany	Pb/Zn veins	0.9 - 1.9
Montcalm, Canada	Magmatic Cu-Ni-sulfide deposit	0.32
Nanisivik, Canada	Zn-Pb deposit in carbonates (various zones)	0.3 - 0.7
Western Australia	Various gold quartz veins	0.8 - 1.6
Cayeli, Turkey	Volcanogenic Cu-Zn-(Pb-Ag-Au) deposit	0.35
Greenbushes, Australia	Massive pegmatite spodumene-zone	0.25
Yeelirrie, Australia	Calcrete-uranium deposit	1.2
Meggen, Germany	Stratiform Zn (Pb)-barite-pyrite deposit in clastic sediments	0.20
	Various iron ore deposits	ca. 0.25
Hungary	Bauxite deposits, Flat, stratabound Flat, lensy „Sink-hole" deposits	0.2 - 0.4 0.3 - 0.5 0.6 - 0.9

Appendix Table 2. Cumulative frequency distribution (cumulative density function) of the standardized normal distribution (standard deviation σ = 1, area under normal distribution F = 1, limiting value x, cumulative requency Ø)

X	φ(x)	X	φ(x)	X	φ(x)	X	φ(x)
-2.00	0.0228	-1.50	0.0668	-1.00	0.1587	-0.50	0.3085
-1.98	0.0239	-1.48	0.0694	-0.98	0.1635	-0.48	0.3156
-1.96	0.0250	-1.46	0.0721	-0.96	0.1685	-0.46	0.3228
-1.94	0.0262	-1.44	0.0749	-0.94	0.1736	-0.44	0.3300
-1.92	0.0274	-1.42	0.0778	-0.92	0.1788	-0.42	0.3372
-1.90	0.0287	-1.40	0.0808	-0.90	0.1841	-0.40	0.3446
-1.88	0.0301	-1.38	0.0838	-0.88	0.1894	-0.38	0.3520
-1.86	0.0314	-1.36	0.0869	-0.86	0.1949	-0.36	0.3594
-1.84	0.0329	-1.34	0.0901	-0.84	0.2005	-0.34	0.3669
-1.82	0.0344	-1.32	0.0934	-0.82	0.2061	-0.32	0.3745
-1.80	0.0359	-1.30	0.0968	-0.80	0.2119	-0.30	0.3821
-1.78	0.0375	-1.28	0.1003	-0.78	0.2177	-0.28	0.3897
-1.76	0.0392	-1.26	0.1038	-0.76	0.2236	-0.26	0.3974
-1.74	0.0409	-1.24	0.1075	-0.74	0.2296	-0.24	0.4052
-1.72	0.0427	-1.22	0.1112	-0.72	0.2358	-0.22	0.4129
-1.70	0.0446	-1.20	0.1151	-0.70	0.2420	-0.20	0.4207
-1.68	0.0465	-1.18	0.1190	-0.68	0.2483	-0.18	0.4286
-1.66	0.0485	-1.16	0.1230	-0.66	0.2546	-0.16	0.4364
-1.64	0.0505	-1.14	0.1271	-0.64	0.2611	-0.14	0.4443
-1.62	0.0526	-1.12	0.1314	-0.62	0.2676	-0.12	0.4522
-1.60	0.0548	-1.10	0.1357	-0.60	0.2743	-0.10	0.4602
-1.58	0.0571	-1.08	0.1401	-0.58	0.2810	-0.08	0.4681
-1.56	0.0594	-1.06	0.1446	-0.56	0.2877	-0.06	0.4761
-1.54	0.0618	-1.04	0.1492	-0.54	0.2946	-0.04	0.4840
-1.52	0.0643	-1.02	0.1539	-0.52	0.3015	-0.02	0.4920

Fig.108. The normal distribution

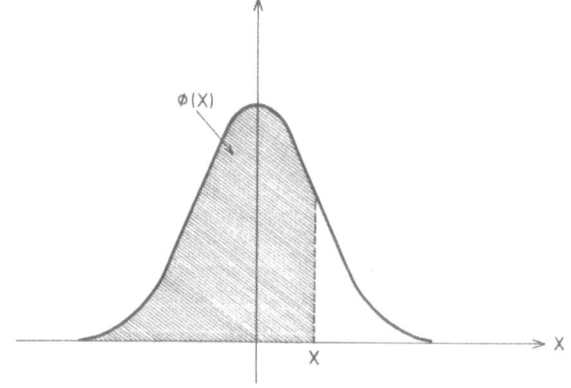

(Table 2 Continuation)

x	$\phi(x)$	x	$\phi(x)$	x	$\phi(x)$	x	$\phi(x)$
+0.02	0.5080	+0.52	0.6985	+1.02	0.8461	+1.52	0.9357
+0.04	0.5160	+0.54	0.7054	+1.04	0.8508	+1.54	0.9382
+0.06	0.5239	+0.56	0.7123	+1.06	0.8554	+1.56	0.9406
+0.08	0.5319	+0.58	0.7190	+1.08	0.8599	+1.58	0.9429
+0.10	0.5398	+0.60	0.7257	+1.10	0.8643	+1.60	0.9452
+0.12	0.5478	+0.62	0.7324	+1.12	0.8686	+1.62	0.9474
+0.14	0.5557	+0.64	0.7389	+1.14	0.8729	+1.64	0.9495
+0.16	0.5636	+0.66	0.7454	+1.16	0.8770	+1.66	0.9515
+0.18	0.5714	+0.68	0.7517	+1.18	0.8810	+1.68	0.9535
+0.20	0.5793	+0.70	0.7580	+1.20	0.8849	+1.70	0.9554
+0.22	0.5871	+0.72	0.7642	+1.22	0.8888	+1.72	0.9573
+0.24	0.5948	+0.74	0.7704	+1.24	0.8925	+1.74	0.9591
+0.26	0.6026	+0.76	0.7764	+1.26	0.8962	+1.76	0.9608
+0.28	0.6103	+0.78	0.7823	+1.28	0.8997	+1.78	0.9625
+0.30	0.6179	+0.80	0.7881	+1.30	0.9032	+1.80	0.9641
+0.32	0.6255	+0.82	0.7939	+1.32	0.9066	+1.82	0.9656
+0.34	0.6331	+0.84	0.7995	+1.34	0.9099	+1.84	0.9671
+0.36	0.6406	+0.86	0.8051	+1.36	0.9131	+1.86	0.9686
+0.38	0.6480	+0.88	0.8106	+1.38	0.9162	+1.88	0.9699
+0.40	0.6554	+0.90	0.8159	+1.40	0.9192	+1.90	0.9713
+0.42	0.6628	+0.92	0.8212	+1.42	0.9222	+1.92	0.9726
+0.44	0.6700	+0.94	0.8264	+1.44	0.9251	+1.94	0.9738
+0.46	0.6772	+0.96	0.8315	+1.46	0.9279	+1.96	0.9750
+0.48	0.6844	+0.98	0.8365	+1.48	0.9306	+1.98	0.9761
+0.50	0.6915	+1.00	0.8413	+1.50	0.9332	+2.00	0.9772

Appendix Table 3. Proposals for quantitative reserve/resource classification systems

Country	Levels of confidence (%)	Block size	Categories of errors				Reference
Germany	90	Production volume of 3 to 4 years	Proven ± 10%	Probable ± 20%	Possible I ± 30%	Possible II ± 50%	Wellmer (1983a)
Austria (national standard)	90	?	reliable estimates				ÖNORM (Anonymous 1989)
			1A ± 20%	1B ± 30%	1C ± 50%		
Germany/ Canada	95	Volume for short-term planning (6-8 month)	A ± 20%				Diehl (1981) Diehl and David (1982)
		Volume for medium-term planning (0,5-2 years)		B ± 40%			
		Volume for long-term planning (2-5 years)			C ± 80%		
Canada	90	Global	Class I < 10% Class II < 20%		Class I < 30% Class II < 40%		Vallée (1986) Vallée et al. (1992) Vallée et al. (1993)
		Local/blocks	Class I < 25% Class II < 50%		Class I > 50% Class II > 50%		
Canada	95	Annual production volume	Proven Class A				Weber and Morgan (1993)
	84	Annual production volume		Proven Class B			
	95	Global			Probable Class C		
	85	Global or annual production volume				Probable Class D	
			Errors not fixed, lower error limit to fulfill minimum economic criteria				
Australia (only for coal)	95	According to) 1 km²) (1x1)	A ± 20%				Whitchurch et al. (1990)
		Queensland) and) New) 4 km² South) (2x2)		B ± 40%			
		Wales) Coal Reserve) 16 km² codes) (4x4)			C ± 60%		

Appendix Table 4. Student's t-distribution

Column 1 No. of samples n	Column 2 90% level of confidence	Column 3 95% level of confidence	Column 4 No. of sample pairs	Column 5[a] No. of degress of freedom generally = F for k lots with n samples (see below)
2	6.13	12.7	3	1
3	2.92	4.30	4	2
4	2.35	3.18	5	3
5	2.13	2.78	6	4
6	2.02	2.57	7	5
7	1.94	2.45	8	6
8	1.90	2.37	9	7
9	1.86	2.31	10	8
10	1.83	2.26	11	9
11	1.81	2.23	12	10
12	1.80	2.20	13	11
13	1.78	2.18	14	12
14	1.77	2.16	15	13
15	1.76	2.15	16	14
16	1.75	2.13	17	15
17	1.75	2.12	18	16
18	1.74	2.11	19	17
19	1.73	2.10	20	18
20	1.73	2.09	21	19
21	1.73	2.09	22	20
23	1.72	2.07	24	22
25	1.71	2.06	26	24
26	1.71	2.06	27	25
29	1.70	2.05	30	28
30	1.70	2.04	31	29
40	1.68	2.02	40	40
50	1.68	2.01	50	50
100	1.66	1.98	100	100
200	1.65	1.97	200	200
∞	1.65	1.96	∞	∞

a)   Column 5: the number of the degrees of freedom F for k lots with n_i samples is

$$F = \left(\sum_{i=1}^{k} n_i \right) - k \quad \text{(Clauß and Ebner 1985)}$$

Appendix Table 5. Table of natural logarithms

	0.0	0.1	0.2	0.3	0.4	0.5	0.6	0.6	0.8	0.9
0	-	-2.30259	-1.60944	-1.20397	-0.91629	-0.69315	-0.51083	-0.35667	-0.22314	-0.10536
1	0.00000	0.09531	0.18232	0.26236	0.33647	0.40547	0.47000	0.53063	0.58779	0.64185
2	0.69315	0.74194	0.78846	0.83291	0.37547	0.91629	0.95551	0.99325	1.02962	1.06471
3	1.09861	1.13140	1.16315	1.19392	1.22378	1.25276	1.28093	1.30833	1.33500	1.36098
4	1.38629	1.41099	1.43508	1.45861	1.48160	1.50408	1.52606	1.54756	1.56862	1.58924
5	1.60944	1.62924	1.64866	1.66771	1.68640	1.70475	1.72277	1.74047	1.75786	1.77495
6	1.79176	1.80829	1.82455	1.84055	1.85630	1.87180	1.88707	1.90211	1.91692	1.93152
7	1.94591	1.96009	1.97408	1.98787	2.00148	2.01490	2.02815	2.04122	2.05412	2.06686
8	2.07944	2.09186	2.10413	2.11626	2.12823	2.14007	2.15176	2.16332	2.17475	2.18605
9	2.19722	2.20827	2.21920	2.23001	2.24071	2.25129	2.26176	2.27213	2.28238	2.29253
10	2.30259	2.31254	2.32239	2.33214	2.34181	2.35138	2.36085	2.37024	2.37955	2.38876
11	2.39790	2.40695	2.41591	2.42480	2.43361	2.44235	2.45101	2.45959	2.46810	2.47654
12	2.48491	2.49321	2.50144	2.50960	2.51770	2.52573	2.53370	2.54160	2.54945	2.55723
13	2.56495	2.57261	2.58022	2.58776	2.59525	2.60269	2.61007	2.61740	2.62467	2.63189
14	2.63906	2.64617	2.65324	2.66026	2.66723	2.67415	2.68102	2.68785	2.69463	2.70136
15	2.70805	2.71469	2.72130	2.72785	2.73437	2.74084	2.74727	2.75366	2.76001	2.76632
16	2.77259	2.77882	2.78501	2.79116	2.79728	2.80336	2.80940	2.81541	2.82138	2.82731
17	2.83321	2.83908	2.84491	2.85071	2.85647	2.86220	2.86790	2.87356	2.87920	2.88480
18	2.89037	2.89591	2.90142	2.90690	2.91235	2.91777	2.92316	2.92852	2.93386	2.93916
19	2.94444	2.94969	2.95491	2.96010	2.96527	2.97041	2.97553	2.98062	2.98568	2.99072
20	2.99573	3.00072	3.00568	3.01062	3.01553	3.02042	3.02529	3.03013	3.03495	3.03975
21	3.04452	3.04927	3.05400	3.05871	3.06339	3.06805	3.07269	3.07731	3.08191	3.08649
22	3.09104	3.09558	3.10009	3.10459	3.10906	3.11352	3.11795	3.12236	3.12676	3.13114
23	3.13549	3.13983	3.14415	3.14845	3.15274	3.15700	3.16125	3.16547	3.16969	3.17388
24	3.17805	3.18221	3.18635	3.19048	3.19458	3.19867	3.20275	3.20680	3.21084	3.21487
25	3.21888	3.22287	3.22684	3.23080	3.23475	3.23868	3.24259	3.24649	3.25037	3.25424
26	3.25810	3.26193	3.26576	3.26957	3.27336	3.27714	3.28091	3.28466	3.28840	3.29213
27	3.29584	3.29953	3.30322	3.30689	3.31054	3.31419	3.31782	3.32143	3.32504	3.32863
28	3.33220	3.33577	3.33932	3.34286	3.34639	3.34990	3.35341	3.35690	3.36037	3.36384
29	3.36730	3.37074	3.37417	3.37759	3.38099	3.38439	3.38777	3.39115	3.39451	3.39786
30	3.40120	3.40452	3.40784	3.41115	3.41444	3.41773	3.42100	3.42426	3.42751	3.43076
31	3.43399	3.43721	3.44042	3.44362	3.44681	3.44999	3.45316	3.45632	3.45947	3.46261
32	3.46574	3.46886	3.47197	3.47507	3.47816	3.48124	3.48431	3.48737	3.49043	3.49347
33	3.49651	3.49953	3.50255	3.50556	3.50856	3.51155	3.51453	3.51750	3.52046	3.52341
34	3.52636	3.52930	3.53223	3.53514	3.53806	3.54096	3.54385	3.54674	3.54962	3.55249
35	3.55535	3.55820	3.56105	3.56388	3.56671	3.56953	3.57235	3.57515	3.57795	3.58074
36	3.58352	3.58629	3.58906	3.59182	3.59457	3.59731	3.60005	3.60278	3.60550	3.60821
37	3.61092	3.61362	3.61631	3.61899	3.62167	3.62434	3.62700	3.62966	3.63231	3.63495
38	3.63759	3.64021	3.64284	3.64545	3.64806	3.65066	3.65325	3.65584	3.65842	3.66099
39	3.66356	3.66612	3.66868	3.67122	3.67377	3.67630	3.67883	3.68135	3.68387	3.68638
40	3.68888	3.69138	3.69387	3.69635	3.69883	3.70130	3.70377	3.70623	3.70868	3.71113
41	3.71357	3.71601	3.71844	3.72086	3.72328	3.72569	3.72810	3.73050	3.73290	3.73529
42	3.73767	3.74005	3.74242	3.74479	3.74715	3.74950	3.75185	3.75420	3.75654	3.75887
43	3.76120	3.76352	3.76584	3.76815	3.77046	3.77276	3.77506	3.77735	3.77963	3.78191
44	3.78419	3.78646	3.78872	3.79098	3.79324	3.79549	3.79773	3.79997	3.80221	3.80444
45	3.80666	3.80888	3.81110	3.81331	3.81551	3.81771	3.81991	3.82210	3.82428	3.82646
46	3.82864	3.83081	3.83298	3.83514	3.83730	3.83945	3.84160	3.84374	3.84588	3.84802
47	3.85015	3.85227	3.85439	3.85651	3.85862	3.86073	3.86283	3.86493	3.86703	3.86912
48	3.87120	3.87328	3.87536	3.87743	3.87950	3.88156	3.88362	3.88568	3.88773	3.88978
49	3.89182	3.89386	3.89589	3.89792	3.89995	3.90197	3.90399	3.90600	3.90801	3.91002
50	3.91202	3.91402	3.91601	3.91800	3.91999	3.92197	3.92395	3.92593	3.92790	3.92986
51	3.93183	3.93378	3.93574	3.93769	3.93964	3.94158	3.94352	3.94546	3.94739	3.94932
52	3.95124	3.95316	3.95508	3.95700	3.95891	3.96081	3.96272	3.96462	3.96651	3.96840
53	3.97029	3.97218	3.97406	3.97594	3.97781	3.97968	3.98155	3.98341	3.98527	3.98713
54	3.98898	3.99083	3.99268	3.99452	3.99636	3.99820	4.00003	4.00186	4.00369	4.00551
55	4.00733	4.00915	4.01096	4.01277	4.01458	4.01638	4.01818	4.01998	4.02177	4.02356
56	4.02535	4.02714	4.02892	4.03069	4.03247	4.03424	4.03601	4.03777	4.03954	4.04130
57	4.04305	4.04480	4.04655	4.04830	4.05004	4.05178	4.05352	4.05526	4.05699	4.05872

	0.0	0.1	0.2	0.3	0.4	0.5	0.6	0.6	0.8	0.9
58	4.06044	4.06217	4.06389	4.06560	4.06732	4.06903	4.07073	4.07244	4.07414	4.07584
59	4.07754	4.07923	4.08092	4.08261	4.08429	4.08598	4.08766	4.08933	4.09101	4.09268
60	4.09434	4.09601	4.09767	4.09933	4.10099	4.10264	4.10429	4.10594	4.10759	4.10923
61	4.11087	4.11251	4.11415	4.11578	4.11741	4.11904	4.12066	4.12228	4.12390	4.12552
62	4.12713	4.12875	4.13035	4.13196	4.13357	4.13517	4.13677	4.13836	4.13995	4.14155
63	4.14313	4.14472	4.14630	4.14789	4.14946	4.15104	4.15261	4.15418	4.15575	4.15732
64	4.15888	4.16044	4.16200	4.16356	4.16511	4.16667	4.16821	4.16976	4.17131	4.17285
65	4.17439	4.17592	4.17746	4.17899	4.18052	4.18205	4.18358	4.18510	4.18662	4.18814
66	4.18965	4.19117	4.19268	419419	4.19570	4.19720	4.19870	4.20020	4.20170	4.20320
67	4,20469	4.20618	4.20767	4.20916	4.21064	4.21213	4.21361	4.21509	4.21656	4.21804
68	4.21951	4.22098	4.22244	4.22391	4.22537	4.22683	4.22829	4.22975	4.23120	4.23266
69	4.23411	5.23555	4.23700	4.23844	4.23989	4.24133	4.24276	4.24420	4.24563	4.24707
70	4.24850	4.24992	4.25135	4.25277	4.25419	4.25561	4.25703	4.25845	4.25986	4.26127
71	4.26268	4.26409	4.26549	4.26690	4.26830	4.26970	4.27109	4.27249	4.27388	4.27528
72	4.27667	4.27805	4.27944	4.28082	4.28221	4.28359	4.28496	4.28634	4.28772	4.28909
73	4.29046	4.29183	4.29320	4.29456	4.29592	4.29729	4.29864	4.30000	4.30136	4.30271
74	4.30407	4.30542	4.30676	4.30811	4.30946	4.31080	4.31214	4.31348	4.31482	4.31615
75	4.31749	4.31882	4.32015	4.32148	4.32281	4.32413	4.32546	4.32678	4.32810	4.32942
76	4.33073	4.33205	4.33336	4.33467	4.33598	4.33729	4.33860	4.33990	4.34120	4.34251
77	4.34381	4.34510	4.34640	4.34769	4.34899	4.35028	4.35157	4.35286	4.35414	4.35543
78	4.35671	4.35799	4.35927	4.36055	4.36182	4.36310	4.36437	4.36564	4.36691	4.36818
79	4.36945	4.37071	4.37198	4.37324	4.37450	4.37576	4.37701	4.37827	4.37952	4.38078
80	4.38203	4.38328	4.38452	4.38577	4.38701	4.38826	4.38950	4.39074	4.39198	4.39321
81	4.39445	4.39568	4.39692	4-39815	4.39938	4.40060	4.40183	4.40305	4.40428	4.40550
82	4.40672	4.40794	4.40916	4.41037	4.41159	4.41280	4.41401	4.41522	4.41643	4.41763
83	4.41884	4.42004	4.42125	4.42245	4.42365	4.42485	4.42604	4.42724	4.42843	4.42963
84	4.43082	4.43201	4.43319	4.43438	4.43557	4.43675	4.43793	4.43912	4.44030	4.44147
85	4.44265	4.44383	4.44500	4.44617	4.44735	4.44852	4.44969	4.45085	4.45202	4.45318
86	4.45435	4.45551	4.45667	4.45783	4.45899	4.46014	4.46130	4.46245	4.46361	4.46476
87	4.46591	4.46706	4.46820	4.46935	4.47050	4.47164	4.47278	4.47392	4.47506	4.47620
88	4.47734	4.47847	4.47961	4.48074	4.48187	4.48300	4.48413	4.48526	4.48639	4.48751
89	4.48864	4.48976	4.49088	4.49200	4.49312	4.49424	4.49536	4.49647	4.49758	4.49870
90	4.49981	4.50092	4.50203	4.50314	4.50424	4.50535	4.50645	4.50756	4.50866	4.50976
91	4.51086	4.51196	4.51305	4.51415	4.51525	4.51634	4.51743	4.51852	4.51961	4.52070
92	4.52179	4.52287	4.52396	4.52504	4.52613	4.52721	4.52829	4.52937	4.53045	4.53152
93	4.53260	4.53367	4.53475	4.53582	4.53689	4.53796	4.53903	4.54010	4.54116	4.54223
94	4.54329	4.54436	4.54542	4.54648	4.54754	4.54860	4.54966	4.55071	4.55177	4.55282
95	4.55388	4.55493	4.55598	4.55703	4.55808	4.55913	4.56017	4.56122	4.56226	4.56331
96	4.56435	4.56539	4.56643	4.56747	4.56851	4.56954	4.57058	4.57161	4.57265	4.57368
97	4.57471	4.57574	4.57677	4.57780	4.57883	4.57985	4.58088	4.58190	4.58292	4.58395
98	4.58497	4.58599	4.58701	4.58802	4.58904	4.59006	4.59107	4.59208	4.59310	4.59411
99	4.59512	4.59613	4.59714	4.59815	4.59915	4.60016	4.60116	4.60217	4.50317	4.60417

Comment: If the natural logarithms for a number larger than 100 has to be calculated, the value $y = 10 \cdot x_i$ or the value $y = 100 \cdot x_2$ is taken, e.g.:

$\ln 160 = \ln(10 \cdot 16) = \ln 10 + \ln 16 = 2.30259 + 2.77259 = 5.07518.$

Appendix Table 6. Sichel's estimator t_{si} for the estimation of the arithmetic mean of a lognormal distribution (logarithmic variance β^2 based on natural logarithms). (South African Institute of Mining and Metallurgy: Sichel 1966)

β^2 \ n	2	3	4	5	6	7	8	9	10	12	14	16	18	20	50	100	1000
0.00	1.000	1.000	1.000	1.000	1.000	1.000	1.000	1.000	1.000	1.000	1.000	1.000	1.000	1.000	1.000	1.000	1.000
0.02	1.010	1.010	1.010	1.010	1.010	1.010	1.010	1.010	1.010	1.010	1.010	1.010	1.010	1.010	1.010	1.010	1.010
0.04	1.020	1.020	1.020	1.020	1.020	1.020	1.020	1.020	1.020	1.020	1.020	1.020	1.020	1.020	1.020	1.020	1.020
0.06	1.030	1.030	1.030	1.030	1.030	1.030	1.030	1.030	1.030	1.030	1.030	1.030	1.030	1.030	1.030	1.030	1.030
0.08	1.040	1.040	1.040	1.040	1.040	1.041	1.041	1.041	1.041	1.041	1.041	1.041	1.041	1.041	1.041	1.041	1.041
0.10	1.050	1.051	1.051	1.051	1.051	1.051	1.051	1.051	1.051	1.051	1.051	1.051	1.051	1.051	1.051	1.051	1.051
0.12	1.061	1.061	1.061	1.061	1.061	1.061	1.061	1.061	1.061	1.062	1.062	1.062	1.062	1.062	1.062	1.062	1.062
0.14	1.071	1.071	1.071	1.072	1.072	1.072	1.072	1.072	1.072	1.072	1.072	1.072	1.072	1.072	1.072	1.072	1.072
0.16	1.081	1.082	1.082	1.082	1.082	1.082	1.082	1.083	1.083	1.083	1.083	1.083	1.083	1.083	1.083	1.083	1.083
0.18	1.091	1.092	1.092	1.093	1.093	1.093	1.093	1.093	1.093	1.094	1.094	1.094	1.094	1.094	1.094	1.094	1.094
0.20	1.102	1.102	1.103	1.103	1.104	1.104	1.104	1.104	1.104	1.104	1.104	1.104	1.104	1.105	1.105	1.105	1.105
0.3	1.154	1.156	1.157	1.158	1.158	1.159	1.159	1.159	1.160	1.160	1.160	1.160	1.160	1.161	1.161	1.162	1.162
0.4	1.207	1.210	1.212	1.214	1.215	1.216	1.216	1.217	1.217	1.218	1.218	1.129	1.219	1.219	1.220	1.221	1.221
0.5	1.260	1.266	1.269	1.272	1.273	1.275	1.276	1.276	1.277	1.278	1.279	1.279	1.280	1.280	1.282	1.283	1.284
0.6	1.315	1.323	1.328	1.332	1.334	1.336	1.337	1.338	1.339	1.341	1.342	1.343	1.344	1.344	1.348	1.349	1.350
0.7	1.371	1.382	1.389	1.393	1.397	1.399	1.401	1.403	1.404	1.406	1.408	1.409	1.410	1.411	1.416	1.417	1.419
0.8	1.427	1.442	1.451	1.457	1.462	1.465	1.468	1.470	1.472	1.475	1.477	1.478	1.480	1.481	1.487	1.490	1.492
0.9	1.485	1.503	1.515	1.523	1.529	1.533	1.537	1.540	1.542	1.546	1.549	1.551	1.552	1.554	1.562	1.565	1.568
1.0	1.543	1.566	1.580	1.591	1.598	1.604	1.608	1.612	1.615	1.620	1.623	1.626	1.628	1.630	1.641	1.645	1.649
1.1	1.602	1.630	1.648	1.661	1.670	1.677	1.682	1.687	1.691	1.697	1.701	1.705	1.708	1.710	1.723	1.728	1.733
1.2	1.662	1.696	1.718	1.733	1.744	1.752	1.759	1.765	1.770	1.777	1.782	1.787	1.790	1.793	1.810	1.816	1.822
1.3	1.724	1.764	1.789	1.807	1.820	1.831	1.839	1.846	1.851	1.860	1.867	1.872	1.876	1.880	1.900	1.908	1.916
1.4	1.786	1.832	1.862	1.884	1.900	1.912	1.922	1.930	1.936	1.947	1.955	1.961	1.966	1.971	1.995	2.004	2.014
1.5	1.848	1.903	1.938	1.963	1.981	1.996	2.007	2.017	2.025	2.037	2.047	2.054	2.060	2.065	2.095	2.106	2.117
1.6	1.912	1.975	2.015	2.044	2.066	2.082	2.096	2.107	2.116	2.131	2.142	2.151	2.158	2.164	2.199	2.212	2.226
1.7	1.977	2.049	2.095	2.128	2.153	2.172	2.188	2.201	2.212	2.229	2.242	2.252	2.260	2.267	2.308	2.323	2.340
1.8	2.043	2.124	2.177	2.214	2.243	2.265	2.283	2.298	2.310	2.330	2.345	2.357	2.367	2.375	2.422	2.440	2.460
1.9	2.110	2.201	2.260	2.303	2.336	2.361	2.382	2.399	2.413	2.436	2.453	2.467	2.478	2.487	2.542	2.563	2.586
2.0	2.178	2.280	2.347	2.395	2.431	2.460	2.484	2.503	2.519	2.545	2.565	2.581	2.594	2.604	2.668	2.692	2.718
2.1	2.247	2.360	2.435	2.489	2.530	2.563	2.589	2.611	2.630	2.659	2.682	2.700	2.714	2.726	2.800	2.827	2.858
2.2	2.317	2.442	2.526	2.586	2.632	2.669	2.698	2.723	2.744	2.778	2.803	2.824	2.840	2.854	2.937	2.969	3.004
2.3	2.388	2.526	2.618	2.686	2.737	2.778	2.811	2.839	2.863	2.900	2.929	2.952	2.971	2.987	3.082	3.118	3.158
2.4	2.460	2.612	2.714	2.788	2.846	2.891	2.928	2.959	2.986	3.028	3.060	3.086	3.108	3.125	3.233	3.274	3.320
2.5	2.533	2.699	2.812	2.894	2.957	3.008	3.049	3.084	3.113	3.160	3.197	3.226	3.250	3.270	3.391	3.438	3.490
2.6	2.607	2.789	2.912	3.003	3.073	3.128	3.174	3.213	3.245	3.298	3.339	3.371	3.398	3.420	3.557	3.610	3.669
2.7	2.682	2.880	3.015	3.114	3.191	3.253	3.304	3.346	3.382	3.441	3.486	3.522	3.552	3.577	3.730	3.791	3.857
2.8	2.759	2.973	3.120	3.229	3.314	3.382	3.437	3.484	3.524	3.589	3.639	3.680	3.713	3.740	3.912	3.980	4.055
2.9	2.836	3.068	3.228	3.347	3.440	3.514	3.576	3.627	3.671	3.743	3.799	3.843	3.880	3.911	4.102	4.178	4.263
3.0	2.914	3.166	3.339	3.469	3.570	3.651	3.718	3.775	3.824	3.902	3.964	4.013	4.054	4.088	4.301	4.387	4.482
3.1	2.994	3.265	3.453	3.593	3.703	3.792	3.866	3.928	3.981	4.068	4.136	4.190	4.235	4.273	4.510		
3.2	3.075	3.366	3.569	3.721	3.841	3.938	4.018	4.086	4.145	4.240	4.314	4.374	4.424	4.465	4.728		
3.3	3.157	3.469	3.688	3.853	3.983	4.088	4.176	4.250	4.314	4.418	4.500	4.566	4.620	4.666	4.956		
3.4	3.240	3.574	3.810	3.988	4.129	4.243	4.338	4.419	4.489	4.603	4.692	4.764	4.824	4.875	5.195		
3.5	3.324	3.682	3.935	4.127	4.279	4.403	4.506	4.594	4.670	4.794	4.892	4.971	5.037	5.092	5.445		
3.6	3.409	3.792	4.063	4.270	4.434	4.568	4.680	4.775	4.858	4.993	5.099	5.186	5.258	5.318	5.706		
3.7	3.496	3.903	4.194	4.416	4.593	4.738	4.859	4.962	5.052	5.198	5.315	5.409	5.488	5.554	5.980		
3.8	3.583	4.017	4.329	4.567	4.757	4.913	5.044	5.156	5.252	5.412	5.538	5.641	5.726	5.799	6.266		
3.9	3.672	4.134	4.466	4.721	4.925	5.093	5.234	5.355	5.460	5.633	5.770	5.882	5.975	6.054	6.566		

n / β²	2	3	4	5	6	7	8	9	10	12	14	16	18	20	50	100	1000
4.0	3.762	4.252	4.607	4.880	5.099	5.279	5.431	5.562	5.675	5.862	6.011	6.132	6.234	6.319	6.879		
4.1	3.853	4.373	4.751	5.042	5.277	5.471	5.634	5.775	5.897	6.099	6.260	6.392	6.502	6.596			
4.2	3.946	4.496	4.898	5.209	5.460	5.668	5.844	5.995	6.127	6.345	6.519	6.662	6.781	6.883			
4.3	4.040	4.622	5.049	5.380	5.649	5.872	6.060	6.223	6.364	6.599	6.788	6.942	7.072	7.182			
4.4	4.135	4.750	5.203	5.556	5.843	6.081	6.283	6.458	6.610	6.863	7.066	7.233	7.373	7.493			
4.5	4.231	4.881	5.361	5.736	6.042	6.297	6.513	6.700	6.863	7.136	7.355	7.536	7.687	7.816			
4.6	4.328	5.014	5.522	5.921	6.247	6.519	6.750	6.950	7.126	7.419	7.655	7.849	8.013	8.152			
4.7	4.427	5.149	5.687	6.111	6.457	6.747	6.995	7.209	7.397	7.711	7.965	8.175	8.351	8.502			
4.8	4.527	5.288	5.856	6.305	6.674	6.983	7.247	7.476	7.677	8.014	8.287	8.512	8.703	8.865			
4.9	4.629	5.428	6.029	6.505	6.896	7.225	7.507	7.751	7.966	8.328	8.620	8.863	9.068	9.243			
5.0	4.732	5.572	6.205	6.709	7.124	7.474	7.774	8.036	8.265	8.652	8.966	9.227	9.447	9.636			
5.1	4.836	5.718	6.386	6.919	7.359	7.731	8.050	8.329	8.574	8.988	9.324	9.604	9.841				
5.2	4.941	5.866	6.570	7.134	7.600	7.995	8.335	8.631	8.893	9.335	9.696	9.996	10.25				
5.3	5.048	6.018	6.759	7.354	7.847	8.266	8.628	8.944	9.222	9.695	10.08	10.40					
5.4	5.156	6.172	6.951	7.579	8.102	8.546	8.930	9.265	9.563	10.07	10.48	10.82					
5.5	5.266	6.329	7.148	7.811	8.363	8.833	9.240	9.598	9.914	10.45	10.89						
5.6	5.376	6.489	7.350	8.048	8.631	9.129	9.561	9.940	10.28	10.85	11.32						
5.7	5.489	6.652	7.555	8.290	8.906	9.433	9.890	10.29	10.65	11.26							
5.8	5.603	6.818	7.766	8.539	9.188	9.745	10.23	10.66	11.04	11.68							
5.9	5.718	6.987	7.980	8.794	9.478	10.07	10.58	11.03	11.44								
6.0	5.834	7.159	8.200	9.054	9.776	10.40	10.94	11.42	11.85								

Appendix Table 7a. Wainstein's table for the estimation of the one-sided lower 95% confidence interval, i.e. the *lower* boundary of the *central* 90% confidence interval, of the arithmetic mean of a lognormal population [logarithmic variance β^2 based on natural logarithms, for n = 5 to 20 (Wainstein 1975) for n = 50 to 1000 (Sichel 1966)] (South African Institute of Mining and Metallurgy) (see Fig. 109a)

β^2	n = 5	n = 10	n = 15	n = 20	n = 50	n = 100	n = 1000
0.00	1.0000	1.0000	1.0000	1.0000	1.0000	1.0000	1.0000
0.02	0.8978	0.9333	0.9458	0.9540	0.9697	0.9782	0.9927
0.04	0.8589	0.9071	0.9246	0.9344	0.9573	0.9692	0.9895
0.06	0.8302	0.8874	0.9079	0.9200	0.9478	0.9622	0.9872
0.08	0.8070	0.8708	0.8943	0.9077	0.9398	0.9564	0.9852
0.10	0.7870	0.8563	0.8821	0.8972	0.9328	0.9512	0.9833
0.12	0.7693	0.8439	0.8716	0.8878	0.9264	0.9464	0.9817
0.14	0.7535	0.8323	0.8617	0.8790	0.9204	0.9420	0.9801
0.16	0.7389	0.8216	0.8527	0.8709	0.9149	0.9380	0.9787
0.18	0.7255	0.8116	0.8442	0.8632	0.9097	0.9341	0.9773
0.20	0.7129	0.8023	0.8360	0.8558	0.9048	0.9304	0.9760
0.30	0.6605	0.7618	0.8008	0.8243	0.8828	0.9139	0.9701
0.40	0.6187	0.7284	0.7717	0.7981	0.8639	0.8996	0.9648
0.50	0.5838	0.6995	0.7462	0.7744	0.8470	0.8867	0.9600
0.60	0.5538	0.6739	0.7270	0.7534	0.8313	0.8741	0.9554
0.70	0.5277	0.6508	0.7020	0.7338	0.8168	0.8632	0.9511
0.80	0.5044	0.6297	0.6825	0.7156	0.8030	0.8525	0.9470
0.90	0.4836	0.6103	0.6646	0.6987	0.7899	0.8421	0.9429
1.00	0.4650	0.5923	0.6476	0.6826	0.7774	0.8322	0.9389
1.10	0.4481	0.5756	0.6317	0.6674	0.7654	0.8226	0.9351
1.20	0.4328	0.5599	0.6165	0.6530	0.7538	0.8133	0.9313
1.30	0.4189	0.5452	0.6023	0.6393	0.7426	0.8042	0.9276
1.40	0.4062	0.5315	0.5888	0.6262	0.7318	0.7954	0.9240
1.50	0.3946	0.5186	0.5760	0.6137	0.7214	0.7868	0.9203
1.60	0.3840	0.5065	0.5637	0.6018	0.7112	0.7784	0.9168
1.70	0.3743	0.4950	0.5521	0.5904	0.7014	0.7702	0.9133
1.80	0.3655	0.4842	0.5410	0.5794	0.6918	0.7622	0.9098
1.90	0.3574	0.4740	0.5305	0.5688	0.6825	0.7544	0.9064
2.00	0.3501	0.4644	0.5203	0.5587	0.6734	0.7466	0.9030
2.10	0.3433	0.4552	0.5106	0.5489	0.6646	0.7391	0.8996
2.20	0.3372	0.4466	0.5014	0.5395	0.6560	0.7317	0.8962
2.30	0.3316	0.4385	0.4925	0.5304	0.6476	0.7245	0.8929
2.40	0.3266	0.4308	0.4840	0.5217	0.6394	0.7173	0.8896

ß²	n = 5	n = 10	n = 15	n = 20	n = 50	n = 100	n = 1000
2.50	0.3220	0.4234	0.4759	0.5133	0.6314	0.7104	0.8864
2.60	0.3179	0.4166	0.4681	0.5044	0.6236	0.7035	0.8831
2.70	0.3142	0.4100	0.4606	0.4974	0.6160	0.6967	0.8799
2.80	0.3110	0.4039	0.4535	0.4899	0.6085	0.6901	0.8767
2.90	0.3081	0.3981	0.4467	0.4826	0.6012	0.6836	0.8736
3.00	0.3055	0.3926	0.4401	0.4756	0.5941	0.6772	0.8704
3.10	0.3033	0.3874	0.4338	0.4689	0.5872		
3.20	0.3014	0.3825	0.4278	0.4624	0.5804		
3.30	0.2999	0.3770	0.4220	0.4561	0.5738		
3.40	0.2986	0.3736	0.4165	0.4500	0.5673		
3.50	0.2976	0.3695	0.4112	0.4442	0.5609		
3.60	0.2969	0.3656	0.4062	0.4385	0.5547		
3.70	0.2964	0.3621	0.4013	0.4331	0.5486		
3.80	0.2962	0.3587	0.3967	0.4278	0.5427		
3.90	0.2963	0.3556	0.3923	0.4228	0.5369		
4.00	0.2965	0.3527	0.3880	0.4179	0.5312		
4.10	0.2971	0.3500	0.3840	0.4132			
4.20	0.2978	0.3475	0.3801	0.4086			
4.30	0.2988	0.3452	0.3765	0.4043			
4.40	0.3000	0.3430	0.3729	0.4001			
4.50	0.3014	0.3411	0.3696	0.3960			
4.60	0.3030	0.3393	0.3664	0.3921			
4.70	0.3048	0.3378	0.3634	0.3883			
4.80	0.3069	0.3363	0.3605	0.3847			
4.90	0.3091	0.3351	0.3578	0.3812			
5.00	0.3116	0.3340	0.3552	0.3778			

Fig.109a-b. The lognormal distribution: a with lower confidence interval; b with upper confidence interval

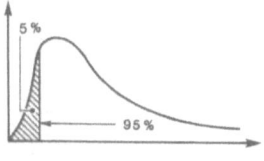

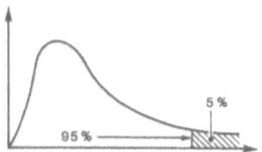

Appendix Table 7b. Wainstein's table for the estimation of the one-sided lower 95% confidence interval, i.e. the *upper* limit of the *central* 90% confidence interval, of the arithmetic mean of a lognormal distribution [logarithmic variance β^2 based on natural logarithms, for n = 5 to 20 (Wainstein 1975) for n = 50 to 1000 (Sichel 1966)] (South African Institute of Mining and Metallurgy) (see Fig. 109b)

β^2	n = 5	n = 10	n = 15	n = 20	n = 50	n = 100	n = 1000
0.00	1.0000	1.0000	1.0000	1.0000	1.000	1.000	1.000
0.02	1.2411	1.1170	1.0845	1.0671	1.038	1.026	1.007
0.04	1.3623	1.1712	1.1221	1.0990	1.055	1.037	1.011
0.06	1.4664	1.2156	1.1538	1.1239	1.069	1.046	1.013
0.08	1.5613	1.2559	1.1812	1.1462	1.080	1.053	1.015
0.10	1.6518	1.2934	1.2070	1.1661	1.091	1.060	1.017
0.12	1.7405	1.3272	1.2301	1.1845	1.100	1.066	1.019
0.14	1.8270	1.3606	1.2527	1.2023	1.109	1.072	1.020
0.16	1.9136	1.3930	1.2741	1.2191	1.118	1.078	1.022
0.18	1.9993	1.4247	1.2951	1.2356	1.126	1.084	1.023
0.20	2.0867	1.4554	1.3160	1.2523	1.135	1.089	1.025
0.30	2.5320	1.6065	1.4147	1.3279	1.172	1.113	1.031
0.40	3.0191	1.7564	1.5090	1.3988	1.207	1.135	1.037
0.50	3.5628	1.9099	1.6031	1.4705	1.240	1.156	1.042
0.60	4.1756	2.0696	1.6824	1.5412	1.273	1.175	1.047
0.70	4.8696	2.2374	1.7984	1.6137	1.306	1.196	1.052
0.80	5.6634	2.4149	1.9007	1.6877	1.338	1.215	1.057
0.90	6.5703	2.6037	2.0061	1.7635	1.371	1.235	1.062
1.00	7.6047	2.8050	2.1169	1.8424	1.404	1.254	1.067
1.10	8.7946	3.0194	2.2328	1.9239	1.437	1.274	1.071
1.20	10.1548	3.2501	2.3551	2.0083	1.471	1.294	1.076
1.30	11.7177	3.4968	2.4827	2.0958	1.506	1.314	1.080
1.40	13.5127	3.7610	2.6168	2.1874	1.540	1.334	1.085
1.50	15.5693	4.0454	2.7581	2.2819	1.576	1.354	1.089
1.60	17.9276	4.3507	2.9069	2.3804	1.613	1.374	1.094
1.70	20.6390	4.6799	3.0640	2.4838	1.650	1.395	1.098
1.80	23.7488	5.0341	3.2289	2.5916	1.688	1.416	1.103
1.90	27.3182	5.4141	3.4032	2.7042	1.728	1.438	1.107
2.00	31.3985	5.8248	3.5883	2.8217	1.767	1.459	1.112
2.10	36.0792	6.2676	3.7832	2.9454	1.808	1.481	1.116
2.20	41.4437	6.7454	3.9895	3.0739	1.850	1.504	1.121
2.30	47.5859	7.2596	4.2078	3.2091	1.893	1.526	1.125
2.40	54.6113	7.8149	4.4382	3.3506	1.937	1.549	1.130

ß²	n = 5	n = 10	n = 15	n = 20	n = 50	n = 100	n = 1000
2.50	62.6606	8.4151	4.6833	3.4979	1.982	1.572	1.134
2.60	71.8613	9.0606	4.9411	3.6700	2.029	1.596	1.139
2.70	82.3661	9.7588	5.2145	3.8156	2.076	1.620	1.144
2.80	94.3775	10.5124	5.5045	3.9860	2.125	1.645	1.148
2.90	108.115	11.3263	5.8112	4.1638	2.175	1.670	1.153
3.00	123.750	12.2057	6.1368	4.3511	2.226	1.695	1.158
3.10	141.632	13.1539	6.4801	4.5477	2.279		
3.20	162.014	14.1793	6.8461	4.7530	2.333		
3.30	185.244	15.2885	7.2324	4.9695	2.388		
3.40	211.722	16.4835	7.6430	5.1958	2.445		
3.50	241.829	17.7800	8.0781	5.4335	2.504		
3.60	276.140	19.1781	8.5394	5.6832	2.564		
3.70	315.198	20.6894	9.0286	5.9445	2.626		
3.80	359.607	22.3201	9.5476	6.2202	2.689		
3.90	410.033	24.0878	10.0983	6.5090	2.754		
4.00	467.417	25.9967	10.6828	6.8126	2.821		
4.10	532.494	28.0586	11.3034	7.1320			
4.20	606.446	30.2911	11.9606	7.4667			
4.30	690.289	32.7042	12.6607	7.8188			
4.40	785.542	35.3133	13.4026	8.1881			
4.50	893.390	38.1351	14.1912	8.5768			
4.60	1015.74	41.1878	15.0297	8.9861			
4.70	1154.22	44.4911	15.9190	9.4172			
4.80	1311.09	48.0599	16.8623	9.8681			
4.90	1488.76	51.9165	17.8687	10.3448			
5.00	1689.53	56.0924	18.9342	10.8454			

Appendix Table 8. Formulas for the regression analysis

In „Economic Evaluations for Exploration" the regression analysis was discussed in Chapter 5.2. The relevant equations are repeated here:

1. Equation for the regression line
   Regression follows the general equation of a straight line:

$$y = a \cdot x + b \tag{1}$$

The straight line is determined in such a way that the distances of individual points (x_i, y_i) to the straight line are minimized, i.e. the straight line is the best fit to the points, with a and b being the regression coefficients. The equations for the regression coefficients are:

$$a = \frac{\sum x_i y_i - \dfrac{\sum x_i y_i}{n}}{\sum x_i^2 - \dfrac{\left(\sum x_i\right)^2}{n}} \tag{2}$$

$$b = \bar{y} - a\bar{x}, \tag{3}$$

with $\bar{x}$ and $\bar{y}$ being the arithmetic mean:

$$\bar{y} = \frac{\sum y_i}{n}$$

$$\bar{x} = \frac{\sum x_i}{n}.$$

To determine the degree of correlation, the correlation coefficient r is calculated. $r = 0$ if there is no correlation at all, and $r = 1$ in case of a perfect correlation, i.e. when all points lie on the regression line.

The square of the correlation coefficient is determined in the following way:

$$r^2 = \frac{\left[\sum x_i y_i - \dfrac{\sum x_i \sum y_i}{n}\right]^2}{\left[\sum x_i^2 - \dfrac{\left(\sum x_i\right)^2}{n}\right]\left[\sum y_i^2 - \dfrac{\left(\sum y_i\right)^2}{n}\right]} = \frac{s_{xy}^2}{s_x^2 \cdot s_y^2}. \tag{4}$$

s_x^2 and s_y^2 are the variances of the values x_i and y_i and s_{xy} is the covariance of the values x_i and y_i.

r^2 is also called the coefficient of determination B. It is a measure of the degree of correlation. It indicates what percentage of the distribution can be explained by linear regression.

2. Confidence interval for the line of regression

If $y = f(x)$ the confidence interval l indicates that for an assumed level of confidence the value y for a given x can lie between $y \pm l$.

$$l = \frac{t \cdot h \sqrt{d}}{\sqrt{n-2}} .$$
(5)

t = factor from Students-t-distribution (see Chap. 7.1 and Appendix Table 4)
n = number of value pairs (see Appendix Table 4, column 4).

$$h^2 = \frac{1}{n} + \frac{(x - \bar{x})^2}{(n-1) \cdot s_x^2}$$
(6)

$$d = (n-1)(s_y^2 - b^2 \cdot s_x^2) .$$
(7)

s_x^2 and s_y^2 are the variances of the values x_i and y_j, b is the slope of the line of regression [see above Eq. (1)].

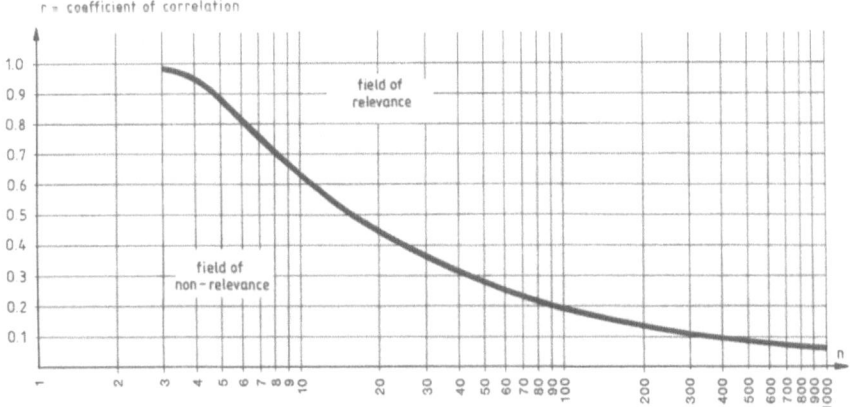

Fig. 110. Minimum coefficients of correlation at a significance level of 5%

3. Examination of the coefficient of correlation

If the correlation coefficient r is small, e.g. close to zero, the question arises whether the correlation between two elements is real or only apparent due to a chance distribution of a limited number of samples. To answer this question, statisticians work with the so-called zero hypothesis. They select a number which is the percentage of the number of times they expect no correlation to exist. This is called the significance level. A common selection for a significance level is 5%, meaning there is a 1 in 20 chance that there is no correlation and a 19 in 20 chance that the correlation is real. Minimum correlation coefficients then can be calculated as a function of the number of data pairs (Fig. 110).

Appendix Table 9. BASIC-program for the calculation of a variogram

```
100  REM Basic-program : estimation of an experimental variogram
150  REM
200  REM X , Y  : cartesian coordinates
210  REM Z      : data
220  REM
1000 INPUT " enter number of data "; N
1100 IF N < 1 THEN PRINT " Error": END
1500 DIM X(N), Y(N), Z(N)
1510 DIM GA(20), NW(20)
2000 FOR I = 1 TO N
2100 INPUT "enter x,y,z : "; X(I), Y(I), Z(I)
2200 NEXT I
3000 REM estimation
3010 REM get direction and lag-size
3020 REM -------------------------
3500 INPUT " enter lag-size and tolerance : "; H, DH
3510 IF (H <= 0) THEN PRINT " enter lag-size > 0 !!! ": GOTO 3500
3520 IF (DH < 0) THEN PRINT " enter tolerance >= 0 !!! ": GOTO 3500
3600 IF (DH > .5 * H) THEN PRINT " tolerance > 0.5*lag-size ": GOTO 3500
3650 INPUT " Enter direction (E-W = 0; N-S = 90) and tolerance : "; A, DA
3700 INPUT " Enter lower and upper bounds for data : "; UZ, OZ
3750 IF (UZ > OZ) THEN PRINT " lower bound > upper bound !!! ": GOTO 3700
4000 FOR I = 1 TO 20
4020 GA(I) = 0
4030 NW(I) = 0
4040 NEXT
4600 T1 = 3.14159 * DA / 360: T1 = COS(T1)
4610 HI = 3.14159 * A / 180: CA = COS(HI): SA = SIN(HI)
5000 FOR I = 1 TO N - 1
5050 IF (Z(I) < UZ) GOTO 8000
5060 IF (Z(I) > OZ) GOTO 8000
5200 FOR J = I * 1 TO N
5250 IF (Z(J) < UZ) GOTO 7000
5260 IF (Z(J) > OZ) GOTO 7000
5300 DI = SQR((X(I) - X(J)) ^ 2 + (Y(I) - Y(J)) ^ 2)
5310 IF (DI < .000001) GOTO 7000
5320 HI = ABS((X(I) - X(J)) * CA / DI + (Y(I) - Y(J)) * SA / DI)
5340 IF (HI < TI) GOTO 7000
5400 A% = (DI / H)
5450 A = A% * H
5500 IF (ABS(A - DI) > DH) GOTO 7000
5600 K = A%
5650 IF (K > N) GOTO 7000
5700 NW(K) = NW(K) * 1
5800 GA(K) - GA(K) + (Z(J) - Z(I)) ^ 2
7000 NEXT
8000 NEXT
8900 PRINT " h", "N(h)", " gamma(h)"
9000 FOR I = 1 TO 20
9010 IF (NW(I) = 0) GOTO 9700
9050 GA(I) = GA(I) * .5 / NW(I)
9600 PRINT I, NW(I), GA(I)
9700 NEXT
9900 INPUT " another estimation ? (Y/N) "; A$
9910 IF (A$ = "Y" OR A$ = "y") GOTO 350
9999 END
```

Appendix Table 10. Examples for range a and relative nugget effect C_0/C ($C=C_0+C_i$=sill)

Deposit	Thickness		Accumulation value GT (grade G x thickness T)		Grade G (in graded variogram i.e. thickness constant)	
	a	$\dfrac{C_0}{C}$	a	$\dfrac{C_0}{C}$	a	$\dfrac{C_0}{C}$
Perseverance - Ni magmatic Australia (Dowd and Milton 1987)	30 m	0.14	15 m (down dip) 30 m (along strike)	0.3		
Au-quartz veins Australia (Siromines) (Guibal 1987)					40-60 m (down dip) 20-30 m (along strike)	0.3
Nanisivik, Pb-Zn in carbonates Canada				10	100 m (along strike) 30 m (perpendicular to strike)	0.4 0.33
Ramsbeck, Pb-Zn in veins, Germany					10 m	0.4
South Yorke'am sedimentary phosphate, Israel (Miller & Gill 1986)	700 m	0			600 m	0
Neves Corvo, Cu volcanogenic, Portugal (Muge 1986)	400 m	0.35	300 m	0.35		
Cayeli, Cu/Zn volcanogenic, Turkey	110 m	0.18	70 m	0.2		
Tar Sands, Athabasca Canada (Zwicky 1975)	1500 m	0	1500 m	0		
					100 m	0.25-0.65
Various Hungarian bauxite deposits (Bardossy 1991, pers. comm.) flat, stratabound, flat, lensy	50-90 m 30-50 m	0.01 0.15				

Appendix Table 11. Solution for a linear system of equations with the method of determining the eigen values

We have a linear system of equations with the unknowns x, y and z, for example:

$$a_1x + a_2y + a_3z = a_r \qquad (1)$$
$$b_1x + b_2y + b_3z = b_r$$
$$c_1x + c_2y + c_3z = c_r.$$

The aim is to produce a matrix, in which we have only the numbers 1 on the diagonal and all other factors are zeros. This is the unit matric, which enables us to read from the vector on the right directly the solutions for the unknows x, y and z:

$$1 \cdot x + 0 \cdot y + 0 \cdot z = a_f \qquad (2)$$
$$0 \cdot x + 1 \cdot y + 0 \cdot z = b_f$$
$$0 \cdot x + 0 \cdot y + 1 \cdot z = c_f$$

or

$$\begin{bmatrix} 1 & 0 & 0 \\ 0 & 1 & 0 \\ 0 & 0 & 1 \end{bmatrix} \longrightarrow \begin{bmatrix} a_f \\ b_f \\ c_f \end{bmatrix}$$

This method will be demonstrated now with the system of Eq. (13) from Chapter 13.4.5.2.2. The unknowns are λ_1, λ_2 and μ.

$$480\,\lambda_1 + 740\,\lambda_2 + \mu = 310 \qquad (4)$$

$$185\,\lambda_1 + 795\,\lambda_2 + \mu = 185$$

$$\lambda_1 + 4\,\lambda_2 = 1$$

1st Step: We realize that in this system of equations the unknown λ_1 occurs already with the factor 1. We therefore rearrange the system in such a way that this factor occurs on the diagonal:

	Column 1	Column 2	Column 3		(5)
line 1:	λ_1 +	$4\lambda_2$ +	0μ	=	1
line 2:	$480\lambda_1$ +	$740\lambda_2$ +	μ	=	1
line 3:	$185\lambda_1$ +	$795\lambda_2$ +	μ	=	1

or simplified in the form of a matrix:

$$
\begin{array}{c}
\text{line 1} \\
\text{line 2} \\
\text{line 3}
\end{array}
\begin{bmatrix}
1 & 4 & 0 \\
480 & 740 & 1 \\
185 & 795 & 1
\end{bmatrix}
\qquad
\begin{bmatrix}
1 \\
310 \\
185
\end{bmatrix}
\qquad (6)
$$

The first value in the 1st line and the 1st column is 1 now. In the next step, we want the 2nd and 3rd value of the 1st line to become zero [cf. Eqs. (2) and (3)]. To transform this matrix, into the unit matrix we proceed by *columns*.

For this transformation we apply the rule that we can multiply any equation in a matrix like the matrix (6) with any factor, without changing the values of the unknowns. Also the rule applies that one line can be subtracted from or added to another line without influencing the values of the unknowns.

2nd Step: In line 2 at the first place we have the number 480. Therefore we multiply the first line with 480 and then we subtract line 2 from line 1:

	line 1	([1	4	0]	[1]) · 480
i.e.	line 1	480	1920	0	480
	line 2	480	740	1	310
	line 2 - line 1	0	-1180 +	1	170

As a consequence, we now have the matrix system:

$$
\begin{array}{c}
\text{line 1} \\
\text{line 2} \\
\text{line 3}
\end{array}
\begin{bmatrix}
1 & 4 & 0 \\
0 & -1180 & 1 \\
185 & 795 & 1
\end{bmatrix}
\qquad
\begin{bmatrix}
1 \\
-170 \\
185
\end{bmatrix}
\qquad (7)
$$

3rd Step: The same procedure we now apply to the 3rd line, i.e. we multiply line 1 with 185 and then subtract line 3 from line 1:

line 1	185	740	0	185
line 2	185	795	1	185
line 2 - line 1	0 -	-55	-1	0

As a consequence, we now have the matrix system:

	Column 1	Column 2	Column 3	

$$
\begin{array}{c}
\text{line 1} \\
\text{line 2} \\
\text{line 3}
\end{array}
\begin{bmatrix}
1 & 4 & 0 \\
480 & -1180 & 1 \\
0 & -55 & -1
\end{bmatrix}
\qquad
\begin{bmatrix}
1 \\
-170 \\
0
\end{bmatrix}
\qquad (8)
$$

4th Step: Comparing now our matrix system (8) with our final goal (2) or (3), we realize that we reached our goal for column 1.

Now we deal with the 2nd column. Again we start by transforming the value on the diagonal to 1, i.e. we divide the line 2 by -1180. The system of equations then has the following form:

$$
\begin{array}{ll}
\text{line 1} \\
\text{line 2} \\
\text{line 3}
\end{array}
\begin{bmatrix}
1 & 4 & 0 \\
0 & 1 & -0.0008 \\
0 & -55 & -1
\end{bmatrix}
\begin{bmatrix}
1 \\
+0.144 \\
0
\end{bmatrix}
\qquad (9)
$$

5th Step: In the 1st line the value in column 2 (9) has to be converted into zero. We therefore multiply the 2nd line with 4 and then subtract line 2 from line 1:

line 1	1	4	0	1
line 2	0	4	- 0.0032	0.576
line 2 - line 1	1	0	+0.0032	+0.424

As a consequence, we now have the matrix system:

$$
\begin{array}{ll}
\text{line 1} \\
\text{line 2} \\
\text{line 3}
\end{array}
\begin{bmatrix}
1 & 0 & +0.0032 \\
0 & 1 & -0.0008 \\
0 & -55 & -1
\end{bmatrix}
\begin{bmatrix}
0.424 \\
0.144 \\
0
\end{bmatrix}
\qquad (10)
$$

6th Step: The same procedure we now apply to the 3rd line. We multiply the 2nd line with -55 and then subtract line 2 from line 3:

line 2	0	-55	+0.0440	-7.92
line 3	0	-55	-1	0
line 3 - line 2	0 -	0	-1.044	+7.92

As a consequence, we now have the matrix system:

	Column 1	Column 2	Column 3	

$$
\begin{array}{ll}
\text{line 1} \\
\text{line 2} \\
\text{line 3}
\end{array}
\begin{bmatrix}
1 & 0 & 0.0032 \\
0 & 1 & -0.0008 \\
0 & 0 & -1.044
\end{bmatrix}
\begin{bmatrix}
0.424 \\
0.144 \\
+7.92
\end{bmatrix}
\qquad (11)
$$

7th Step: Comparing now our matrix system (11) with our goals (2) and (3) we realise that in column 1 and column 2 we reached the final goal. Now we deal with the 3rd column. To obtain the value 1 on the diagonal we divide the 3rd line by -1.044. We therefore then have the matrix:

$$
\begin{array}{l}
\text{line 1} \\
\text{line 2} \\
\text{line 3}
\end{array}
\begin{bmatrix}
1 & 0 & 0.0032 \\
0 & 1 & -0.0008 \\
0 & 0 & 1
\end{bmatrix}
\begin{bmatrix}
0.424 \\
0.144 \\
-7.59
\end{bmatrix}
\tag{12}
$$

8th Step: We now want to transform the 3rd number in line 1 to zero. Therefore we multiply line 3 with 0.0032 and then subtract line 3 from line 1:

line 1	1	0	0.0032	0.424
line 3	0	0	0.0032	-0.024
line 3 - line 1	1	0	0	0.448

As a consequence, we now have the matrix system:

$$
\begin{bmatrix}
1 & 0 & 0 \\
0 & 1 & -0.0008 \\
0 & 0 & 1
\end{bmatrix}
\begin{bmatrix}
0.448 \\
0.144 \\
-7.59
\end{bmatrix}
\tag{13}
$$

9th Step: We now apply the same procedure to line 2. We multiply line 3 by -0.0008 and then subtract line 3 from line 2:

line 2	0	1	-0.0008	0.144
line 3	0	0	-0.0008	0.006
line 2 - line 3	0	1	0	0.138

As a consequence, we now have the matrix system:

$$
\begin{bmatrix}
1 & 0 & 0 \\
0 & 1 & 0 \\
0 & 0 & 1
\end{bmatrix}
\begin{bmatrix}
0.488 \\
0.138 \\
-7.59
\end{bmatrix}
\tag{14}
$$

If we now compare our matrix system (14) with our goal (2) and (3), we realize that we have achieved our final goal. System (14) can now be transcribed into:

$$
\begin{bmatrix}
1 \cdot \lambda_1 & 0 & 0 \\
0 & 1 \cdot \lambda_2 & 0 \\
0 & 0 & 1 \cdot \mu
\end{bmatrix}
\begin{bmatrix}
0.488 \\
0.138 \\
-7.59
\end{bmatrix}
$$

or

$$
\lambda_1 = 0.448
$$
$$
\lambda_2 = 0.138
$$
$$
\mu = -7.59
$$

The final result, of course, is the same – taking into account rounding errors – as the one in Chapter 13.4.5.2.2.

Appendix Table 12. Spheric model: auxiliary function 1-H(H1, H2)

H1/H2	0.000	.250	.500	.750	1.000	1.250	1.500	1.750	2.000	2.250	2.500	2.750	3.000	3.250	3.500	3.750	4.000
.000	1.000	.814	.641	.490	.375	.300	.250	.214	.187	.167	.150	.136	.125	.115	.107	.100	.094
.100	.925	.790	.626	.479	.367	.293	.244	.210	.183	.163	.147	.133	.122	.113	.105	.098	.092
.200	.851	.744	.594	.456	.349	.279	.233	.199	.174	.155	.140	.127	.116	.107	.100	.093	.087
.300	.778	.690	.555	.426	.326	.260	.217	.186	.163	.145	.130	.118	.109	.100	.093	.087	.081
.400	.708	.633	.511	.392	.299	.239	.200	.171	.150	.133	.120	.109	.100	.092	.086	.080	.075
.500	.641	.575	.465	.357	.272	.217	.181	.155	.136	.121	.109	.099	.091	.084	.078	.072	.068
.600	.577	.519	.420	.321	.245	.196	.163	.140	.122	.109	.098	.089	.082	.075	.070	.065	.061
.700	.518	.467	.377	.287	.219	.175	.146	.125	.109	.097	.087	.080	.073	.067	.062	.058	.055
.800	.464	.418	.337	.256	.195	.156	.130	.111	.098	.087	.078	.071	.065	.060	.056	.052	.049
.900	.416	.375	.302	.229	.174	.140	.116	.100	.087	.078	.070	.063	.058	.054	.050	.047	.044
1.000	.375	.338	.272	.207	.157	.126	.105	.090	.079	.070	.063	.057	.052	.048	.045	.042	.039
1.100	.341	.307	.247	.188	.143	.114	.095	.082	.071	.063	.057	.052	.048	.044	.041	.038	.036
1.200	.312	.281	.226	.172	.131	.105	.087	.075	.065	.058	.052	.048	.044	.040	.037	.035	.033
1.300	.288	.260	.209	.159	.121	.097	.081	.069	.060	.054	.048	.044	.040	.037	.035	.032	.030
1.400	.268	.241	.194	.148	.112	.090	.075	.064	.056	.050	.045	.041	.037	.035	.032	.030	.028
1.500	.250	.225	.181	.138	.105	.084	.070	.060	.052	.047	.042	.038	.035	.032	.030	.028	.026
1.600	.234	.211	.170	.129	.098	.079	.065	.056	.049	.044	.039	.036	.033	.030	.028	.026	.025
1.700	.221	.199	.160	.121	.092	.074	.062	.053	.046	.041	.037	.034	.031	.028	.026	.025	.023
1.800	.208	.188	.151	.115	.087	.070	.058	.050	.044	.039	.035	.032	.029	.027	.025	.023	.022
1.900	.197	.178	.143	.109	.083	.066	.055	.047	.041	.037	.033	.030	.028	.025	.024	.022	.021
2.000	.187	.169	.136	.103	.079	.063	.052	.045	.039	.035	.031	.029	.026	.024	.022	.021	.020
2.100	.179	.161	.129	.098	.075	.060	.050	.043	.037	.033	.030	.027	.025	.023	.021	.020	.019
2.200	.170	.153	.124	.094	.071	.057	.048	.041	.036	.032	.029	.026	0.24	.022	.020	.019	.018
2.300	.163	.147	.118	.090	.068	.055	0.46	.039	.034	.030	.027	.025	.023	.021	.020	.018	.017
2.400	.156	.141	.113	.086	.065	.052	.044	.037	.033	.029	.026	.024	.022	.020	.019	.017	.016
2.500	.150	.135	.109	.083	.063	.050	.042	.036	.031	.028	.025	.023	.021	.019	.018	.017	.016
2.600	.144	.130	.105	.079	.060	.048	.040	.035	.030	.027	.024	.022	.020	.019	.017	.016	.015
2.700	.139	.125	.101	.076	.058	.047	.039	.033	.029	.026	.023	.021	.019	.018	.017	.016	.015
2.800	.134	.121	.097	.074	.056	.045	.037	.032	.028	.025	.022	.020	.019	.017	.016	.015	.014
2.900	.129	.116	.094	.071	.054	.043	.036	.031	.027	.024	.022	.020	.018	.017	.015	.014	.014
3.000	.125	.113	.091	.069	.052	.042	.035	.030	.026	.023	.021	.019	.017	.016	.015	.014	.013
3.100	.121	.109	.088	.067	.051	.041	.034	.029	.025	.023	.020	.018	.017	.016	.014	.014	.013
3.200	.117	.106	.085	.065	.049	.039	.033	.028	.025	.022	.020	.018	.016	.015	.014	.013	.012
3.300	.114	.102	.082	.063	.048	.038	.032	.027	.024	.021	.019	.017	.016	.015	.014	.013	.012
3.400	.110	.099	.080	.061	.046	.037	.031	.026	.023	.021	.018	.017	.015	.014	.013	.012	.012
3.500	.107	.096	.078	.059	.045	.036	.030	.026	.022	.020	.018	.016	.015	.014	.013	.012	.011
3.600	.104	.094	.075	.057	.044	.035	.029	.025	.022	.019	.017	.016	.015	.013	.012	.012	.011
3.700	.101	.091	.073	.056	.042	.034	.028	.024	.021	.019	.017	.015	.014	.013	.012	.011	.011
3.800	.099	.089	.072	.054	.041	.033	.028	.024	.021	.018	.017	.015	.014	.013	.012	.011	.010
3.900	.096	.087	.070	.053	.040	.032	.027	.023	.020	.018	.016	.015	.013	.012	.012	0.11	.010
4.000	.094	.084	.068	.052	.039	.031	.026	.022	.020	.017	.016	.014	.013	.012	.011	.010	.010

Appendix Table 13. Produced gold in ounces per 1 m development. (Hester 1982)

Porcupine district, Ontario, Canada

Hallnor	48.2
Dome	41.6
McIntyre	33.2
Hollinger	31.7
Pamour	25.2
Preston East Dome	16.4

Red Lake, Ontario, Canada

Campbell	44.8
Dickenson	38.4
Madsen	34.3
Cochenour	19.6
McKenzie	15.9

Cripple Creek, Colorado, USA

Portland	29.9

Kirkland Lake, Ontario, Canada

Toburn	20.2
Sylvanite	18.1
Wright-Hargreaves	45.8
Lake Shore	71.9
Kirkland Lake Gold	25.1
Macassa	27.4

Other Ontario mines, Canada

Little Long Lac	41.4
Pickle Crow	35.4
Renabie	35.4
Central Patricia	32.8
Upper Canada	21.4

Appendix Table 14

A: P (probability, completeness of search) for a linear target with angle δ between strike direction and line direction.

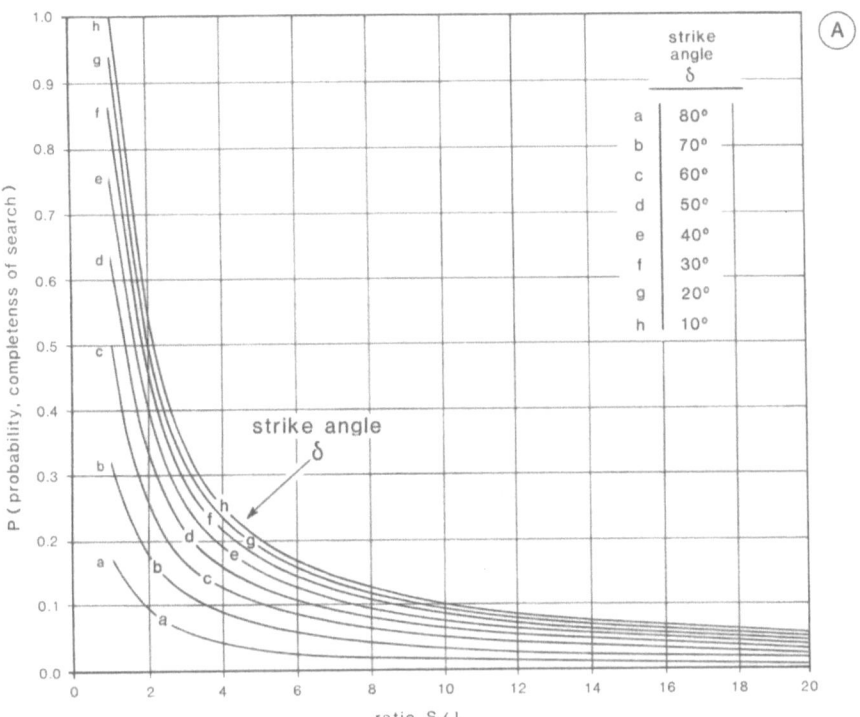

Fig. 111. Appendix Table 14, A: P (probability, completeness of search) for a linear target with angle δ between strike direction and line direction

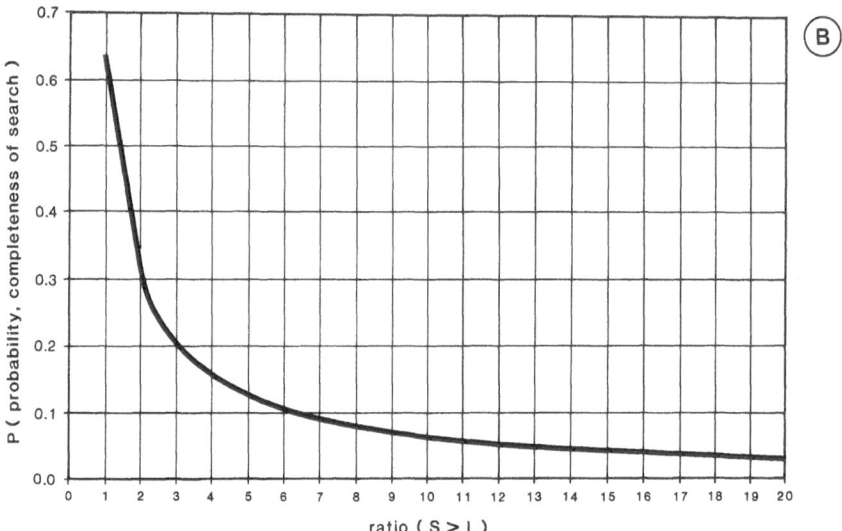

Fig. 112. Appendix Table 14, B: P (probability, completeness of search) for a random-oriented target for S>L

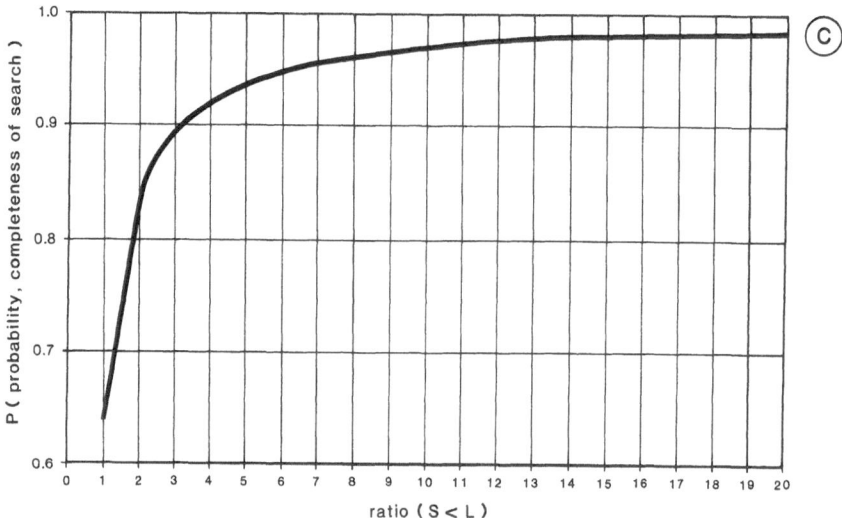

Fig. 113. Appendix Table 14, C: P (probability, completeness of search) for a random-oriented target for S<L

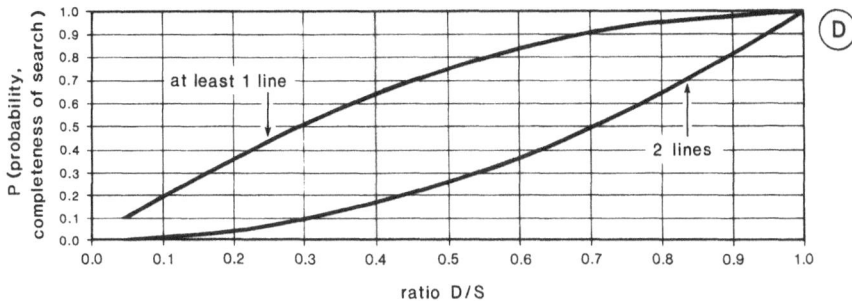

Fig.114. Appendix Table 14, D: P (probability, completeness of search) for a grid with survey lines perpendicular to each other

Appendix Table 15. Table for the generation of random numbers

Row no.	Column no.									
	0	1	2	3	4	5	6	7	8	9
0	7559	2521	7011	5411	4602	3855	5384	2697	5179	4821
1	5095	6202	7216	0940	9232	6432	9077	7937	0850	3981
2	8747	1275	3096	9648	4414	6737	9202	3851	4238	9962
3	2256	7331	6233	6440	5196	0798	8617	1869	1402	4583
4	9462	1915	3991	9690	6927	2099	5815	6390	9027	8146
5	0282	5629	4096	5520	0420	7737	0089	0121	9345	3600
6	5167	7352	0144	3609	7458	2822	5990	8590	5066	2345
7	0031	9991	6296	1471	5263	5583	8956	5721	5008	8987
8	1528	0819	2205	9681	1762	2346	0236	8464	8082	5850
9	1862	5754	8529	7205	0524	2408	1118	1785	9161	8903
10	4067	5403	1034	5921	5564	3491	9144	4115	3653	1806
11	6693	3989	6463	5955	6739	6256	2097	1517	2427	1597
12	8625	4161	6990	0430	4443	9522	3397	3613	7329	3211
13	4941	1739	5618	0806	4531	5256	7383	8092	5518	5052
14	0783	0472	0871	1513	3832	3671	3648	5353	4865	3270
15	7228	7251	1720	5052	4610	7745	2791	4420	8880	7456
16	5974	7101	3096	7097	9604	8790	0637	7231	1754	5039
17	4267	8510	0348	1484	7425	1559	3753	3851	3167	6971
18	0068	7486	4522	5311	7249	8289	4723	7348	3690	1781
19	6739	3705	7170	3571	3862	4859	6141	9355	9119	1609
20	8558	4683	5627	4491	2465	5194	6588	4010	6225	6494
21	5179	0995	7822	4282	4309	6009	5330	4106	7329	9589
22	7023	1605	7208	2475	0917	2504	7120	5261	5112	2266
23	0210	8510	7797	6674	9478	2877	7132	1250	8245	9100
24	1130	1601	9566	7540	8128	3542	6588	9113	5961	2387
25	2427	4352	2908	4010	0051	3768	7099	3901	2753	8067
26	6635	7757	3678	7933	9704	1246	5823	3901	4234	9886
27	2511	6595	8588	9288	6563	6724	0867	5265	1361	4738
28	9027	3249	0018	1522	1227	8878	5505	7465	5832	7879
29	0997	3395	0591	5829	9755	9761	3928	8280	4660	7331
30	3887	0986	9152	6612	8596	9556	6053	8197	6041	9451
31	8625	2885	4527	6457	0102	1212	2214	6210	3335	3801
32	6998	4320	1733	1246	7099	6248	8207	4228	0821	7896
33	6108	3282	6091	4550	8287	9150	1039	2170	8036	2308
34	3883	6218	8500	3554	2578	2245	2678	3425	5518	6122
35	3774	2542	0717	3224	4962	1618	6158	9246	9269	1099
36	0148	9857	9930	1756	0612	0693	9913	4094	5915	7364
37	2511	8941	4041	5449	0750	6745	3180	0614	7747	8159
38	1373	8498	6706	3805	6869	4591	0683	4395	7651	4709
39	2824	0493	1908	4399	6246	2634	7801	9301	4150	1676

NORMAL PROBABILITY PAPER

(EUROPEAN x/y - AXES)

APPENDIX A1

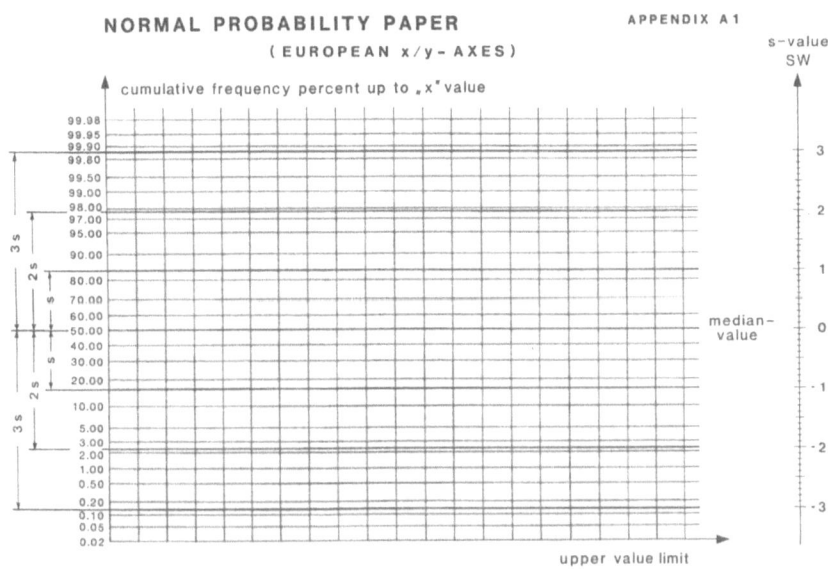

s-value
SW

NORMAL PROBABILITY PAPER

(NORTH AMERICAN x/y-AXES)

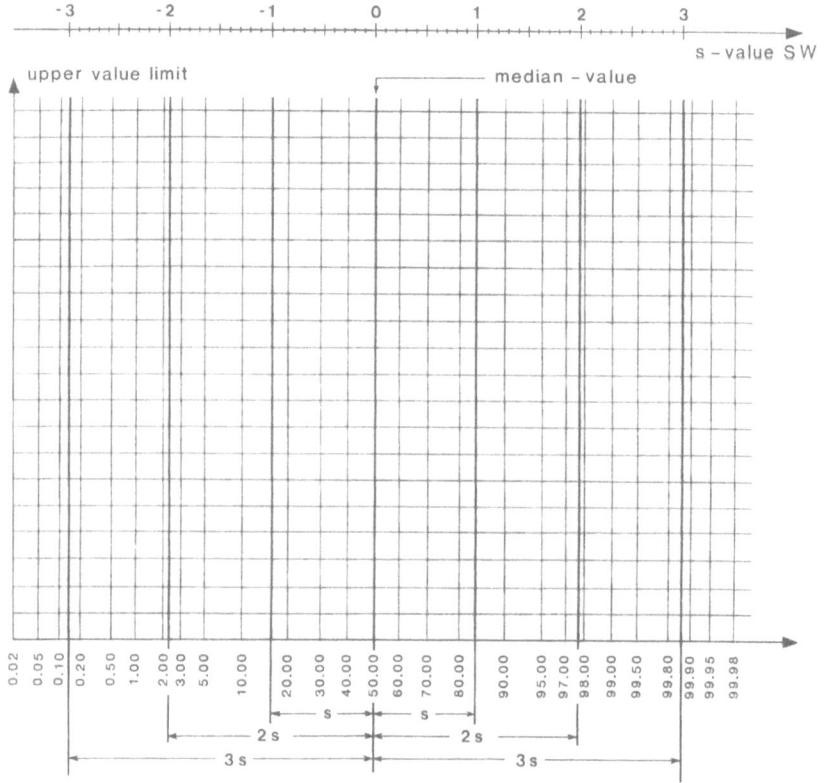

cumulative frequency percent up to „x" value

LOGARITHMIC PROBABILITY PAPER

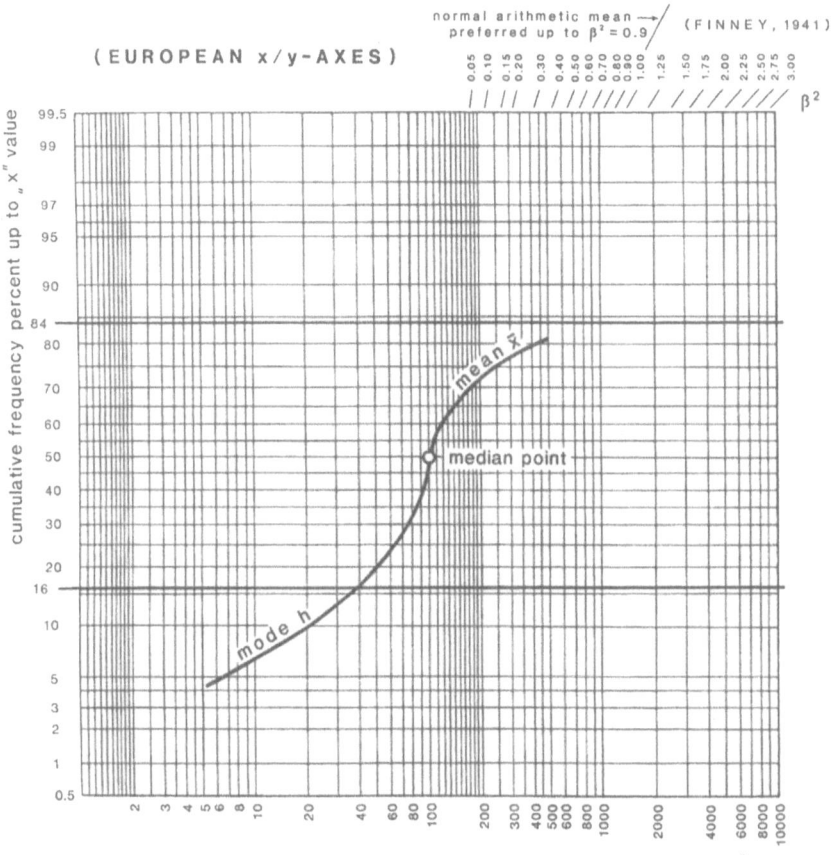

normal arithmetic mean →
preferred up to $\beta^2 = 0.9$ (FINNEY, 1941)

(EUROPEAN x / y − AXES)

β^2

cumulative frequency percent up to „x" value

mean x̄

median point

mode h

upper value limit „x"

LOGARITHMIC PROBABILITY PAPER
(NORTH AMERICAN x/y-AXES)

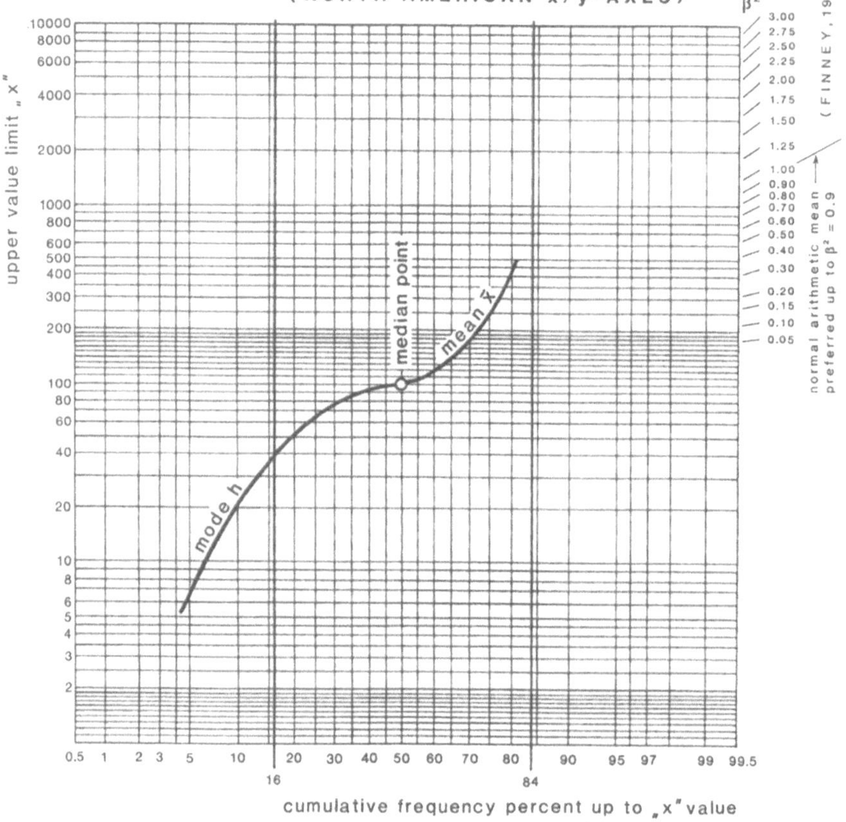

cumulative frequency percent up to „x" value

NORMAL PROBABILITY PAPER

(EUROPEAN x/y - AXES)

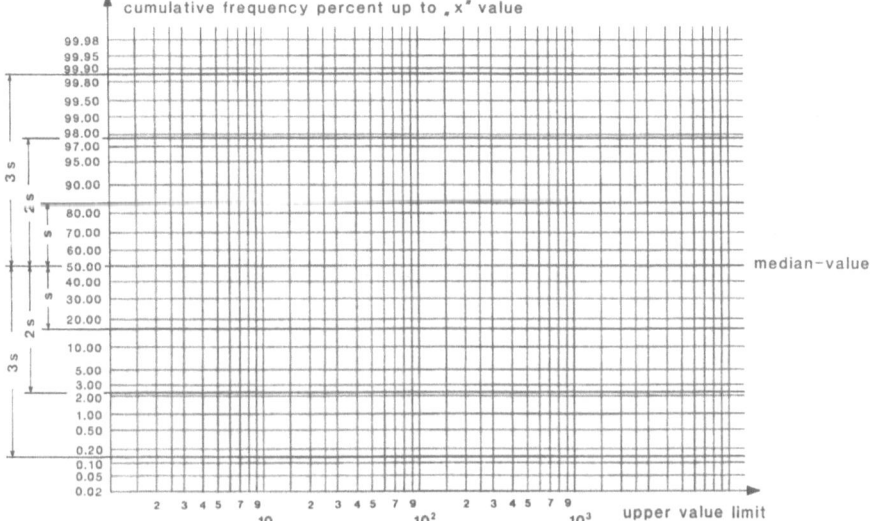

NORMAL PROBABILITY PAPER

(NORTH AMERICAN x/y - AXES)

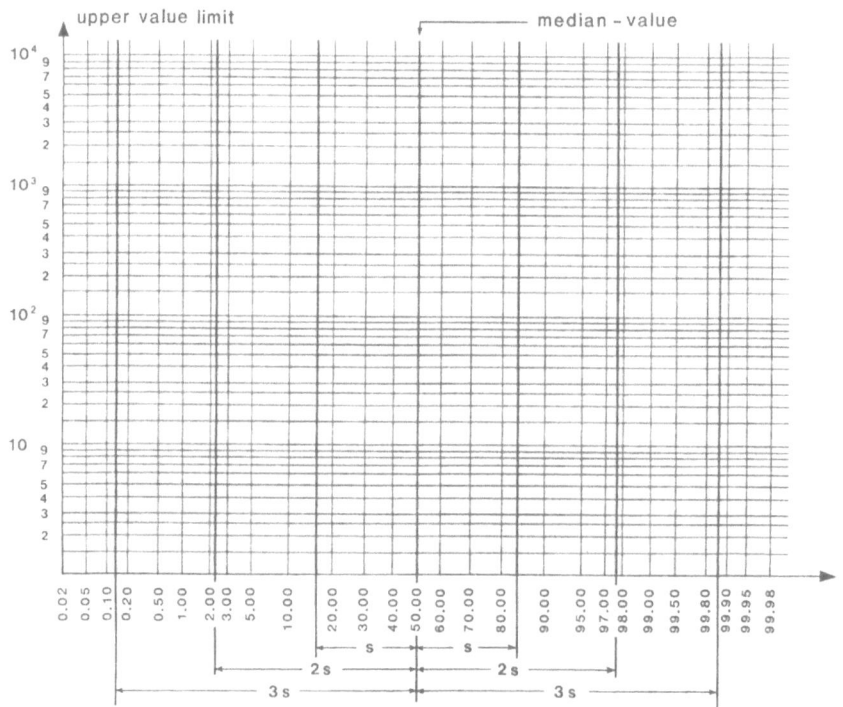

cumulative frequency percent up to „x" value

References

Agocs WB (1955) Line spacing effect and determination of optimum spacing illustrated by Marmora, Ontario, magnetic anomaly. Geophysics 20(4):871-885

Agterberg FP (1975) Statistical methods for regional occurrence of mineral deposits. Schr Operations Res Datenverarbeitung Bergbau 4:C11-C15

Aitchison J, Brown JAC (1969) The lognormal distribution. Cambridge University Press, London, 176 pp

Akin H (1983) Geostatistische Aspekte bei der Vorratsermittlung. In: Klassifikation von Lagerstättenvorräten mit Hilfe der Geostatistik. GDMB Schriftenr 39:83-100

Akin H, Siemes H (1988) Praktische Geostatistik. Springer, Berlin Heidelberg New York, 304 pp

Annels A (ed) (1992) Case histories and methods in mineral resource evaluation. Geol Soc Lond, Spec Publ 63, 313 pp

Anonymus (1981) The Marquarie dictionary. NSW, St Leonards, 2049 pp (Marquarie Library)

Anonymus (1989) Klassifikation von Vorkommen fester mineralischer Rohstoffe (Classification of occurrences of solid mineral resources). ÖNORM, Vienna (Austrian National Standard Series)

Armstrong M (1983) Comparing drilling patterns for coal reserve assessment. Proc Aust Inst Min Metall 288:1-5

Armstrong M (ed) (1989) Geostatistics. Proc 3rd Int Geostatistics Congr, Avignon 1988, 2 vols

Attanasi ED, Drew LJ (1985) Lognormal field size distribution as a consequence of economic truncation. Math Geol 17(4):335-351

Australasian Institute of Mining and Metallurgy (ed) (1982) Field geologist's manual. Parkville, Victoria, (Monogr. Ser. 9), 301 pp

Barnes RJ (1991) Teacher's aide: the variogram sill and the sample variance. Math Geol 23(4):673-678

Barnett V, Lewis T (1984) Outliers in statistical data. Wiley series in probability and mathematical statistics. Wiley, New York, 463 pp

Beckmann P (1971) The history of pi. Golem, Press New York

Bronstein IN, Semendjajew KA (1965) Taschenbuch der Mathematik. Deutsch, Frankfurt, 584 pp

Burgin LB (1976) Time required in developing selected Arizona copper mines. US Bureau of Mines, Information Circular 8702, Washington, DC, 144 pp

Caldwell RH, Johnston D (1985) Drilling statistics influence most appropriate risk technique. Oil Gas J (Technol) 83(52):136-144

Cameron MA (1983) Analysis of geochemical data. In: Smith RE (ed) Geochemical exploration in deeply weathered terrain. CSIRO Division of Mineralogy, Floreat Park/Perth, pp 185-193

Carras S (1986) Concepts for calculating recoverable reserves for selective mining in open pit gold operations. Selective Open Pit Gold Mining Seminar, May 9. Australasian Institute of Mining and Metallurgy, Perth

Carswell JT, Schofield NA (1993) Estimation of high-grade copper stope grades in QTS North, Cobar Mine, Cobar NSW. Aust Inst Min Metall Proc 2:19-26

Cassie RM (1954) Some uses of probability paper in the analysis of size frequence distributions. Aust J Mar Freshw Res 5:513-522

Champigny N, Armstrong M (1989) Effect of high grades on geostatistical estimation of gold deposits. 28th Int Geological Congr, July 18th, Washington, DC

Champigny N, Armstrong M (1993) Geostatistics for the estimation of gold deposits. Miner Deposita 28:279-282

Chung CF, Fabbri AG, Sinding-Larsen R (eds) (1988) Quantitative analysis of mineral and energy resources. Reidel, Dordrecht, 738 pp

Clark I (1979) Practical geostatistics. Applied Science, London, 129 pp

Clarke PJ (1956) Grouping in spatial distributions. Science 123:373-374

Clauß G, Ebner E (1985) Statistik für Soziologen, Pädagogen, Psychologen und Mediziner, vol 1. Thun, Frankfurt, 530 pp

Clifton HE, Hunter RE, Swanson FJ, Phillips RL (1969) Sample size and meaningful gold analysis. US Geol. Surv. Prof Pap 625-C, Washington, DC, 17 pp

Cressie NAC (1991) Statistics for spatial data. Wiley, New York, 900 pp

Cressie N, Hawkins DM (1980) Robust estimation of the variogram, I. Math Geol 12(2):115-125

David M (1977) Geostatistical ore reserve estimation. Developments in geomathematics 2. Elsevier, Amsterdam, 364 pp

David M (1988) Dilution and geostatistics. Can Inst Min Metall Bull 81(914):29-35

Dean RB, Dixon WJ (1951) Simplified statistics for small numbers of observations. Anal Chem 23(1):636-638

Deming WE (1950) Some theory of sampling. Wiley, New York, 571 pp

De Geoffroy JG, Wignall TK (1985) Designing optimal strategies for exploration. Plenum, New York , 364 pp

De Wijs HJ (1951) Statistics of ore distribution. Part 1, Frequency distribution of assay values. Geol en Mijnbouw [Nieuw Ser] 13(11):365-375

De Wijs HJ (1953) Statistics of ore distribution. Part 2. Theory of binomial distribution applied to sampling and engineering problems. Geol Mijnbouw [Nieuw Ser] 15(1): 12-24

Didyk M, Tulcanaza E (1970) La determination de errores en el calculo de reservas. Minerales (Inst Ing Min Chile) 25(3):6-19

Diehl P (1981) Eine Klassifikation der Erzvorräte auf geostatistischer Grundlage. Glückauf Forschungsh 42(1):39-48

Diehl P, David M (1982) Classification of ore reserves/resources based on geostatistical methods. Can Inst Min Metall Bull 75(838):127-136

Dimitrakopoulos R (ed) (1994) Geostatistics for the next century – an international forum in honour of Michel David's contribution to geostatistics, Montreal 1993. Kluwer, Dordrecht, 497 pp

Dixon WJ, Massey FJ Jr (1957) Introduction to statistical analysis. McGraw-Hill, New York, 370 pp

Doerffel K (1962) Beurteilung von Analysenverfahren und -ergebnissen. Fresenius Z Anal Chem 185:1-98

Doerffel K (1967) Die statistische Auswertung von Analysenergebnissen. In: Schormüller I (ed) Handbuch der Lebensmittelchemie, vol 2/2. Springer, Berlin Heidelberg New York, pp 1194-1246

Dowd PA, Milton DW (1987) Geostatistical estimation of a section of the Perseverance nickel deposit. In: Matheron G, Armstrong M (eds) Geostatistical case studies. Reidel, Dordrecht, pp 39-67

Dutter R (1985) Geostatistik. Reihe Mathematische Methoden in der Technik, no 2. Teubner, Stuttgart, 159 pp

Finney DJ (1941) On the distribution of a variate whose logarithm is normally distributed. J R Stat Soc Suppl 8(2):155-161

François-Bongarçon D (1991) Geostatistical determination of sample variances in the sampling of broken gold ores. Can Inst Min Metall Bull 84(950):46-57

François-Bongarçon D (1993) The practice of the sampling theory of broken ores. Can Inst Min Metall Bull 86(970):75-81

Fraser DC (1971) VLF-EM data processing. Can Inst Min Metall Bull 64(705):39-41

Fraser DC (1981) A review of some useful algorithms in geophysics. Can Inst Min Metall Bull 74(828):76-83

Friedrich G, Plüger W, Hilmer E (1972) Die Anwendung geochemischer Methoden bei der Prospektion im Lagerstättengebiet von Meggen. GDMB Schriftenr 24:211-225

Froidevaux R (1982) Geostatistics and ore reserve classification. Can Inst Min Metall Bull 75(843):77-83

Gries JP (1979) Providing new sources of mineral supply. US Bureau of Mines, Washington, Information Circular 8789, 42 pp

Guibal D (1987) Recoverable reserve estimation of an Australian gold project. In: Matheron G, Armstrong M (eds) Geostatistical case studies. Reidel, Dordrecht, pp 149-168

Günther M, Kelter D (1994) Geostatistische Fehlerabschätzung der geologischen Vorratsklassen bei der Kreismethode. Z Angew Geol 40(1):21-25

Gy PM (1967) Théorie générale, vol 1. Bur Rech Geol Min Mem 56, 186 pp

Gy PM (1979) Sampling of particulate materials-theory and practice. Developments in geomathematics, 4. Elsevier, Amsterdam, 431 pp

Hammond JS (1967) Better decisions with preference theory. Harvard Business Rev 45(6):123-141

Harbaugh JW, Dovetan JH, Davis JC (1977) Probability methods in oil exploration. Wiley, New York, 269 pp

Hawkes HE, Webb JS (1962) Geochemistry in mineral exploration. Harper and Row, New York, 415 pp

Hawkins DM (1980) Identification of outliers. Monographs on applied probability and statistics. Chapman and Hill, London, 188 pp

Hazen A (1914) Trans Ame Soc Civ Eng 77:1539

Hazen SW Jr (1967) Some statistical techniques for analysing mine and mineral deposit sample and assay data. US Bureau of Mines, Bull 621, Washington, 223 pp

Herzfeld UC (1993) A method for seafloor classification using directional variograms, demonstrated for data from the Western flank of the mid-Atlantic ridge. Math Geol 25(7):901-924

Hester B (1982) Problems in estimating ore reserves in some current exploration targets. In: Metals Economics Group (ed) Mine developments in the 80's. Seminar, Feb 4-5, Denver

Hewlett RF (1965) Design of drill-hole grid spacings for evaluating low-grade copper deposits. US Bureau of Mines, Washington, Report of Investigation 6634, 46 pp

International Atomic Energy Agency (IAEA) (1985) Methods for the estimation of uranium ore reserves. An instruction manual. IAEA Tech Rep Ser 255, 92 pp

Isaaks EH, Srivastava RM (1989) Applied geostatistics. Oxford University Press, Oxford 561 pp

James C (1986) Geochemisches Arbeitspapier. Kalgoorlie School of Mines, Kalgoorlie

Journel AG (1983) Nonparametric estimation of spatial distribution. Math Geol 15(3):445-468

Journel AG, Huijbregts C (1978) Mining geostatistics. Academic Press, New York, 600 pp

Karsten KG (1925) Charts and graphics. 2nd edn. Prentice Hall, New York

Kelter D, Wellmer F-W (1993) Is the introduction of a 3-dimensional coal reserves/resources classification desirable? Z Förderer des Bergbaus und des Hüttenwesens an der TU Berlin 1:11-13

Kesler SE (1994) Mineral resources, economics and the environment. Macmillan, New York, 389 pp

King HF, McMahon DW, Bujtor GJ (1982) A guide to the understanding of ore reserve estimations. Proc Aust Inst Min Metall Suppl 281

Knudsen HP, Kim YK, Mueller E (1978) Comparative study of the geostatistical ore reserve estimation method over the conventional method. Mining Eng 30(1):54-58

Koch GS, Link RF (1970) Statistical analysis of geological data vol 1. Wiley, New York, 375 pp

Koch GS, Link RF (1971) Statistical analysis of geological data vol 2. Wiley, New York, 438 pp

Kraft G (1981) Theoretische Grundlagen der Probenahme. GDMB Schriftenr. 36

Kreyszig E (1968) Statistische Methoden und ihre Anwendungen. Vandenhoeck and Ruprecht, Göttingen, 422 pp

Krige DG (1951) A statistical approach to some basic mine valuation problems on the Witwatersrand. J Chem Metall Min Soc S Afr 52:119-139

Krige DG (1960) On the departure of ore valuation distributions from lognormal models in South African gold mines. J S Afr Inst Min Metall 61:231-224

Krige DG (1962) Statistical applications in mine valuation. J Inst Min Surv S Afr 12:2, 3

Krige DG (1966) Two-dimensional weighted moving average trend surfaces for ore valuation. In: Symposium on mathematical statistics and computer applications in ore valuation. South African Institute of Mininig and Metallurgy, Johannesburg, pp 13-19

Krige DG (1978) Lognormal-de Wijsian geostatistics for ore evaluation. Monogr Ser, Geostatistics 1. South African Institute of Mining and Metallurgy, Johannesburg, 51 pp

Krige DG (1994) A statistical approach to some basic mine valuation problems on the Witwatersrand. J S Afr Inst Mining Metal 94(3):95-111

Krige DG, Magri EJ (1982a) Geostatistical case studies of the advantages of lognormal-de Wijsian kriging with mean for a base metal mine and a gold mine. Math Geol 14(6):547-555

Krige DG, Magri EJ (1982b) Studies of the effects of outliers and data transformation on variogram estimates for a base metal and a gold orebody. Math Geol 14(6):557-564

Krige DG, Kleingeld WJ, Osterveld MM (1990) A-priori parameter distribution patterns for gold in the Witwatersrand basin to be applied in borehole evaluation of potential new mines using Bayesian geostatistical techniques. Proc 22nd Int Symp APCOM, Sept 17-21, Berlin, vol 2, pp 715-726

Lasky SG (1950) How tonnage and grade relations help to predict ore reserves. Eng Min J 151(4):81-85

Lepeltier C (1969) A simplified statistical treatment of geochemical data by graphical representation. Econ Geol 64:538-550

Levy IW (1982) Exploration costs, risks, decisions and returns. Australian Development Assistence Bureau Exploration Course, Nov 23, Perth

Link RF, Koch GS (1975) Some consequences of applying lognormal theory to pseudolognormal distributions. Math Geol 7:117-128

Mackenzie BB (1973) Corporate exploration strategies. In: Proc 10th Symp. on Application of Computer Methods in the Mineral Industry (APCOM). South African Institute of Mining and Metallurgy, Johannesburg, pp 1-7

Mackenzie BB. Bilodeau ML (1984) Economics of mineral exploration in Australia. Australian Mineral Foundation, Adelaide, 171 pp

Mackenzie B, Woodall R (1988) Economic productivity of base metal exploration in Australia and Canada. In: Tilton JE, Eggert RG, Landsberg HH (eds) World mineral exploration. Trends and economic issues. Resources for the Future, Washington, DC, 464 pp

Matheron G (1971) The theory of regionalized variables and its application. Centre de Morphologie Mathématique de Fontainebleau, Cahier 5, 211 pp

Matheron G, Armstrong M (eds) (1987) Geostatistical case studies. Quantitative geology and geostatistics. Reidel, Dordrecht, 248 pp

Miller E, Gill D (1986) Geostatistical ore-reserve estimation of South York'am phosphate deposit, Zin Valley, southern Israel. Trans Inst Min Metall [A] 95:A1-A7

Miskelly N (1994) A comparison of international definitions for reporting mineral resources and reserves. Miner Ind Int 1019 (July):28-36

Moroney MJ (1970) Facts from figures. Penguin, Hammondsworth, 412 pp

Muge FH (1986) Case studies on modelling complex sulphide orebodies for ore reserve estimation using geostatistical methods. In: Chung CF, Fabbri AG, Sinding-Larsen R (eds) Quantitative analysis of mineral and energy resources. Reidel, Dordrecht, pp 341-358

Müller A (1986) Kleine Anleitung zur graphischen Auswertung geochemischer Daten. BGR, Hannover

Nowak D (1994) Ore reserves and grade control of the Broken Hill deposit, Aggeneys, Proc 15th CMMI Congr, Johannesburg, vol 3, pp 233-238

Olea RA (1977) Measuring spatial dependence with semivariograms. No 3 Series on Spatial analysis. Kansas Geol Surv, Lawrence, Kansas, 29 pp

Olea RA (1991) Geostatistical glossary and multilingual dictionary. International Association for Mathematical Geological Studies in Mathematical Geology, vol 3. Oxford University Press, Oxford, 177 pp

Parker HM (1991) Statistical treatment of outlier data in epithermal gold deposit reserve estimation. Math Geol 23(2):157-199

Paterson NR, Ronka V (1971) Five years of surveying with the very low frequency-electromagnetic method. Geoexploration, 9(7):7-26

Perillo GME, Marone F (1986) Determination of optimal numbers of class intervals using maximum entropy. Math Geol 18(4):401-407

Raymond G (1979) Ore estimation problems in an erratically mineralized orebody. Can Inst Min Metall Bull 72(806):90-98

Rehder S (1988) Manuscript about the separation of two populations. BGR, Hannover

Rendu JM (1978) An introduction to geostatistical methods of mineral evaluation. S Afr Inst Min Metall Monogr Ser 2:100 pp (2nd edn, 1981)

Rose AW, Keith ML (1976) Reconnaissance of geochemical techniques for detecting uranium deposits in sandstone of northeastern Pennsylvania. J Geochem Explor 6:119-137

Royle AG (1977) Global estimating of ore reserves. Trans Inst Min Metall [A] 86:A1-A39

Royle AG (1988) The volume-variogram relationships in gold mines. Leeds University Mining Association, Luma

Rudenno V (1985) Technical note-polygonal blocks of influence intriangular grids. Bull Proc Austr Inst Min Metall 290(6):67

Sabourin RL (1983) Geostatistics as a tool to define various categories of resources. Math Geol 15:131-143

Sames CW, Wellmer F-W (1981) Exploration I: Nur wer wagt, gewinnt. Risiken, Strategien, Aufwand, Erfolg. Glückauf 117(10):580-589

Savinskii ID (1965) Probability tables for locating elliptical underground masses with a rectangular grid. Consultants Bureau, New York, 110 pp

Sichel HS (1952) New methods in the statistical evaluation of mine sampling data. Inst Min Metall Trans 61:261-288

Sichel HS (1966) The estimation of means and associated confidence limits for small samples from lognormal population. Symp on Mathematical statistics and computer application in ore evaluation. South African Inst Min Metall, Proc, Johannesburg, pp 106-123

Sichel HS (1968) Development of large sample estimates for the 3-parameter lognormal distribution. Operational Research Bureau, Johannesburg

Sinclair AJ (1974) Selections of threshold values in geochemical data using probability graphs. J Geochem Explor 3(2):121-149

Sinclair AJ (1975) Some considerations regarding grid orientation and sample spacing. In: Elliot IL, Fletcher WK (eds) Geochemical exploration 1974. Elsevier, Amsterdam, pp 133-140

Sinclair AJ (1976) Applications of probability graphs in mineral exploration. Assoc Explor Geochem Spec 4, Richmond, British Columbia, 95 pp

Slichter LB (1955) Geophysics applied to prospecting for ores. Econ Geol 50:885-969

Slichter LB (1960) The need of a new philosophy of prospecting. Min Eng 12:570-576

Taylor HK (1972) General background theory of cut-off grades. Inst Min Metall Trans [A]81:A160-A179

Thompson M, Howarth RJ (1973) The rapid estimation and control of precision by duplicate determination. Analysist 98:153-160

Tukey JW (1962) The future of data analysis. Ann Math Stat 33:1-67

Turner GW (ed) (1987) The Australian concise Oxford dictionary of current English. Oxford University Press, Melbourne, 1340 pp

Vallée MJ (1986) Mineral inventory, from resource reconnaissance and evaluation to ore reserve. In: David M, Froidevaux R, Sinclair AJ, Vallée M (eds) Proceedings of Canadian Institute of Mining and Metallurgy Symposium 1986. Ore reserve estimation, methods, models and reality. Montreal, pp 10-31

Vallée MJ, Belisle JM, David M (1977) Kriging as a tool to avoid overestimation of grade in sulphide orebodies. 14th Symp Application of Computer Methods in the Mineral Industry (APCOM), Proc vol AIME, pp 1013-1025

Vallée MJ, David M, Dagbert M, Desrochers C (1992) Guide to the evaluation of gold deposits.Can Inst Min Metall Bull, Spec Pap 45, 299 pp

Vallée MJ, Dagbert M, Côte D (1993) Quality control requirements for more reliable mineral deposit and reserve estimates. Can Inst Min Metall Bull 86 (969):65-75

Wainstein BM (1975) An extension of lognormal theory and its application to risk analysis models for new mining ventures. J S Afr Inst Min Metall 75:221-238

Weber E (1956) Grundlagen der biologischen Statistik, 2nd edn. Fischer, Jena, 456 pp

Weber HW, Morgan PJ (1983) Classification of ore reserves based on geostatistical and economic parameters. Can Inst Min Metall Bull 86(966):73-76

Wellmer F-W (1976) Der Einfluß von Lagerstättenanisotropien und -inhomogenitäten auf die wirtschaftliche Optimierung von Lagerstättenprojekten. Erzmetall 29(6):270-276

Wellmer F-W (1983a) Klassifikation von Lagerstättenvorräten mit Hilfe der Geostatistik. GDMB Schriftenr 39:9-43

Wellmer F-W (1983b) Classification of ore reserves by geostatistical methods - results to date of the discussions of the GDMB-working group. Erzmetall 36(7/8): 315-321

Wellmer F-W (1983c) New genetic and geochemical concepts as guides to mineral exploration. In: Bender F (ed) New paths to minerals exploration. Nägele, Stuttgart, pp 135-153

Wellmer F-W (1983d) Neue Entwicklungen in der Exploration I: Kosten, Reserven, Technologien. Erzmetall 36(1):7-13

Wellmer F-W (1983e) Neue Entwicklungen in der Exploration II: Kosten, Erlöse, Technologien. Erzmetall 36(3):124-131

Wellmer F-W (1984) Kosten-Nutzen-Analyse von Explorationsmethoden. Teil I Erzmetall. 37(2):52-57; Teil II 37(3):144-150

Wellmer F-W (1986) Risk elements characteristic of mining investments. 13th Congr Council of Mining and Metallurgy Institutions, May 11-16, Singapore. Congr Vol 5, Aust Inst Min Metall, Melbourne

Wellmer F-W, Greinwald S (1982) Optimale Methodenkombination in der Exploration. Z Dtsch Geol Ges 133:509-533

Wellmer F-W, Atmaca T, Günther M, Kästner H, Thormann A (1994) The economics of sediment-hosted zinc-lead deposits. In: Fontboté L, Boni M (eds) Sediment-hosted Zn-Pb-Ores. Springer, Berlin, Heidelberg, New York, pp 429-462

Whateley MKG, Harvey PK (eds) (1994) Mineral resource evaluation II. Methods and case histories. Geol Soc Lond, Spec Publ 79, 271 pp

Whitchurch KD, Gillies ADS, Just GD (1990) Coal resource classification and geostatistics. Aust Inst Min Metall Proc 295(2):7-17

Wonnacott RJ, Wonnacott TH (1985) Introductory statistics, 4th edn. Wiley, New York, 649 pp

Wood GH, Kehn TM, Carter ME, Culberton WC (1983) Coal resource classification system of the US Geological Survey. US Geol Surv Circ 891, 65 pp

Zwicky RW (1975) Preliminary geostatistical investigations of tar bearing sands. Lease – 13, Athabasca Tar Sands, Canada. Schr Operation Research Datenverarbeitung im Bergbau 4(1/I-III):1-15

Glossary

(own definitions and definitions drawing on David 1977; Isaaks and Srivastava 1989; Kesler 1994; Olea 1991; Macquarie Dictionary, Anonymous (1981) and the Australian Concise Oxford Dictionary (Turner 1987))

accumulation	the product out of grade and thickness (=intensity factor)
accuracy	degree to which an analysis approaches the correct number
anisotropic	having different statistical properties in different directions, meaning an anisotropic property is only revealed in a spatial analysis
anisotropy	the state of property of being anisotropic
anomalous	samples that differ significantly from all others in a group or population
anomaly	a distinctive local deviation from average values a=anomalous values) determined in mineral exploration survey (e.g. geochemical or geophysical anomaly)
approximation	a result that is not exact, but is sufficiently so for a given purpose
assay	analysis of the grade of ore or concentrate
background	a term used to refer to values characteristic of the average or most common sample in a population (see anomalies)
bias	distortion of result by neglected factor in general, in statistics the mean of the error distribution
block estimate	the estimate of grade or tonnage of a block within an ore deposit (contrary to the global estimate, see)
category	interval in a histogram (see histogram)
core recovery	the amount of drill core gained by core drilling, expressed as the percentage of drill core in relation to the total length drilled

confidence limits	a pair of numbers used to estimate a characteristic of a population from a sample, which are such that it can be stated with a specified probability that the pair of numbers calculated from a sample will include the value of the population characteristic between them
correlation	standardized covariance with values between -1 and +1 (see covariance)
covariance	a measurement of mutual relationship between pairs of random variables
cross-validation	any of the validation methods, in which observations are dropped one at a time from a sample and estimates are computed using all or most of the remaining measurements or observations
cumulative	increasing or growing by successive additions
cumulative frequency distribution	a set of accumulated frequencies (either by successive addition from the lowest or highest value of the frequency distruction) associated with different categories, intervals or values to which items of a statistical group belong
cut-off grade	minimum grade of ore that can be extracted profitably
detection limit	the level below which an analytical method cannot detect the element
distribution	a graphical or analytical representation showing the frequency of occurrence for outcome of random variables
error	the deviation of a computed observed or measured quantity from the time, specified, or theoretically correct values of such a variety. The error is called positive or an overestimation if the result is greater than the true value, and negative or underestimation if it is less than the true value
estimator	a value used for an approximate calculation
extrapolation	the estimation process of finding a value of a function at a point x based on information provided by values, whereby x lies outside the hull of the values providing the information (see interpolation)

feasibility study	a geological, technical and economical study made to determine whether a deposit can be exploited economically
frequency diagram	a graph of a frequency distribution (see frequency distribution)
frequency distribution	set of frequencies associated with different categories, intervals or values to which items of a statistical group belong
geochemical exploration	mineral exploration based on analysis of the chemical composition of rocks, soils, water, gas and living organism
geophysical exploration	mineral exploration by measurement of magnetic gravity, electrical, radioactive or other physical features of rocks and minerals
geostatistics	statistical treatment of spatially arranged data or processes with continuous spatial indices such as assay values from a mineral deposit for example
global estimate	the estimate of grade or tonnage of the total ore deposit (contrary to block estimate, see)
grade tenor	the order of magnitude of the grade of a mineralization or an ore deposit
gradient	1. the degree of inclination, 2. change in a variable quantity per unit distance
goodness of fit	the measure of how closely a set of observed values approximates those derived from a theoretical model
grid	a systematic array of points or lines at or along which field observations are made and samples taken during mineral exploration
histogram	graph of a frequency distribution, in which equal intervals of values are marked on a horizontal axis and the frequency corresponding to each interval is indicated by the height of a rectangle having the interval as its base
intensity factor	the product out of grade and thickness (=accumulation)

interpolation	the estimation process of finding a value at a point x based on information provided by surrounding known values
isotropic	a state of having one or more statistical properties that are the same in all directions, meaning that an isotropic property is only revealed in a spatial analysis
isotropy	the state of a property of being isotropic
kriging	a linear regression technique for minimizing the estimation variance of a spatial variance (see variance) from a defined model, the variogram
kurtosis	the shape or a measure indicating the shape, of the curve of a frequency distribution near its mean or the flatness of a distribution
lag	distance between observation or sample points
lognormal distribution	a skewed distribution which can be transformed into the normal distribution by converting the values of the distribution to logarithms (see skewed, normal distributions)
mean	a quantity having a value intermediate between the values of other quantities; an average as of the arithmetic mean by summing of all values in a group divided by number of values
mean deviation	the average of the absolute values of a set of deviations from the mean (see mean) in a statistical distribution
median	central value in a group with the same number of values above and below
mode	the most common number in a population
noise	irregular fluctuation in a series of measurements accompaning but not relevant to a signal to be searched for in a survey
normal distribution	a group of samples or population having a symmetric distribution, meaning the mean, median and mode are the same and fulfilling mathematically exactly defined properties, sometimes also called the Gaussian distribution

normalization	mathematical procedure to make a value or group of values conform to a standard (=standardization)
nugget	unusually large piece of natural metal, usually found in a placer deposit
nugget effect	a geostatistical term describing in a quantitative way the small-scale variation of regionalized variables like grades
ore deposit	mineral deposit that can be exploited at a profit
outlier	a value that lies outside the normal range of a frequency distribution
partitioning	separation of two or more subpopulations from a population
polygon	a closed plane figure with several edges and angles, usually more than four
population	in statistics, a set of samples
precision	agreement between replicate analysis of the same sample by the same method
prefeasibility study	a geological, technical and economic study made to determine whether a deposit can be exploited
probable error	a value such that the error in an error distribution is equally likely to be greater or smaller than it
probability	the relative frequency of the occurrence of an event as measured by the ratio of the number of cases of alternatives favourable to the event to the total number of cases or alternatives
probability curve	a curve which describes the distribution of probability over the values of a variable
quantile	any number (n-1) that partitions a set of values of a random variable into n classes each containing the same number of data. The quantile that divides the random variable into two classes is called the median (see median), into 4 classes the quartile etc.
random	not according to a pattern or method

random drilling	mineral exploration conducted by drilling holes of locations that have been selected randomly in a specified area
random sampling	the drawing of a sample from a statistical population in which all members of the population have equal probabilities of being included in the sample
random variable	a variable which can take one of several possible values
range	a) the limits between which variation is possible, the difference between the largest and the smallest value; b) in a transitive type variogram, the distance in which the variogram value (gamma value) increases, indicating a regional interdependence of the properties analysed by the variogram
regionalised variable	a variable used to describe spatial natural phenomena that are characterised by smooth variations on a global scale, but erratic at a local scale
regression analysis	any of several statistical methods to investigate relationships between variables
reserve	that part of the reserve base that can be extracted at a profit at the time of determination whether or not facilities are available on the property, meaning a reserve figure is not a constant, but varies with time, influenced by the commodity prices applicable at the time of the reserve determination
residual	a) the deviation of one of a set of observations or numbers from the mean of the set b) the deviation between an empirical and a theoretical set
resource	concentration of natural material that can be extracted now or in the future. This includes reserves (see). Resources are divided into demonstrated resources that have been measured to some degree, and inferred resources that are only thought to be present.
risk	a concept used to describe the chance or probability of an adverse outcome
risk assessment	a scientific undertaking to quantify the risk

robust	the characteristic of a model or a method to persist in yielding correct predictions or estimations in spite of errors in the collection of data
semivariogram	see variogram
sill	the maximum value that the transitive variogram function approximates with increasing lag (see variogram, lag). The sill normally has the value of the variance of the data considered.
simulation	the use of an analogue model in order to study the properties of a system
skewed	refers to a population or a frequency distribution that is not symmetric (mean, mode and median are not the same).
skewness	a measure for a frequency distribution of the degree of deviation for symmetry
source	in geochemistry, a reservoir that releases or emits a substance of interest
standard	in chemistry, a sample against which other samples are compared
standard deviation	the square root of the average of the squares of a set of deviations about an arithmetic mean, the root mean square of the deviations of a set of values
standardization	mathematical procedure to make a value or a group of values conform to a standard (normalisation)
stationarity	the property of being stationary (see)
stationary	the statistical properties are independent of location, meaning if the field of analysis is changed, the statistical properties do not change
support	the volume, the mass or the shape of the sample
tenor	(see: grade tenor)
threshold	the lowest value of any signal which will produce the effect of being recognized as an anomaly

trace metal	metal present in very small (trace) amounts
transitive type variogram	a variogram in which the variogram value first increases with increasing lag (see) and then (in some cases only asymptotically) reaches a limiting value (the sill, see)
trend	the large-scale variation of a regionalized variable
variance	the square of the standard deviation (see)
variate	the numerical value of an attribute belonging to a statistical item (=random variable)
variogram	graph of the measurement of degree of dissimilarity in the outcome of sampling as the distance between the observation sites increases
weighting	multiplying components by factors to take into account their different importance

Index

Springer
and the
environment

At Springer we firmly believe that an international science publisher has a special obligation to the environment, and our corporate policies consistently reflect this conviction.

We also expect our business partners – paper mills, printers, packaging manufacturers, etc. – to commit themselves to using materials and production processes that do not harm the environment. The paper in this book is made from low- or no-chlorine pulp and is acid free, in conformance with international standards for paper permanency.

 Springer